Library of
Davidson College

APPLIED PROBABILITY

A Series of the Applied Probability Trust

Editors
J. Gani   C.C. Heyde

The Craft of Probabilistic Modelling

A Collection of Personal Accounts

Edited by J. Gani

With Contributions by

N. T. J. Bailey   J. W. Cohen   R. L. Disney   W. J. Ewens
E. J. Hannan   M. Iosifescu   J. Keilson   D. G. Kendall
M. Kimura   M. F. Neuts   K. R. Parthasarathy
N. U. Prabhu   H. Solomon   R. Syski   L. Takács
R. L. Tweedie   D. Vere-Jones   G. S. Watson   P. Whittle

With 21 Illustrations

Springer-Verlag
New York Berlin Heidelberg Tokyo

Series Editors

J. Gani
Statistics Program
Department of Mathematics
University of California
Santa Barbara, CA 93106
U.S.A.

C. C. Heyde
Department of Statistics
Institute of Advanced Studies
The Australian National University
GPO Box 4, Canberra ACT 2601
Australia

519.2
C 885

87-8314

AMS Classifications: 60-03, 01A65

Library of Congress Cataloging in Publication Data
Main entry under title:
The craft of probabilistic modelling.
  (Applied probability)
  Bibliography: p.
  1. Probabilities—Addresses, essays, lectures.
  I. Gani, J. M. (Joseph Mark)   II. Bailey, Norman T. J.
  III. Series.
  QA273.18.C73   1986   519.2   85-32060

© 1986 by Applied Probability Trust.
All rights reserved. No part of this book may be translated or reproduced in any form without written permission from the copyright holder.

Typeset by Asco Trade Typesetting Ltd, Hong Kong.
Printed and bound by R. R. Donnelley & Sons, Harrisonburg, Virginia.
Printed in the United States of America.

9 8 7 6 5 4 3 2 1

ISBN 0-387-96277-8 Springer-Verlag New York Berlin Heidelberg Tokyo
ISBN 3-540-96277-8 Springer-Verlag Berlin Heidelberg New York Tokyo

To Ruth, my wife,
on whose support I have always relied

Preface

This book brings together the personal accounts and reflections of nineteen mathematical model-builders, whose specialty is probabilistic modelling.

The reader may well wonder why, apart from personal interest, one should commission and edit such a collection of articles. There are, of course, many reasons, but perhaps the three most relevant are:
(i) a philosophical interest in conceptual models; this is an interest shared by everyone who has ever puzzled over the relationship between thought and reality;
(ii) a conviction, not unsupported by empirical evidence, that probabilistic modelling has an important contribution to make to scientific research; and finally
(iii) a curiosity, historical in its nature, about the complex interplay between personal events and the development of a field of mathematical research, namely applied probability.

Let me discuss each of these in turn.

Philosophical

Abstraction, the formation of concepts, and the construction of conceptual models present us with complex philosophical problems which date back to Democritus, Plato and Aristotle. We have all, at one time or another, wondered just how we think; are our thoughts, concepts and models of reality approximations to the truth, or are they simply functional constructs helping us to master our environment?

Nowhere are these problems more apparent than in mathematical modelling, where idealized concepts and constructions replace the imperfect realities for which they stand. Using mathematics, a symbolic language of great power, modellers work out the relations between these concepts, as well as their consequences; they then test their deductive results against the real world. If these results are in reasonable agreement with reality, the mathematical models are deemed to "explain" natural phenomena; their predictive capacity then proves extremely useful. I have never ceased to marvel at the complexity of the process, a complexity which the contributors to this volume illustrate in the variety of problems which they outline.

Scientific

Applied probability is the field of mathematical research in which probabilistic methods are applied to real-life phenomena. Probabilistic modelling is possibly the most basic element of applied probability. While one might not be able to claim particular intellectual pre-eminence for the subject, its uses have spread to almost every area of human endeavour: its contributions to scientific research are far too numerous to list.

Most natural phenomena in the biological, physical, social and technological sciences have random components. These may not always be as clearly apparent as in a game of roulette or in life insurance, but they are sufficiently important to make probabilistic models essential in "explaining" reality, and in predicting the likely results of changed conditions. If, for example, one wished to alter the rules of play of a game of roulette, one could predict fairly accurately the results of such changes by constructing a simple probabilistic model of the game.

How has the field of applied probability developed? It clearly existed long before its name was coined: Pascal (1623–1662), Jacob Bernoulli (1654–1705), Laplace (1749–1827), Gauss (1777–1855), Maxwell (1831–1879), Gibbs (1839–1903), Boltzmann (1844–1906), Markov (1856–1922), Planck (1858–1947), Einstein (1879–1955), Lévy (1886–1971), Khinchin (1894–1959), Kolmogorov (b. 1903), Feller (1906–1970) and Kac (1914–1984), among many others, were concerned with applications of probability to a wide range of real-life problems. The term "applied probability" first appeared in the title of an American Mathematical Society proceedings in 1955, and was popularized by its use after 1959 in the Methuen Monographs on Applied Probability and Statistics, edited by M. S. Bartlett.

The past thirty years have witnessed a blossoming of the field. There is a growing list of journals partially or exclusively devoted to applied probability: *Teoriya Veroyatnostei i ee Primeneniya* (1956), *Zeitschrift für Wahrscheinlichkeitstheorie* (1962), *Journal of Applied Probability* (1964), *Advances in Applied Probability* (1969), *Stochastic Processes and their Applications* (1973), *Annals of Probability* (1973), *Stochastics* (1973), *Stochastic Analysis and Applications* (1983), *Stochastic Models* (1985), *Applied Stochastic Models and Data Analysis* (1985), and most recently *Queueing Systems: Theory and Applications* (1986).

One can also find considerable discussion of applied probability models in several biological, chemical, engineering, operations research, physical and psychological journals; in fact, there is hardly an area of research in which applied probability has not made a contribution. Our modellers' accounts attest to the wide range of problems in which probabilistic models have been applied.

Historical

Most of the authors invited to contribute to this volume willingly agreed to write an article for it. They have distinguished themselves in such areas as biological and genetic models, control and optimization, epidemic theory, geometric probability, jurimetrics and learning theory, physical, quantum theoretical and seismic models, population and migration models, probabilistic algorithms, and queueing theory. Many of them have worked in more than one area.

This has made it very difficult to classify them in appropriate groupings; no sooner had someone been allocated to a particular category (queueing theory or biological modelling, for example) than it became apparent that this modeller could equally well be placed in another (physical models or probabilistic algorithms). In the end, I opted for a chronological ordering, with natural if slightly arbitrary breakpoints; the first four contributors were early in the field, the following nine were directly involved in the formal organization of applied probability, while the last six belong to the newer generation of researchers.

In a very real sense, however, the development of probabilistic modelling has been a continuous process, and I would not wish to claim any special virtue for the subdivision of the contributors into the present three groups.

How have the efforts of individuals contributed to the development of applied probability? And how have personal events as apparently random as those described in the following accounts led to so ordered a body of scientific work? This is a historical problem on which the contributors throw much light. Some started off as engineers and physicists, others as biologists, yet others as mathematicians and statisticians; all were eventually led by their scientific questioning to some aspect of probabilistic modelling.

While this activity clearly involves mathematical ability, it is also a craft in the sense that mathematics plays only a partial role in the way the problems are tackled. Every one of the contributors responded to the lure of very real scientific questions which they wished to answer; all, in addition to their mathematical skills, carefully selected the fundamental characteristics of their problems to build into appropriate models. In this lay their craftsmanship: the feel for which structures and concepts were most relevant to their problems, and which others could safely be disregarded. Their accounts, describing as they do both personal and intellectual histories, provide a documented record of their endeavours which should help fellow modellers and students in future investigations and studies.

It has been my good fortune to be associated with the development of applied probability since the mid-1950s. I have been lucky to assist the progress of the field through the foundation of the Applied Probability Trust in 1964, and the subsequent editing of the *Journal of Applied Probability* and *Advances in Applied Probability*. The Trust now brings to readers a new series of books and monographs in Applied Probability, edited by Professor

C. C. Heyde and myself, and published by Springer-Verlag. I hope that this, the first volume of the series, will give readers an insight into probabilistic modelling and its place in scientific research.

My thanks are due to Mrs Kathleen Lyle and Miss Mavis Hitchcock of the Applied Probability Trust Office for their editorial assistance, and to Springer-Verlag for publishing the book on behalf of the Trust.

University of California J. GANI
Santa Barbara
January 1986

Contents

1
EARLY CRAFTSMEN

David G. Kendall	3
Crafty Modelling	3
Herbert Solomon	10
Looking at Life Quantitatively	11
E. J. Hannan	31
Remembrance of Things Past	31
G. S. Watson	43
A Boy from the Bush	43

2
THE CRAFT ORGANIZED

Norman T. J. Bailey	63
An Improbable Path	64
J. W. Cohen	88
Some Samples of Modelling	89
Ryszard Syski	109
Markovian Models—An Essay	110
N. U. Prabhu	126
Probability Modelling Across the Continents	127
Lajos Takács	139
Chance or Determinism?	140
Motoo Kimura	150
Diffusion Models of Population Genetics in the Age of Molecular Biology	151
Julian Keilson	166
Return of the Wanderer: a Physicist Becomes a Probabilist	167

Peter Whittle 186
    In the Late Afternoon 186

Ralph L. Disney 196
    The Making of a Queueing Theorist 197

3
THE CRAFT IN DEVELOPMENT

Marcel F. Neuts 213
    An Algorithmic Probabilist's Apology 214

D. Vere-Jones 222
    Probability, Earthquakes and Travel Abroad 223

K. R. Parthasarathy 235
    From Information Theory to Quantum Mechanics 235

Marius Iosifescu 250
    From Real Analysis to Probability: Autobiographical Notes 251

W. J. Ewens 276
    The Path to the Genetics Sampling Formula 276

R. L. Tweedie 291
    In and Out of Applied Probability in Australia 292

    Index 309

Contributors

Norman T. J. Bailey
    Division d'Informatique, Hôpital Cantonal Universitaire, 1211 Genève 4, Switzerland

J. W. Cohen
    Mathematical Institute, State University of Utrecht, Budapestlaan 6, P.O. Box 80.010, 3508 TA Utrecht, The Netherlands

Ralph L. Disney
    Department of Industrial Engineering and Operations Research, Virginia Polytechnic Institute and State University, Blacksburg, VA 24061, USA

W. J. Ewens
    Department of Mathematics, Monash University, Clayton, VIC 3168, Australia

E. J. Hannan
    Department of Statistics, I.A.S., The Australian National University, GPO Box 4, Canberra ACT 2601, Australia

Marius Iosifescu
    Str. Dr. N. Manolescu 9-11, 76222 Bucharest 35, Romania

Julian Keilson
    Graduate School of Management, University of Rochester, Rochester, NY 14267, USA

David G. Kendall
    Statistical Laboratory, University of Cambridge, Cambridge CB2 1SB, England

Motoo Kimura
    National Institute of Genetics, Yata 1,111, Mishima, Shizuoka-ken, 411 Japan

Marcel F. Neuts
    Department of Systems and Industrial Engineering, University of Arizona, Tucson, AZ 85721, USA

K. R. Parthasarathy
    Indian Statistical Institute, Delhi Centre, 7, S. J. S. Sansanwal Marg, New Delhi-110016, India

N. U. Prabhu
  School of Operations Research and Industrial Engineering, Cornell University, Upson Hall, Ithaca, NY 14853, USA

Herbert Solomon
  Department of Statistics, Stanford University, Sequoia Hall, Stanford, CA 94305, USA

Ryszard Syski
  Department of Mathematics, University of Maryland, College Park, MD 20742, USA

Lajos Takács
  Department of Mathematics and Statistics, Case Western Reserve University, Cleveland, OH 44106, USA

R. L. Tweedie
  SIROMATH Pty Ltd, St Martins Tower, 31 Market St, Sydney, NSW 2000, Australia

D. Vere-Jones
  Department of Mathematics, Victoria University of Wellington, Private Bag, Wellington, New Zealand

G. S. Watson
  Princeton University, Fine Hall, Princeton, NJ 08544, USA

Peter Whittle
  Statistical Laboratory, University of Cambridge, Cambridge, CB2 1SB, England

1
EARLY CRAFTSMEN

David G. Kendall

David George Kendall was born in Ripon, Yorkshire, England on 15 January 1918. He was educated at Ripon Grammar School and the Queen's College, Oxford. During the war he worked on rocketry under L. Rosenhead CBE FRS and M. S. Bartlett FRS, in association with F. J. Anscombe and P. A. P. Moran. This transformed him from a mathematician with a taste for astronomy into a statistician with a taste for mathematics. After the war he returned to Magdalen College, Oxford as fellow and lecturer in mathematics, and in 1962 he was elected to the newly founded chair of mathematical statistics in Cambridge, and to a fellowship of Churchill College.

In 1970 he was joint organizer (with F. R. Hodson and P. Tautu) of the Anglo-Romanian Conference on Mathematics in the Archaeological and Historical Sciences held at Mamaia in Romania. In 1973–4 he was joint editor (with E. F. Harding) of the books *Stochastic Analysis* and *Stochastic Geometry* which, together with the work of the Rollo Davidson Trust, constitute a memorial to a famous colleague and former pupil.

He was president of the London Mathematical Society 1972–4, and the first president of the Bernoulli Society at its formation in 1975.

In 1952 he married Diana Louise Fletcher. They have six children.

Crafty Modelling

1. Initiation in Modelling

Trendy phrases generate confusion. This is because everyone thinks it fun to use them, but not everyone uses them in the same way. "The art of modelling" is no exception. I shall try to demonstrate this by releasing a few skeletons from my own cupboard.

A paper like [2] illustrates one sort of modelling. The "models" in that paper were never intended for use in exact calculations with real biological populations. The intention there (as in earlier work by Feller, by Arley, and by Bartlett) was to formulate the simplest nontrivial mathematical structures which capture the essentials of multiplicative growth, death, ageing, immigration, and emigration, and to study these in order to surmount the irreducible minimum of difficulty which surrounds the infinitely more complex task of studying such processes as they actually arise in biology, physics, and other fields. A phrase often used at the time was "we are here at the stage of solving the problem of two bodies; only when (and if) we have succeeded in doing that, will we start to think about the problem of three, or n, bodies". In such an activity as this it would perhaps be better to describe the product as a *caricature* rather than a model. A caricature is not a faithful portrait, but it catches enough of its subject to make it clear to all that the Honourable Samuel Slumkey of Slumkey Hall is intended, and not Horatio Fizkin Esq of Fizkin Lodge.

The paper [2] was my contribution to an afternoon Symposium on Stochastic Processes organized by the Research Section of the Royal Statistical Society in 1949. Joe Moyal gave a paper on stochastic processes and statistical physics, while Maurice Bartlett gave one on evolutionary stochastic processes. I had intended writing a contribution on stochastic epidemics, but I was depressed by the lack of interest shown by medical colleagues and the excess of interest shown by some concerned with military applications, and so I suppressed that work for about five years, until I felt that there were realistic prospects of constructive applications, and spent a busy summer writing a replacement paper on stochastic population growth. Bartlett was really responsible for interesting me in such things; just before leaving for the United States in 1946 he had advised me to get hold of the now famous thesis by Niels Arley (who was to become a good friend) on the stochastic analysis of cosmic radiation. This approach Bartlett and I transferred to and extended in the context of human and other biological populations, thus following up an initiative taken by Feller in 1939. One topic which especially interested me was the application of stochastic techniques to describe the chance fluctuations in age-distributions. Quite recently I had the satisfaction of seeing Russell Gerrard develop that topic yet further in the context of a population of manuscripts subject to random processes of copying and loss.

It would be a sufficient justification for this approach to modelling if it led naturally on to the construction of more nearly faithful models, or *portraits* as we might call them. But in fact it is a matter of record (see Whittle's paper [16], and also [5]) that theorems about the behaviour of such a caricature can lead to very valuable insights even at the portrait level. Similar remarks might be made about the early work on queueing, and on epidemic spread, and certainly when writing [3], [4], [5], [6] I felt myself to be working within the same framework and with the same intentions. The huge literature which has since evolved around birth-and-death processes, queueing processes, and epidemic processes suggests that those primitive caricatures had an in-built plasticity and capacity for generalization which was not at all envisaged at the time. So the art of caricature has been known to be fruitful.

2. Queueing Theory

My work on queueing was suggested by the congestion problems associated with the famous Berlin airlift. It involved an exhausting survey of the earlier work by writers like Erlang, much of it in the Danish language, and mostly published in technological journals rather difficult of access. I was extremely lucky in the fortunate coincidence that the first volume of Will Feller's book appeared at just that moment. This enabled me to grab the first chance of attacking realistic queueing problems with the Markovian techniques that we all now take for granted. Further Scandinavian associations followed this work; I came to know (and later to collaborate with) Arne Jensen, and also Conny Palm (though we two were never to meet). Palm essentially gave me the idea of "the imbedded Markov chain". Feller and Sam Wilks invited me to spend an academic year in Princeton at this time, and Jerzy Neyman extended this by a summer invitation to Berkeley. As Diana and I had met and married a few weeks before, this was an opportunity of which we were delighted to make the most.

Meetings with Marc Kac, with Kai Lai Chung, and with Joe Doob, together with Feller's influence, now pushed me firmly into theoretical probability theory, and for some years after this Harry Reuter and I thought of little else but what Chung has called "those damned little p_{ij}'s". This led on to further adventures in stochastic analysis in the company of David Williams, John Kingman, and Rollo Davidson, and in stochastic geometry in the company of Davidson, Ruben Ambartzumian, and Klaus Krickeberg.

3. Archaeology and Bird Navigation

But this total absorption in mathematical probability was not to last. Towards the end of my 16 years in Oxford a Magdalen colleague, R. W. Hamilton, introduced me to the atrocious problem of archaeological seriation. I racked my brains over this for several years, happy in the knowledge that Flinders Petrie's 800 × 900 matrix had been destroyed, so was not available for reanalysis. But on moving to Cambridge in 1962 I was introduced by Morris Walker to an archaeologist, Roy Hodson, who had been a gunner and so knew some mathematics, while at about the same time service on a Medical Research Council committee brought me into touch with Donald Michie who first told me about "nonmetric" (i.e. ordinal) multidimensional scaling. Now Hodson had a first-class data-set, and a possible seriation technique suggested itself almost immediately, so I was sucked into the maelstrom of data analysis at the outset of my Cambridge career. This was to change my ideas about models and their uses.

At about the same time an old friendship with David Lack (who had been an Oxford colleague) and a new one with Geof Watson (whom I met monastery-hunting in Yugoslavia) made me interested in bird navigation, and prompted me to write [8], which could be taken to illustrate an entirely different sort of activity, though this too has been regarded as "model-building". The factual puzzle to be explained was the extraordinary faculty many birds have for navigating back from an unfamiliar point of release to their "home". Some writings at the time were concerned with the actual mechanism that might be used in the navigation process, but I felt unhappy with these detailed explanations in the absence of confirmatory physiological evidence. It seemed better to attempt to answer a much more loosely formulated question: let the bird be supposed to possess an unspecified navigational aid which directs it continually, *but with massive random angular errors*, in the direction towards home. Would this, when combined with observational values for all the other parameters, produce the very remarkable successes scored by such birds as the Manx Shearwater? If the answer had been negative, that would have required a complete change in the direction of research into such matters. In fact the answer proved to be affirmative, thus encouraging further work into the question of the kinds of very poor navigational aids that might be employed, whether in the way in which we ourselves would use them, or in coded form as "rules of claw".

The new feature in this second sort of "model-building" is the use of observationally determined values for the residual parameters such as flight speed, orientation performance, and so on. But the "model" is still a caricature, and not a portrait.

The writing of that paper involved learning many new techniques. The diagnostic study of empirical distributions seemed to me insufficiently structured, and I had recently been involved with Liliana Boneva and Ivan Stefanov in an attempt to amend this situation by introducing the use of spline techniques into density estimation. This proved very useful in the bird navigation work, but a serious gap on the mathematical side was revealed by David Williams, who pointed out that much of my very clumsy asymptotics could be handled painlessly by means of the Itô calculus. Obviously I had to learn this, and did; it has been immensely useful to me ever since. This led eventually to the conviction that the Itô calculus, once mastered, should be used in an explicit and globally conceived geometrical setting, and so I took a sabbatical year and attended all my colleagues' lectures on geometry and topology. This propelled me into an attack on the statistics of "shape", from which I have not yet emerged, and also has subtly qualified my latest adventures in data analysis.

4. Random Lateral Perturbations of the Data

In recent years consulting work on my part [9]–[15] has been in the rather different fields of archaeology and astronomy—sometimes called the "historical" sciences because of the great difficulty of testing hypotheses in the laboratory. Just occasionally models of the caricature variety suggest themselves as worthy of study, but in archaeology that is only very rarely the case, while in astronomy one does not often meet it as a statistician because if the problem could be so formulated then the astronomers with their magnificent mathematical heritage would have solved it already.

So, as one has to use some model or other, it will normally be one which it is only too easy to criticize, and therefore one tries to make use of it only in the most tentative way: (i) in suggesting what may be useful as at any rate a moderately powerful test (and of course one checks the power by studying faked situations with a genuine effect present); and (ii) in suggesting what may be the relevant ancillaries (of which more below). Typically in carrying out (i) and (ii) one will be fairly relaxed about calling in asymptotic procedures and associated limit theorems, even though one knows that one will never find oneself to be "at the limit", and this will not be objectionable because such appeals are made only at the stages (i) and (ii) of the investigation, where one is choosing a kind of procedure, rather than actually executing it. The object is to keep model and limit theorems locked away in another room when one is making the test itself. The test in fact is to be "data based", using simulations constructed by random corruptions of the data modulated by conditioning on the relevant ancillaries.

There is of course something of the Efron "bootstrap" philosophy about this, but as here no subsampling procedures are involved it has seemed preferable to employ a different terminology hinting at the way in which one would like to see these rather different techniques developed. Indeed the stimulus which produced these ideas was not the "bootstrap" work at all, but a justly famous throw-away remark by George Barnard [1], the great importance of which was pointed out to me by Brian Ripley, then a research student at Cambridge.

I do not know of any attempt to construct a general theory of "random corruption", though I think that one is much overdue. I like to imagine that something of this sort will emerge in a natural way out of current work marrying statistical inference to global differential geometry, and for that reason Wilfrid Kendall and I [15] refer to random corruption conditioned on the relevant ancillaries as "random *lateral* perturbation of the data". Here the force of the important word "lateral" is that there are many possible "directions" of perturbation, but we are to choose those which are tangent to the submanifold on which the relevant ancillaries are constant. The definition of a relevant ancillary is that a change in its level will alter the propensity of the data to yield a spurious effect of the sort we are interested in. The data-corruption must therefore be free of distortions in these dangerous directions, and "lateral" in our terminology means "in that subspace of the tangent-space that is orthogonal to them".

Of course one has also to decide on the extent of the allowable perturbations, and here there are two rules which must be observed; first, the perturbation should be strong enough to wreck if not eliminate the actual observed effect under test, and second, it must not be so strong that qualitative features of the data (such as the existence of a particular pattern of clusters, etc.) are changed beyond recognition. This second rule might perhaps be seen as maintaining the value of some "relevant ancillaries" not readily described in terms of the parameters we are working with.

The task of performing the random corruption a suitably large number of times to provide a data-based simulation-experience in terms of which (following Barnard [1]) a significance test is to be formulated can be extremely laborious, and so it will always be worth considering whether it can be adapted to serve in other structurally similar circumstances where the ancillaries are to be preserved at some other levels. In at least one case [10], [14] this has proved possible, and here again limit theorems play an important role, but as always they are prevented from intruding into the testing explicitly.

5. Concluding Remarks

I hope that this essay will not be read as a recommendation that one be slapdash in statistics. That is not at all my intention. My text is rather that models and limit theorems both have a place, *and that they should be kept in it*. As will perhaps be sufficiently evident from the papers I have quoted, the process of doing this will often involve all the most exotic resources of modern probability theory. To illustrate that, it is enough to note that the transition from [15] to [13] (two successive papers in the "lateral perturbation" tradition) went via the highly mathematical article [11]. It will often be so.

On reading through all this again, I see that perhaps there is another moral to be drawn. Consulting can be hard work, but it is a two-way process. It is fine to be in a position to help one's "client" over a temporary obstacle, but do we sufficiently appreciate all that he does for us? The effects may lie unperceived for several years, but at length they can change the whole direction of one's work.

References

[1] Barnard, G. A. (1963) In discussion of paper by M. S. Bartlett. *J. R. Statist. Soc.* B 25, 294.

[2] Kendall, D. G. (1949) Stochastic processes and population growth. *J. R. Statist. Soc.* B 11, 230–264.

[3] Kendall, D. G. (1951) Some problems in the theory of queues. *J. R. Statist. Soc.* B 13, 151–173.

[4] Kendall, D. G. (1953) Stochastic processes occurring in the theory of queues etc. *Ann. Math. Statist.* 24, 338–354.

[5] Kendall, D. G. (1956) Deterministic and stochastic epidemics in closed populations. *Proc. 3rd Berkeley Symp. Math. Statist. Prob.* 4, 149–165.

[6] Kendall, D. G. (1964) Some recent work and further problems in the theory of queues. *Teor. Veroyatnost. i Primenen.* 9, 3–15.

[7] Kendall, D. G. (1965) Mathematical models of the spread of infection. In *Mathematics and Computer Science in Biology and Medicine*, HMSO, London, 213–225.

[8] Kendall, D. G. (1974) Pole-seeking brownian motion and bird navigation. *J. R. Statist. Soc.* B 36, 365–417.

[9] Kendall, D. G. (1975) The recovery of structure from fragmentary information. *Phil. Trans. R. Soc. London* A 279, 547–582.

[10] Kendall, D. G. (1977) Hunting quanta. In *Proc. Symp. to Honour Jerzy Neyman*, ed. R. Bartoczynski et al., Polish Scientific Publishers, Warsaw, 111–159.

[11] Kendall, D. G. (1984) Shape manifolds, procrustean metrics, and complex projective spaces. *Bull. London Math. Soc.* 16, 81–121.

[12] Kendall, D. G. (1984) Statistics, geometry, and the cosmos (The Milne Lecture). *Quart. J. R. Astronom. Soc.* 25, 147–156.

[13] Kendall, D. G. (1985) Exact distributions for shapes of random triangles in convex sets. *Adv. Appl. Prob.* 17, 308–329.

[14] Kendall, D. G. (1986) Quantum hunting. In *Encyclopedia of the Statistical Sciences*, ed. S. Kotz and N. L. Johnson, Wiley, New York.

[15] Kendall, D. G. and Kendall. W. S. (1980) Alignments in two-dimensional random sets of points. *Adv. Appl. Prob.* 12, 380–424.

[16] Whittle, P. (1955) The outcome of a stochastic epidemic: a note on Bailey's paper. *Biometrika* 42, 116–122.

Herbert Solomon

Herbert Solomon was born in New York City on 13 March 1919. His parents had arrived in the USA at rather young ages around the turn of the century, and like many Jewish immigrants from Russia they came to find a better life. He profited from the New York City public education system, receiving a B.Sc. from City College in 1940 with mathematics as his major subject. In 1941, he completed a master's degree in mathematical statistics under Harold Hotelling and Abraham Wald at Columbia. The Second World War intervened at this point, delaying a Ph.D. in statistics until 1950 at Stanford.

Through Hotelling he secured an appointment with the Mathematical Research Group and subsequently, the Statistical Research Group at Columbia, both of which were engaged in military research during the Second World War. From 1948–52, he served in the newly established Office of Naval Research, where he was named the first head of a newly created statistics branch. Some 25 years later he was invited to serve as Chief Scientist for the two-year period 1978 and 1979 for the Office of Naval Research in London.

In 1952, he accepted an associate professorship at Teachers College, Columbia University and was promoted to professor in 1957. This position provided him with opportunities for research in statistics in the behavioral sciences, and an affiliation with the Department of Mathematical Statistics kept him in touch with theoretical and methodological issues in statistics and probability. After a sabbatical year at Stanford, 1958–9, he was invited to serve there as chairman of the Department of Statistics. He held this post for five years and continues as Professor of Statistics in the department. During his chairmanship the number of master's and doctoral students grew dramatically.

Professor Solomon has enjoyed a wide variety of research interests in statistical and probabilistic methodology and in their applications to engineering, the behavioral and social sciences, marketing, law, education, health, and military issues. He is a fellow of both the American Statistical Association and the Institute of Mathematical Statistics, which he served as president in 1964–5. In 1975, the American Statistical Association awarded him the Wilks Medal for his contributions to statistics, and in 1977, the City College of New York presented him with the Townsend Harris Medal for his contributions to knowledge. The Secretary of the Navy awarded Professor Solomon the Navy Department Distinguished Public Service Medal in 1978 for his research contributions and for his leadership in furthering basic research in the academic community for Navy Department programs. This is the highest civilian award offered by the Navy Department to an individual not employed by the department.

He married Lottie Lautman, a violinist, on 1 January 1947. Their daughter, Naomi, is a vice-president in database management for the Bank of America in San Francisco, and their two sons, Mark and Jed, are lawyers in the San Francisco area.

Looking at Life Quantitatively

1. Early Training

At City College in New York City, I took one course in mathematical statistics and three courses in what today we term noncalculus statistics. The former was given by the department of mathematics in which I was enrolled to fulfill the requirement for a major subject. Professor Selby Robinson was my instructor, as he was for generations of students. The noncalculus statistics courses were not listed in any specific department and, in fact, were labeled Unattached 15.1, 15.2, and 15.3. They were offered by Professor John Firestone, who also provided statistical instruction to countless students over the years. Among my contemporaries at City College, (1936–40) were a number of individuals such as Kenneth J. Arrow, Herman Chernoff and Milton Sobel who made their mark in statistics and allied subjects in later years.

At the time, statistics and probability courses, especially at the undergraduate level, were not widely available. Moreover, employment opportunities were exceedingly limited in general and particularly limited in these specialities. The United States, like the rest of the industrial world, was in the second half of the Great Depression, and the only realistic purposes of a college education seemed to be the pursuit of scholarship and possibly a teaching job. The National Youth Administration, a federal government agency, supplied funds for needy students, which at City College and later in graduate work at Columbia University translated into small stipends for me.

I was fortunate that City College provided some instruction and experience in statistics and probability. The pre-Second World War national defense period in the United States and the new social and economic government programs provided professional opportunities that had not existed before. The City College graduates of the late 1930s found that a four-hour train ride to Washington was the prelude to jobs in a wide variety of fields in a number of federal agencies.

At De Witt Clinton High School in the Bronx, mathematics also featured in my program and I had become a member of the school's mathematics team. Contests with other New York City high schools took place over the school year and in the competition we fared rather well. The public high schools and public colleges of New York City in the 1930s were populated by many first-generation American offspring of immigrants mainly from eastern, southern and central Europe, and large numbers of these students were eager and avid learners.

At this point, I should mention I was born on 13 March 1919 in the Harlem section of Manhattan. My father had arrived in New York City in 1905 at the age of 17 after spending some time in England and Scotland. His family had migrated to the British Isles, escaped would be a better word—from that province of Russia known at one time or another as Latvia, to avoid conscription of their four sons into the Russian Army. The oldest son had served during the Russo-Japanese War at the turn of the century, but the plight of Jewish families in Russia was deplorable and was not ameliorated by military service. My mother was also born in Russia, arriving in New York City as a one-year-old infant in 1898. This brief history may explain why my education through college took place in New York City. The munificence of the taxpayers of that metropolis provided an excellent public education for which I will be eternally grateful.

While at City College, I found myself favoring statistics and probability more than pure mathematics. For graduate work, I chose Columbia University because of the reputation of Harold Hotelling in mathematical statistics and because it was in New York City. The latter made it possible for me to live at home and become a "subway" student—a student who traveled long distances daily by underground railway to attend classes and use libraries. I had considered attending Iowa State at Ames, Iowa, but in those days, it seemed too far away and financially out of reach. In 1940, I enrolled in mathematics at Columbia for a master's degree in statistics and probability—no statistics department existed. I completed the degree in one year after an intensive program of 10 courses that included instruction by Harold Hotelling, Abraham Wald, and B. O. Koopman. My master's thesis was supervised by Koopman, whom I did not see after my graduation until we both attended a NATO Advanced Research Seminar on Search Theory held on the Algarve coast of Portugal in 1979. Upon completion of my master's degree in 1941, I had studied, but not as yet really worked in, statistics and probability.

2. The Second World War and Its Impact

The bombing of Pearl Harbor and the entry of the United States into the war changed my life dramatically. I found myself doing time series analyses on prices for the Army Quartermaster Corps in New Jersey, then radio instruction in the Army Air Force in South Dakota, and in 1943, through Hotelling's good offices, I joined the Mathematical Research Group (MRG) at Columbia University, a sister organization to the Statistical Research Group (SRG) to which I transferred in 1944. Both groups were hard at work on a number of problems faced by the military which were filtered through the Applied Mathematics Panel, a unit of the Office of Scientific Research and Development headquartered in Washington. In this agency was Dr Mina S. Rees, with whom I was to work later in the initial and developmental period of the Office of Naval Research.

I could not envision at that time that this effort was to be the beginning of my long affiliation with the research components of the military services, especially the navy. Nor could I imagine that I was serving as a contemporary or a junior colleague to a large number of mathematicians and statisticians who would play prominent roles in our profession over the next 40 years. The Columbia groups contained, in addition to Hotelling and Wald, such statistical scholars as Jack Wolfowitz, M. A. Girschick, W. Allen Wallis, Milton Friedman, George Stigler, Churchill Eisenhart, J. Kenneth Arnold, Millard Hastay, L. J. Savage, Edward Paulson, Albert H. Bowker, and under a related aegis, Frederick Mosteller. Wallis [90] has given an account of this period. Readers may also be interested in a history prepared by Mina Rees [87].

My major work in the Columbia groups, at first, was not statistical. Eisenhart and I worked on the development of aiming rules to be taught to flexible gunners in bombers for use against oncoming fighter planes. Since the fighter's path was such that a tangent to it at any point in time intersected the bomber, the classical pursuit curve model was adopted. This problem was essentially deterministic, but probability could enter it through perturbations in the bomber's path or ballistic error in the projectile's path or aiming error in the gunner's sighting. At any rate, the major issue here was to determine the aiming points for gunners in the bomber as a function of direction of oncoming fighter (angle off), speeds of bomber and fighter, and muzzle velocity of bullets. These analyses suggested that at times the gunner should aim to the rear of the fighter (due to the forward velocity of the bomber), which led initially to training problems but finally resulted in the abandonment of the Army Air Force doctrine that the gunner should always fire at the front of the fighter, which in many cases would lead to zero hits if the doctrine were executed perfectly. The British had already recognized this as early as 1942. My memos on this topic are found in the Summary Technical Report, Applied Mathematics Panel, NDRC, 1946 which is deposited in the National Archives and Record Service, Center for Polar and Scientific Archives in Washington, DC.

At this time, a number of colleagues in the Statistical Research Group were busy on the development of sequential analysis plans. The motivation for this was the development of acceptance sampling models for procurement to insure compliance with quality standards. Earlier, single-sampling and double-sampling techniques had been developed for attribute (defective–nondefective) inspection. The Statistical Research Group began to develop multiple-sampling techniques and sequential procedures for acceptance sampling. These acceptance sampling plans were codified (different sampling procedures with the same operating characteristic (OC) curves) for use by the military for procurement purposes. I recall that by the end of the war, prototype sequential analyzers had been developed in which sequential inspection schemes were embedded. The inspector would receive a green light for acceptance, a red light for rejection, and no light for continuation of sampling as he or she recorded an item as defective or nondefective into the analyzer's memory. The codification of the different sampling plans through the OC curve index followed from the work of Neyman and Pearson done some 15 years earlier. While I was not involved in any major way in acceptance sampling during the war period, I did engage in it and contribute to it for a few years after the war, as I shall explain.

During the war, statistical analyses of experimental data on height finders and range finders occupied my time. One problem was to test whether the largest of a set of observed variances was significantly larger than the others. Our efforts led to extensions of Cochran's test for homogeneity of variances and tables of percentage points over a range of sample sizes and number of variances. This appears in an article coauthored with Eisenhart [15] in one of the volumes that emerged from the work of the Statistical Research Group.

3. Postwar Days

The end of the Second World War provided my generation with a period for returning to studies or seeking peacetime employment, or both. I was a First World War baby whose teens coincided with the Great Depression and whose early twenties fell during the Second World War. For about five years, a period in which most present-day students complete their doctorates, I had concentrated on war preparation or war. I completed some work beyond my master's degree, including the Ph.D. oral comprehensives and language examinations at Columbia in early 1946, but family needs required that I earn some funds. In the summer of 1946, I worked in a newly formed Operations Analysis group in the Air Force and looked into the vulnerability of bombers to anti-aircraft fire under varying conditions based on data from the war. Then I taught college courses in mathematics, but I knew that for me, teaching calculus and differential equations would never replace working in statistics and probability. During this period, I courted Lottie Lautman, a very attractive and bright young music student at Columbia who was to become my wife.

In January 1947, two weeks after our marriage, we went west to Stanford University. There I joined an SRG colleague, Albert Bowker, in the development of a statistics program and to complete a Ph.D. Bowker had trained at Columbia and then at Chapel Hill and was also in the throes of completing his degree. The focus of our work was an Office of Naval Research (ONR) project on acceptance sampling that extended work done during the war. This project was proposed by W. Allen Wallis, who left Stanford for the University of Chicago before work began. Our effort gave initial impetus to the beginning of the Department of Statistics at Stanford. It came into being in 1948 with the addition of M. A. Girschick, who had been working at the Rand Corporation in Santa Monica. One can see here the genesis of Stanford's east–west ties.

I left Stanford before 1948 (to return again permanently in 1958) and became a member of an Air Force–Navy Intelligence group in Washington which then led to a post with the ONR. It was during my stint in the Pentagon that I encountered geometric probability problems such as coverage models. This led to a paper [34] on distribution of coverage of a fixed circle by random circles when there is a fixed offset between the center of the fixed circle and the mean of the points of impact of the center of the random circles. A similar problem while at ONR led to a paper coauthored with Arthur Grad [21] on the distribution of quadratic forms as a model to secure hit probabilities over circular areas or spherical volumes when both location and aiming points were subject to random errors in two or three dimensions, respectively. This differed from the prior paper where aiming error alone provided the stochastic factor. Grad was a colleague in the Mathematics Program at ONR who had worked in complex variables, particularly Schlicht functions. We had also known each other at Stanford in 1947. I like to believe that I influenced his thinking on statistics and probability, for when he assumed the leadership of the Mathematics Program at the National Science Foundation (NSF), financial support for research in statistics and probability flourished.

My efforts at ONR focused on the development of basic research programs in statistics and probability at universities. While the support of federal agencies for such development is quite common today, ONR was the first and set the pattern for the other military agencies as well as the National Science Foundation (NSF). It may have been an accident of history that cast ONR rather than NSF as the pioneer in university funding, but the ONR program did so well that there was no major effort to phase it out when NSF began. A number of my ONR colleagues did transfer to NSF at its inception in 1951.

Under my aegis, two rather large projects were initiated about 1950–1. One was to continue and extend the work in acceptance sampling and quality control; the second was to revive in microcosm the Applied Mathematics Panel groups of the Second World War to help in the Korean conflict. Since a nucleus of the SRG (Bowker, Girshick) was already at work on acceptance sampling for variables inspection at Stanford, their project was expanded to include work not yet initiated on continous sampling (rectifying inspection), surveillance sampling, and other topics that could arise. To accomplish the second task, several universities—Chicago, North Carolina State and Princeton in addition to Stanford—were engaged. Because of the size and importance of these programs, the three services contributed funds and coordination and the Joint Services Advisory Committee (JSAC) in Applied Mathematics and Statistics came into being with me as its first chairman. At about this time, a Statistics and Probability Branch was established at ONR and I was appointed its head. It continues today some 35 years later. The JSAC continued for many years under joint service sponsorship, but this structure ended

4. Columbia University Days

In 1952 I left ONR to join the Columbia University faculty, returning some 25 years later to serve as Chief Scientist for ONR London for a two-year period. In that post, I guided a group of scientists who reported on work in their various disciplines in Europe (west and east) and the Middle East. This gave me a splendid opportunity to see how statistics and probability were organized, taught, and studied in many countries.

At Columbia, my primary appointment was in Teachers College (essentially Columbia's school of education) but I always enjoyed close relations with the Department of Mathematical Statistics; my activities included guiding their participation in funded project activity. One project in particular was initiated by the desire of the School of Aviation Medicine in Texas to look into models of achievement design or mental tests. Briefly, we can imagine a pool of N items (questions or statements) from which n are to be selected to yield a test. How large should n be? What should the characteristics of the items be, e.g., item difficulty, item intercorrelation, higher-order item intercorrelations in order to achieve a "good" test? Those questions led to a wide variety of investigations in which several models were explored. At one extreme, the multivariate normal served as the structure for responses to the n items, and at the other, few or no assumptions were made. On the other hand, those favoring the former began to loosen their assumptions while those favoring the other began to add assumptions; both to achieve some reasonable resolution of the problems. I coordinated the work of this group, some of which appears in a volume I contributed to and edited [43]. Some of this work I presented at the 1955 Berkeley Symposium [36]. I was fortunate to participate in all subsequent Berekeley Symposia until the last in 1970. This will become apparent in subsequent discussion.

Some of the chapters in the 1961 volume examine the problem where the multivariate normal is not assumed and responses to items are dichotomous. This problem motivated a resurgence of interest in dichotomous algebra. The sociologist Paul F. Lazarsfeld, through his latent structure models developed to measure attitudes, also employed dichotomous algebra and had intuitively developed a representation of the joint density of n responses as a series expansion of terms whose coefficients were the higher order interactions. This was formalized by Bahadur [77] in one of the essays in the volume. In another article in this volume, I applied this representation to some data on high-school students' attitudes toward science, and their IQ scores [43]. This data has been employed recently by investigators in multidimensional contingency table analysis. Electrical engineers whose modeling often included 0–1 processes also required the development of dichotomous algebra, and the

Bahadur–Lazarsfeld series representation is now employed and appears from time to time in their literature. However log-linear models for 0–1 responses, which are quite similar, have displaced this application.

While at Columbia, I also maintained a close relationship with the Bureau of Applied Social Research—an interdisciplinary research group initiated and developed by Lazarsfeld. This unit served as the vehicle for expression of another interest of mine, namely, mathematics in the behavioral sciences. In 1952, I secured funds from the Group Psychology Branch of the ONR to do research in the area through the Bureau and with scholars like Lazarsfeld and Theodore W. Anderson from the Department of Statistics. Other Columbia faculty involved were Ernest Nagel (Philosophy) William Vickrey (Economics) and Howard Raiffa (Statistics). Duncan Luce was employed as a postdoctoral research associate and James Coleman as a predoctoral research assistant. The volume by Luce and Raiffa [84] was the result of this enterprise. Two other volumes appeared; Solomon [38] and Luce [83]. I wrote the chapter on factor analysis models in the former. Coleman, who was a pioneer in mathematical sociology and who is famous for his work on achievement in high schools associated with socio-economic factors, wrote on small-group models, also in the volume I edited.

I also served as a statistical consultant to the faculty of Teachers College and to the large number of Ph.D. students whose theses were based on design and analyses of experiments in learning and behavior. Much of the consulting derived from research in psychology and educational measurement. One question raised by my colleague, Irving Lorge, about 1953, was how the efficacy of a group depended on its size, and more specifically how did group performance compare with individual performance. At that time we were working on a project funded by the Air University at Maxwell Field, Alabama. A specific question was how groups of size 3 and larger compared with an individual in resolving a tactical problem of how to get across a mined road. Interestingly, the problem was posed in different ways; for example, a photograph was shown or the officers were taken to the site that had been photographed. In this problem there were only two possible outcomes: success or failure.

There had been a period in social psychology when the theory was put forward that individuals working together could produce, for each individual, a performance that would exceed the performance of the individual working alone. I formulated a simple probabilistic model, namely: the probability of success by a group of n individuals, P_G, was the probability that it contained at least one individual who could solve the problem; or

$$P_G = 1 - (1 - P_I)^n,$$

where P_I was the probability that an individual working alone could solve the problem. The experimental data available at the time for $n = 3, 4, 5$ did not cause rejection of this model or a similar model that allowed for pooling of partial solutions to achieve closure. If a problem required success in two or more independent stages for total solution, the simple model alone could be applied to each stage. Both of these models were based on no personal interaction in the group. Even if they are violated, they serve as baselines through which the effect of personal interaction could be examined. A paper on this by Lorge and Solomon [26] was the first in a series we prepared; see Lorge and Solomon [27], [28]. Over the past 30 years, a large number of articles and a monograph by Davis [79] on group performance have appeared, many of which refer to the original work. I could not have realized in the early 1950s that some 25 years later I would once again investigate group performance in terms of group size and interpersonal behavior when I analyzed jury decision-making—more on this later.

I returned to Stanford for the summer of 1954. At that time, a number of topics on acceptance sampling and quality control that I had helped assemble several years before were receiving attention. One of these was "continuous sampling", a rectification inspection procedure in which sampling is employed to weed out a defective product and in this way guarantee an average outgoing quality limit (AOQL). Harold Dodge at Bell Labs had developed several plans in which essentially a sampling level was employed, say 1 in k items was inspected. On the basis of whether it was nondefective or defective, sampling continued at this level or 100% sampling over i consecutive, nondefective items took place after which the sampling level was resumed. Thus two parameters, k and i, together with AOQL could determine a procedure. Every inspected defective product was removed from the assembly line and replaced with a good item.

It seemed that an inspection scheme that tolerated less sampling if the product was good or more sampling if the product was bad might be more reasonable. This led to the multilevel continuous sampling models developed with Lieberman [25] and later with Derman and Littauer [12]. These models developed the AOQL for levels in which the sampling rate decreased geometrically for a good product and increased geometrically if the product was poor. An elaboration of these multilevel plans appears in a Department of Defense Handbook, H-107, that is available to manufacturers. Harold Dodge informed me in the late 1950s or early 1960s that Western Electric (the manufacturing affiliate of AT&T) was using some multilevel plans as well as his plans.

Another topic that came under the acceptance sampling and quality control setting was surveillance sampling. The motivation here was the development of an inspection scheme to be applied to an inventory of stored items on the basis of which items would either be replaced or kept in storage. Naturally the life distribution of the item in storage would play a role in these schemes. The goal here is to maintain a specified quality of product in an inventory in the presence of item deterioration. Repeated application of sampling inspections together with replacement policy are employed to maintain a desired quality level. In a paper jointly authored with Derman [11], we analyzed sampling plans commonly used in acceptance sampling and a replacement policy where lots judged defective by the sampling procedure are replaced. Markov processes are used to evaluate the effectiveness of sampling plans and replacement policy.

Briefly, we considered the following type of problem. Suppose we have k different lots, each of size N, which are stored. At regular intervals these lots are inspected. If a lot passes inspection, it is kept until its next inspection; if it does not pass inspection, it is replaced by a new lot of acceptable quality. The only way that a lot can leave the system and be replaced is for it to be rejected on inspection. This situation obtains if the lots stored are to be used only in cases of emergency, as in the stockpiling of vital material. The quality of the lots can be expected to deteriorate with age. The aim of the surveillance plan is to maintain the quality of the k lots (Nk items) so that, for example, if an emergency arises, there will be on hand sufficient stock of good quality to respond. We showed how to measure the effectiveness of the surveillance plan which accomplishes this purpose. Once we have an index of effectiveness of a plan, a catalogue of plans can be constructed. A plan can then be available for selection which will meet some desired cost or other relevant criteria. The question of a catalogue of plans I explored in another paper [42]. The paper by Derman and Solomon [11] served as a springboard for a number of works by Derman and other coauthors.

5. Stanford Days

For the 1960 Berkeley Symposium, I returned to the distribution of quadratic forms in normal variables. I wanted to extend the Grad–Solomon tables [21] for two and three dimensions to higher dimensions. Harold Ruben had proposed a recursive scheme that I used to obtain table values for three dimensions from values for two dimensions. The values obtained were checked against known values for three dimensions. This led to my Berkeley Symposium paper [42] which unfortunately the editor chose to publish without the tables. The newly developed two and three-dimensional tables were much more extensive and appeared only in a Stanford University technical report (1960) for which over the years I have received a number of requests. The latest occurred in 1984 when Mary Ellen Bock suggested that the tables were invaluable for a report she prepared on percentiles of more general quadratic forms [78].

In the Berkeley paper I highlighted a rather interesting phenomenon gleaned from the tables for percentiles for the quadratic form in two variables. Let $F_2(t) = P\{a_1\chi^2 + a_2\chi^2 \leq t\}$ where $a_i > 0$, $a_1 + a_2 = 1$, and χ^2 is chi-squared with one degree of freedom. Examination of the Grad–Solomon table of F_2 shows that, as (a_1, a_2) take values from $(\frac{1}{2}, \frac{1}{2})$ to $(1, 0)$, F_2 is strictly increasing for fixed t when $0 < t \leq 1$, strictly decreasing for fixed t when $t \geq 2$, and has a maximum for fixed t when $1 < t < 2$. It would be good to have more extensive tables of F_2 to get a better picture of this situation. This phenomenon has an interesting application. We can view (a_1, a_2) as representing the variation along each coordinate axis. Thus if we fix the total variation in the system, we can for fixed t manipulate the probability by suitable allocation of the total variation along each axis. For small t we get maximum probability by splitting the total variation in half, for large t by allocating the total variation along one axis. But for $1 < t < 2$ there is a different pair of values (a_1, a_2) between $(\frac{1}{2}, \frac{1}{2})$ and $(1, 0)$ for fixed t which produces the maximum probability. This has important implications.

We can show analytically that this phenomenon is true and that the range for t is a consequence of the fact that the sum of a_1 and a_2 is 1. If the sum had been standardized to another constant, the phenomenon would occur but for a different range of t. The analytical relationship between a_1, a_2, and t that produces the maximum of $F_2(t)$ when $1 < t < 2$ is given by

$$a_1 - \frac{1}{2} = -\frac{1}{2} \frac{I_1\left(\frac{\lambda}{2}\right)}{I_0\left(\frac{\lambda}{2}\right)},$$

where

$$\lambda = \frac{(1 - 2a_1)t}{2a_1(1 - a_1)}$$

and I_1 and I_0 are modified Bessel functions of order 1 and 0.

In the late 1950s and early 1960s, I published some follow-up papers to my previous efforts on group versus individual behavior [27], [28], [45], test design [16], [37], [38], and an isolated paper on survey sampling techniques applied to disarmament inspection [37]. The five-year period (1959–64) of my chairmanship of the Department of Statistics at Stanford led to a diminution of published research but a heightening of diverse statistical projects in the department for which I served as a principal agent in securing funds, reaching the sum of $500,000 per year in 1964.

In the second half of the 1960's, my research expanded to include statistics in law, geometrical probability, and multivariate analysis. These topics dominate my published research, but along the way there have been papers on other subjects. I will briefly discuss some but not all of these other subjects and then return to the three dominant topics. A procedure for selecting the largest multiple correlation coefficients appears in papers with Rizvi [31] and with Alam and Rizvi [1]. Estimating the parameters of zero-one processes appears in papers with B. Brown and M. A. Stephens [8], [72]. There was a large literature in model formulation of 0–1 processes, but these papers were the first to consider the problem of estimation.

I have worked with M. A. Stephens on distribution approximation, and our efforts appear in several papers [63], [64], [65]. In this work we resurrect and expand the efforts of Karl Pearson on the problem of approximating percentiles when some of the moments of a distribution are given. The Pearson approach requires the first four moments, but we have also looked at "chi-squared type" approximations where only the first three moments are required.

With Mark Brown, I looked at some probabilistic models in inventory analysis [4], Poisson processes [5], and renewal reward processes [6], and how to combine pseudo-random number generators [7]. In 1976, S. Zacks and I published a paper [75] on clinical trials. The thrust was the determination of confidence intervals for the interaction between treatments and environmental or geographic locations. This article considered only the case of two locations. Much more recently this was extended to the case of three or more locations by Heilbron [82] employing the approach we developed for two locations.

The question of quantizing the univariate normal has come up repeatedly over the past 50 years. Motivation has come from different fields of application, including psychometrics where an optimal partitioning of the normal distribution into $k + 1$ sections employing k points is required. The electrical engineers have provided other motivation. A number of papers have appeared in the journals of these two fields. More recently the question of quantizing bivariate and higher-order normals has been investigated. In a paper with S. Iyengar [23], we give some results for the bivariate case. Iyengar is a recent student of mine who worked in multivariate analysis; over 20 years ago, another student, P. Zador, also looked into this problem for his dissertation. Unfortunately, Zador's results [91] were not published until very recently, although they served as an unpublished reference for a number of investigators in electronics and information theory. I should mention that over the past 25 years, 17 students have received their Ph.D.'s under my sponsorship and I have served as a committee member for many others awarded the degree.

6. Multivariate Data Analysis

Previously, I mentioned that I would return to work in multivariate analysis, statistics in law, and geometrical probability. Over the past 20 years, about half of my 45 published papers have been in these fields. There is also a monograph in geometrical probability [6], and a book on applied multivariate analysis [30].

In the 1950s, my interest in multivariate data analysis was stimulated by a number of problems arising in psychology. While I was at Columbia in those years, I was concerned with psychological measurement which gives rise to multivariate problems. Problems in classification and clustering abounded in the many databases being collected. These subjects, of course, had been present when modern statistics began or at least thrived under the British anthropometricians, biologists, and statisticians at the beginning of the twentieth century. At that time neither computer technology nor enough discrete mathematics was available for the analysis of multivariate data. There had been some improvement by the 1950s, but not enough. Until then there were some papers on the subject employing the multivariate normal assumption. A. Wald and T. W. Anderson, among others, had looked into a classification statistic and, of course, R. A. Fisher had published his paper on linear discriminant analysis and immortalized the four-dimensional data on three iris flowers.

For the Third Berkeley Symposium in 1954–5, I presented and published a paper [36], in which I discussed classification in two categories: the assignment problem—an element is to be assigned to one of k groups where k is known, but information on the probability distribution of observables for the groups runs the entire gamut from complete ignorance of the functional form of the distribution to the situation where the functional form and all parameters are known. The second category is similar except that the value of k is unknown, and we can refer to this as the "clustering" problem.

It required the computer developments of the 1960s to make clustering feasible, and it is during this period that we see a literature beginning to develop on this subject. For the First International Symposium on Multivariate Analysis in 1965, J. J. Fortier and I [17] presented some results on an attempt to cluster 19 variables by looking at all possible configurations. It was immediately obvious that our task would be gargantuan and impossible, given the total number of possibilities. By invoking the fact that we were statisticians we decided to sample from all possible configurations of $k = 2, 3, 4, \ldots, 19$ clusters, choose the best two-cluster arrangement, three-cluster arrangement, etc. from our sample (using a clustering algorithm) and then make some probability statement about how good the best sample partition was. We employed a sample size of 10,000 for each value of k and did not feel too successful in our endeavor. Obviously the simple idea of looking at all possibilities, even through sampling, was not going to be viable.

In the late 1960s, with improving computer technology, I employed clustering techniques on a number of databases—clustering of 82 aphasic children into diagnostic categories on the basis of 25 tests; clustering 590 nouns in the Russian language based on frequency of usage in each of six grammatical cases; clustering of the R. A. Fisher 150 iris flowers for which four characteristics are measured on each; and clustering of 238 individuals convicted of first-degree murder in California on the bases of personal characteristics, characteristics of the murder, characteristics of defense and prosecuting attorneys and judges' characteristics. These results were presented at a meeting in Romania in 1970 and published in the proceedings [50]. Subsequent work in multivariate date analysis appears in [52], [54], [55], [60]. A colleague in psychiatry, Juan E. Mezzich and I published a book [30] that describes and evaluates clustering algorithms proposed by a number of investigators by trying them on each of four databases.

7. Geometrical Probability

In the late 1940s, I became interested in coverage problems that had initially been motivated by hit probability problems. Prompted by Second World War queries, H. E. Robbins and J. Neyman published papers on this topic in connection with hitting targets. Robbin's paper [89] serves as a benchmark for obtaining moments of the coverage random variable. The question of percentiles of this random variable had not been attacked directly. The moments, of course, could give some information but they could be obtained only with much analytical effort, and difficulty increased with the order of the moment. Today, one might be tempted to forego this approach and have the computer provide answers by simulation. In [34], I was able to obtain, for a special case, the distribution of the measure of a random two-dimensional set.

It was to be some time before I engaged in geometrical probability again, but in the early 1960s the random packing problem was brought to my attention. The one-dimensional version of this, called the parking problem, was investigated by Rényi [88]. Subsequently it received attention from Dvoretzky and Robbins [80]. Briefly, unit intervals are placed at random on a line segment until none can be added and the segment approaches infinity. The portion of the line covered is the random variable of interest. Rényi as well as Dvoretzky and Robbins obtained its mean; Mannion [85] found its variance. The problem in three dimensions had originated in a question asked by the crystallographer J. D. Bernal about the random packing of unit spheres in space as an analogue to the random placement of molecules of the inert gases. Bernal had posed this problem to Rényi who found it analytically intractable and retreated to the one-dimensional version. A colleague of Rényi, Ilona Palásti [86], conjectured that the mean packing ratio in n dimensions was the nth power of the mean in one dimension.

The Palásti conjecture has led to a number of papers and was instrumental in my looking into the problem. For the Fifth Berkeley Symposium in 1965, I presented an analysis of random packing density and Palásti's work. In a symposium paper [48] I also gave an analytical account of restricted packing in one dimension. In physical simulation in three dimensions, it was noted that ball bearings in a flask provided a packing that could be altered with shaking, after which more ball bearings could be added to produce a higher packing density. A one-dimensional analogue would permit a unit interval to come along at random and if it could not fit, it would be permitted to move up to a unit length in each direction to achieve a place. If it could then occupy a place, it would remain. I showed the mean packing density in that case was 0.809 as compared with 0.748 in the unrestricted packing case.

After this, I became interested in other geometrical probability problems and initiated an advanced graduate course in the subject at Stanford. In 1978, the Society for Industrial and Applied Mathematics published a monograph on the subject which I prepared [57]. This was based on a series of ten lectures given at the University of Nevada in Las Vegas. The monograph covered my work, results of earlier writers, e.g., Crofton, Buffon, Sylvester, and results obtained by my students.

One set of results I developed with Peter C. C. Wang was given at the Sixth Berkeley Symposium in 1970 and published in the proceedings [69]. In that paper we indicated how random lines in the plane and some idealized traffic models can be viewed to provide some results on traffic flow. There are a number of results for isotropic Poisson random lines in the plane, for the moments and the distribution of the number of sides of polygons formed by the lines, and for the perimeter and areas of the polygons formed by the lines. The trajectory of a car on a highway can be viewed as a random line in a plane. Edward I. George, a former student, developed a number of results on polygons formed by random lines in the plane. In his dissertation [81] moments and distributions of the number of sides and areas of these polygons are presented.

The intersections of random lines in the plane can represent time and space coordinates where an overtaking occurs. Let one of the vehicles be an observer car (arbitrary line). The number of intersections of the arbitrary line (observer car) by the other lines determines the number of overtakings of slower cars by the observer car plus the number of times it was overtaken by faster cars. The distribution of the number of intersections can be determined from anisotropic lines in the plane as a function of the Poisson parameter producing the lines in the plane and the distribution of car velocities. All the details of this geometric probability solution are given in [69].

8. Statistics in Law

In the early 1960s, I was invited to be an expert witness in some pinball cases before the courts in San Francisco. At issue was whether a pinball game was a game of skill or a game of chance. The California Code indicated that where a game was not specifically mentioned in the legislation (such as pinball) it would have to be a game where skill predominates if money could be paid to a player who was considered a winner. Several shopkeepers had been arrested for violating this law by paying policemen (in plain clothes) who at times won while playing the pinball device in their shop.

One obvious question here is to decide whether skill predominates, and for this an experiment was designed. A robot (graduate student) played the game not employing any skills and an expert player was told to play as best he could. Each played 1000 games so that the standard deviation of the proportion of wins would be small. The number of games was designed to be large but not to exhaust the players. The proportion of wins for the robot was 0.05 while for the skilled player, it was 0.10. There are two quantitative issues here, one statistical and the other not. Trials of 1000 games were based on statistical considerations; the effect of the difference between 0.10 and 0.05 is central to the case but not easily resolved.

This was the first of a large number of legal cases in which I served as consultant or expert witness over the past 25 years. In each case there were statistical issues and nonstatistical quantitative issues that in recent times have come under the formal rubric of risk analysis. The French probabilists of the late eighteenth and early nineteenth century gave considerable attention to the question of when a number is small enough to be, with propriety, treated as 0. In the pinball case above, the question is whether the 0.05 difference suggests that skill predominates.

The pinball cases and several others were discussed in an article published in the Neyman Festschrift volume [47]. Other litigation discussed there were "driving under the influence" cases (a criminal misdemeanor like the gambling violations), and some civil and administrative cases. As far as I know, this was one of the earlier papers in the field and I entitled it jurimetrics. I learned subsequently that a paper with this title had appeared in the Minnesota Law Review in 1949. Shortly after my paper appeared, a journal published by the American Bar Association changed its name to *Jurimetrics Journal* and reprinted my paper in its first issue. Between 1966 and 1980, I served in a number of other legal situations. Some information on this I presented at a talk at the 200th anniversary of the Academy of Sciences, Lisbon and the paper for this talk appears in the anniversary proceedings [58]. Another very recent article in which I feature the role of confidence intervals in legal settings appears in a volume on statistics and law, [61].

When I first appeared in some criminal misdemeanor cases, I became curious about jury size and how a jury of size 12 came about in Anglo-Saxon law. While I never fully resolved this issue to my satisfaction it led me to the work of Poisson and other French probabilists who gave attention to jury size and jury behavior. Prominent among them were Condorcet, Cournot, LaPlace, and Bienaymé. This led to several papers with Alan Gelfand who had been my student and had written an interesting thesis on mathematical seriation models in archeology. The first of these papers [18], reopened the Poisson jury model for possible use in present day jury behavior. Two subsequent papers [19], [20] began to incorporate American jury experience.

Beginning about 1970 the U.S. Supreme Court began to make life interesting for jury modelers in a series of decisions that permitted juries of size 6 or more and removed unanimity as a requisite for a jury decision. In Poisson's time, a jury decision could be reached by a majority of 7 out of 12. This suggested essentially a first ballot decision. Unanimity in the American system required deliberation after a first ballot. The Gelfand–Solomon papers introduced social decision schemes to take the jury from initial ballot to final verdict. This permitted an examination of jury size in terms of the probabilities of the two kinds of error—conviction of an innocent defendant and acquittal of a guilty defendant. Quite helpful here was the work of James Davis at Illinois, a social psychologist who has written extensively on deliberation processes. In a paper prepared for a conference celebrating the 200th anniversary of Poisson's (1781–1840) birth, I have recapitulated and discussed all our jury results [59].

9. Concluding Remarks

I have listed or discussed a wide variety of subjects in statistics and probability in which I have been interested. Essentially, each arose as a result of some specific problem in a field of application, although this may be clouded at times in my papers where revision has been performed to satisfy an editor. For me, listening to a problem discussed by a colleague in some field, formulating it, and then achieving some resolution has always provided the excitement and effort that led to results, published or otherwise. Much of the research undertaken was supported by federal agencies, in particular the Office of Naval Research. Other supporting agencies were the Office of Education, the Department of Transportation, the Law Enforcement Assistance Administration, and the Army Research Office. I am very appreciative of their support and the skilful administration of these research programs by statistical colleagues in these agencies, particularly those in the Office of Naval Research.

Another source of enjoyment and achievement has been the development of statistical research programs. I was fortunate enough to have opportunities to do this in government and in academic life. My four years with the Office of Naval Research shortly after the Second World War and the programs developed there have received some attention in an article in the *Encyclopedia of Statistical Sciences* [70] and need not be listed here. At Columbia University, I initiated and developed several programs but my five years as chairman of the Department of Statistics at Stanford (1959–64) saw the fuller development and burgeoning of many statistical enterprises and I am delighted that this has continued.

Publications

[1] Alam, K., Rizvi, M. H. and Solomon. H. (1976) Selection of largest multiple correlation coefficients. *Ann. Statist.* 4, 614–620.

[2] Blaisdell, B. E. and Solomon, H. (1970) On random sequential packing in the plane and a conjecture of Palásti. *J. Appl. Prob.* 7, 667–698.

[3] Blaisdell, B. E. and Solomon, H. (1982) Random sequential packing in Euclidean spaces of dimension three and four and a conjecture of Palásti. *J. Appl. Prob.* 19, 382–390.

[4] Brown, M. and Solomon, H. (1973) Optimal issuing policies under stochastic field lives. *J. Appl. Prob.* 10, 761–768.

[5] Brown, M. and Solomon, H. (1974) Some results for secondary processes generated by a Poisson process. *Stoch. Proc. Appl.* 2, 337–348.

[6] Brown, M. and Solomon, H. (1975) A second order approximation for the variance of a renewal reward process. *Stoch. Proc. Appl.* 3, 310–314.

[7] Brown, M. and Solomon, H. (1979) On combining pseudo-random number generators. *Ann. Statist.* 7, 691–695.

[8] Brown, M., Solomon, H. and Stephens, M. A. (1979) Estimation of parameters of zero-one processes by interval sampling: an adaptive strategy. *Operat. Res.* 27, 606–615.

[9] Brown, M., Solomon, H. and Stephens, M. A. (1981) Monte Carlo simulation of the renewal function. *J. Appl. Prob.* 18, 426–434.

[10] Criswell, J., Solomon, H. and Suppes, P. (eds) (1962) *Stanford Symposium on Mathematical Methods and Small Group Processes*. Stanford University Press, Stanford, CA.

[11] Derman, C. and Solomon, H. (1958) Development and evaluation of surveillance sampling plans. *Management Sci.* 5, 72–88.

[12] Derman, C., Littauer, S. B. and Solomon, H. (1957) Tightened multi-level continuous sampling plans. *Ann. Math. Statist.* 28, 395–404.

[13] Do, Kim-Anh and Solomon, H. (1986) A simulation study of Sylvester's problem in three dimensions. *J. Appl. Prob.* 23, 509–513.

[14] Dolby, J. and Solomon, H. (1975) Information density phenomena and random packing. *J. Appl. Prob.* 12, 364–370.

[15] Eisenhart, C. and Solomon, H. (1947) Significance of the largest of a set of sample estimates of variance. Chapter 15 of *Techniques of Statistical Analysis*, ed. M. Hastay, C. Eisenhand and W. A. Wallis. McGraw-Hill, New York.

[16] Elfving, G., Sitgreaves, R. and Solomon, H. (1959) Item selection procedures for item variables with known factor structure. *Psychometrika* 24, 189–205.

[17] Fortier, J. J. and Solomon, H. (1966) Clustering procedures. In *Multivariate Analysis*, ed. P. R. Krishnaiah, Academic Press, New York, 493–506.

[18] Gelfand, A. and Solomon, H. (1973) A study of Poisson's models for jury verdicts in criminal and civil trials. *J. Amer. Statist. Assoc.* 68, 271–278.

[19] Gelfand, A. and Solomon, H. (1974) Modeling jury verdicts in the American legal system. *J. Amer. Statist. Assoc.* 69, 32–37.

[20] Gelfand, A. and Solomon, H. (1975) Analyzing the decision-making process of the American jury. *J. Amer. Statist. Assoc.* 70, 305–310.

[21] Grad, A. and Solomon, H. (1955) Distribution of quadratic forms and some applications. *Ann. Math. Statist.* 26, 464–477.

[22] Itoh, Y. and Solomon, H. (1986) Random sequential coding by Hamming distance. *J. Appl. Prob.* 23 (to appear).

[23] Iyengar, S. and Solomon, H. (1983) Selecting representative points in normal populations. In *Recent Advances in Statistics: Papers in Honor of Herman Chernoff on his Sixtieth Birthday*, ed. M. H. Rizvi, J. Rustagi and D. Siegmund, Academic Press, New York, 579–591.

[24] Jensen, D. and Solomon, H. (1972) Gaussian approximation to distribution of a quadratic form. *J. Amer. Statist. Assoc.* 67, 898–902.

[25] Lieberman, G. and Solomon, H. (1955) Multi-level continuous sampling plans. *Ann. Math. Statist.* 26, 686–704.

[26] Lorge, I. and Solomon, H. (1955) Two models of group behaviour in the solution of eureka type problems. *Psychometrika* 20, 139–148.

[27] Lorge, I. and Solomon, H. (1959) Individual performance and group performance in problem solving related to group size and previous exposure to the problem. *J. Psychol.* 48, 107–114.

[28] Lorge, I. and Solomon, H. (1960) Group and individual performance in problem solving related to previous exposure to problem, level of aspiration, and group size. *Behavioral Sci.* 5, 28–38.

[29] MacNeill, I. and Solomon, H. (1967) Spelling ability: A comparison between computer output based on a phonemic-graphemic algorithm and actual student performance in elementary grades. *Res. Teaching of English* 1 (2).

[30] Mezzich, J. E. and Solomon, H. (1980) *Taxonomy and Behavioral Science: Comparative Performance of Grouping Methods*. Academic Press, New York.

[31] Rizvi, M. H. and Solomon, H. (1973) Selection of the largest multiple correlation coefficients. *J. Amer. Statist. Assoc.* 68, 184–188.
[32] Rosner, B. and Solomon, H. (1954) Factor analysis. *Rev. Educational Res.* 24, 421–438.
[33] Sitgreaves, R. and Solomon, H. (1957) Status studies and sample surveys. *Rev. Educational Res.* 27, 460–471.
[34] Solomon, H. (1953) Distribution of the measure of a two-dimensional random set. *Ann. Math. Statist.* 24, 650–656.
[35] Solomon, H. (1955) Trends in statistics and probability in psychology. Chapter 33 of *Present Day Psychology*. Philosophical Library Publishing Co., New York.
[36] Solomon, H. (1956) Statistics and probability in psychometric research: item analysis and classification techniques. *Proc. 3rd Berkeley Symp. Math. Statist. Prob.* 5, 169–184.
[37] Solomon, H. (1958) The use of sampling in disarmament inspection. In *Inspection for Disarmament*, Columbia University Press, New York, 225–230.
[38] Solomon, H. (ed) (1960) *Mathematical Thinking in the Measurement of Behavior*. Free Press, Chicago, IL.
[39] Solomon, H. (1960) Classification procedures based on dichotomous response vectors. Chapter 36 of *Contributions to Probability and Statistics*. Stanford University Press, Stanford, CA.
[40] Solomon, H. (1960) Measures of worth in item analysis and test design. Chapter 22 of *Mathematical Methods in the Social Sciences, 1959*. Stanford University Press, Stanford, CA.
[41] Solomon, H. (1960) Analytical survey of mathematical models in factor analysis. Chapter 3 of *Mathematical Thinking in the Measurement of Behavior*. Free Press, Chicago, IL.
[42] Solomon, H. (1961) On the distribution of quadratic forms. *Proc. 4th Berkeley Symp. Math. Statist. Prob.* 1, 645–653.
[43] Solomon, H. (ed) (1961) *Studies in Item Analysis and Prediction*. Stanford University Press, Stanford, CA.
[44] Solomon, H. (1962) Selection of surveillance sampling plans. *Bull. Inst. Internat. Statist., Proc. 33rd Session ISI, Paris* 39, 59–65.
[45] Solomon, H. (1962) Group and individual behavior in verbal recall. [10], pp. 221–231.
[46] Solomon, H. (1965) Effect of group size on group performance. In *Mathematical Explorations in Behavioral Science*, ed. F. Massarik and P. Ratoosh, J. D. Irwin and Co., Homewood, IL., 201–213.
[47] Solomon, H. (1966) Jurimetrics. In *Research Papers in Statistics: Festschrift for J. Neyman*, ed. F. N. David, Wiley, London, 319–350.
[48] Solomon, H. (1967) Random packing density. *Proc. 5th Berkeley Symp. Math. Statist. Prob.* 3, 119–134.
[49] Solomon, H. (1968) How quantitative is education? *Socio-Economic Planning Sciences* 2.
[50] Solomon, H. (1971) Numerical taxonomy. In *Mathematics in the Archaeological Historical Sciences*, Edinburgh University Press, Edinburgh, 62–81.
[51] Solomon, H. (1971) Statistics in legal settings in federal agencies. In *Federal Statistics: Report of the President's Commission* 2, 497–525.
[52] Solomon, H. (1975) Multivariate data analysis. *Proc. 20th Conf. Design of Experiments in Army Research Development and Testing*, ARO Report 75–2, 609–645.
[53] Solomon, H. (1975) A highway traffic model. In *Perspectives in Probability and Statistics: Papers in Honour of Professor M. S. Bartlett*, ed. J. Gani, Applied Probability Trust, Sheffield, 303–312.
[54] Solomon, H. (1976) Parole outcome: A multidimensional contingency table analysis. *J. Res. Crime And Deliquency* 13, 107–126.

[55] Solomon, H. (1977) Data dependent clustering techniques. In *Classification and Clustering*, ed. J. Van Ryzin, Academic Press, New York, 155–174.

[56] Solomon, H. (1977) Applied statistics. In *Science, Technology and the Modern Navy*, ed. E. Salkovits, ONR-37, 129–141.

[57] Solomon, H. (1978) *Geometrical Probability*. Regional Conference Series in Applied Mathematics 28, Siam, Philadelphia.

[58] Solomon, H. (1982) Measurement and burden of evidence. In *Recent Advances in Statistics*, ed. B. Epstein and J. Tiago de Oliveira, Academic Press, London, 1–22.

[59] Solomon, H. (1983) Jury size and jury verdicts. In *Commun. Statist. A Statist. Rev.* 12, 2179–2215.

[60] Solomon, H. (1984) Log-linear model applications. *Proc. Pacific Area Statistics Conf.*, ed. K. Matusita, North-Holland, Amsterdam.

[61] Solomon, H. (1985) Confidence intervals in legal settings. In *Statistics in Law*, ed. M. DeGroot, S. Feinberg and J. Kadane, Wiley, New York.

[62] Solomon, H. and Gelfand, A. (1977) An argument in favour of 12 member juries. *Jurimetrics J.* 17, 292–313.

[63] Solomon, H. and Stephens, M. A. (1977) Distribution of a sum of weighted chi-squared variables. *J. Amer. Statist. Assoc.* 72, 881–885.

[64] Solomon, H. and Stephens, M. A. (1978) Approximations to density functions using Pearson curves. *J. Amer. Statist. Assoc.* 73, 153–160.

[65] Solomon, H. and Stephens, M. A. (1980) Approximations to densities in geometric probability. *J. Appl. Prob.* 17, 145–153.

[66] Solomon, H. and Stephens, M. A. (1983) An approximation to the distribution of the sample variance. *Canad. J. Statist.* 11, 149–154.

[67] Solomon, H. and Stephens, M. A. (1983) On Neyman's statistic for testing uniformity. *Commun. Statist.* B 12, 127–134.

[68] Solomon, H. and Sutton, C. (1986) A simulation study of random caps on a sphere. *J. Appl. Prob.* 23 (to appear).

[69] Solomon, H. and Wang, P. C. C. (1972) Non-homogeneous Poisson fields of random lines with applications to traffic flow. *Proc. 6th Berkeley Symp. Math. Statist. Prob.* 3, 383–400.

[70] Solomon, H. and Wegman, E. J. (1985) Military statistics. In *Encyclopedia of Statistical Sciences* 5, Wiley, New York.

[71] Solomon, H. and Zacks, S. (1970) Optimal design of sampling from finite populations: A critical review and indication of new research areas. *J. Amer. Statist. Assoc.* 65, 653–677.

[72] Solomon, H., Brown, M. and Stephens, M. A. (1977) Estimation of parameters of zero-one processes by interval sampling. *Operat. Res.* 25, 493–505.

[73] Suranyi-Unger, T. and Solomon, H. (1984) Toward a behavior-specific measure of well-being. *J. Psychol. Marketing* 1, 59–67.

[74] Zacks, S. and Solomon, H. (1975) Lower confidence limits for the impact probability within a circle in the normal case. *Naval Res. Logist. Quart.* 22, 19–30.

[75] Zacks, S. and Solomon, H. (1976) On testing and estimating the interaction between treatments and environmental conditions in binomial experiments, I.: The case of two stations. *Commun. Statist.* A 5, 197–223.

[76] Zacks, S. and Solomon, H. (1981) Bayes and equivariant estimators of the variance of a finite population. *Commun. Statist.* A 10, 407–426.

References

[77] Bahadur, R. R. (1961) A representation of the joint distibution of responses to n dichotomous items. In *Studies in Item Analysis and Prediction*, ed. H. Solomon, Stanford University Press, Stanford, CA, 158–168.

[78] Bock, M. E. (1984) Distribution results for positive definite quadratic forms with repeated roots. Technical Report No. 347, Department of Statistics, Stanford University, Stanford, CA.

[79] Davis, J. H. (1969) *Group Performance*. Addison-Wesley, Reading, MA.

[80] Dvoretzky, A. and Robbins, H. (1964) On the 'parking' problem. *Publ. Math. Inst. Hungar. Acad. Sci.* 9, 209–224.

[81] George, E. I. (1981) Sequential Stochastic Construction of Random Polygons. Ph. D. Dissertation, Department of Statistics, Stanford University, Stanford, CA.

[82] Heilbron, D. C. (1981) The analysis of ratios of odds ratios in stratified contingency tables. *Biometrics* 37, 55–66.

[83] Luce, R. D. (1960) *Developments in Mathematical Psychology*. Free Press, Glencoe, Ill.

[84] Luce, R. D. and Raiffa, H. (1957) *Games and Decisions: Introduction and Critical Survey*. Wiley, New York.

[85] Mannion, D. (1964) Random space-filling in one dimension. *Publ. Math. Acad. Hungar. Sci.* 9, 143–154.

[86] Palásti, I. (1960) On some random space filling problems. *Publ. Math. Inst. Hungar. Acad. Sci.* 5, 353–359.

[87] Rees, M. (1980) The mathematical sciences and World War II. *Amer. Math. Monthly.* 87, 607–621.

[88] Rényi, A. (1958) On a one-dimensional problem concerning space-filling. *Publ. Math. Inst. Hungar. Acad. Sci.* 3, 109–127.

[89] Robbins, H. E. (1944) On the measure of a random set. *Ann. Math. Statist.* 15, 70–74.

[90] Wallis, W. A. (1980) The statistical research group. *J. Amer. Statist. Assoc.* 75, 320–330.

[91] Zador, P. L. (1982) Asymptotic quantization error of continuous signals and the quantization dimension. *IEEE Trans. Inf. Theory* 28, 139–148.

E. J. Hannan

Edward J. Hannan was born in Melbourne, Australia, on 29 January 1921. He was educated at Xavier College, Melbourne, leaving school towards the end of his fifteenth year to commence work as a bank clerk. After war service he studied economics and commerce at the University of Melbourne from 1946 to 1948. He subsequently worked in the Australian Central Bank until 1953 and then went to The Australian National University where he took his Ph.D. in 1956.

Professor Hannan has held visiting appointments in the USA (at the University of North Carolina, Brown University, The Johns Hopkins University, Yale University, MIT and Princeton University) and in Europe (at the Technical University of Vienna, the Politecnico di Milano and the University of Sheffield). At present he is head of the Department of Statistics in the Institute of Advanced Studies at The Australian National University, where he has been a faculty member since 1953 and a professor since 1959.

Professor Hannan's predominant research interest is in time series analysis, and his publications include three books and many papers on this subject. He is a member of the International Statistical Institute, and a fellow of the Econometric Society, the Australian Academy of Sciences and the Academy of Social Sciences in Australia. He is married; he and his wife Irene have two sons, Patrick and David, and two daughters, Christine and Jennifer. His interests include walking and reading, mostly history and literature.

Remembrance of Things Past

> When to the sessions of sweet silent thought
> I summon up remembrance of things past
> I sigh for lack of many a thing I sought, ...
>
> Shakespeare: *Sonnet XXX*

1. Early Years and Lost Opportunities

There are a great many ways by which a scholar may come to choose his field of study. The story-book account where a man sees some great human need and devotes his life to providing for it is probably rarely true, and could hardly be true for a statistician. Who would believe a man who claimed to have been seized by a passionate desire to relieve human suffering through the provision of good methods of summarizing data? Some pure mathematicians believe that statisticians are failed mathematicians, and there is a grain of truth in this. However this is basically a superficial judgement, for science calls for different combinations of abilities and men naturally move to those fields best suited to their nature. Not an uncommon progression would be from the more abstract to the more concrete, because it is when we are young that we are most idealistic and only later that we begin to perceive how incredibly difficult it is to achieve anything worthwhile in a very theoretical way. To be a good statistician certainly requires some mathematical ability, but it also requires other qualities.

In any case I did not come to statistics that way. Having left school in Melbourne at 15, I worked as a bank clerk until early 1941. I was employed in the Commonwealth Bank of Australia, which at that time was both a commercial bank and the central bank. I was sent to the savings bank part of the commercial bank. A bank clerk's was a dreary job in those days, especially in a savings bank. The standard of the new employees was high and all of the (rather few) young men who began work with me in 1937 were intelligent. The older men seemed dull and sometimes foolish. The evolutionary process leading to this state is not difficult to understand. I recall an examiner at whom all were casting pitying glances because he had failed to notice a small error (a few pence at most) which had caused trouble at the annual balance. Why an error of a few pence had to be accounted for is incomprehensible, but it was certain that the poor fellow's career was blighted. I recall, also, being told by the chief superintendent's secretary that if I put on a stamp (bearing the likeness of the king) upside down (I was postage boy), I would have committed the crime of *lèse majesté* and could be put in prison.

In 1941 I entered the Australian Army, and eventually saw active service as a lieutenant in northern New Guinea. Before the war I had learnt some economics in an accountancy course. When the war finished those with low priorities for early repatriation (for example, those who were not married) were gathered together at Wewak and I was asked to help to entertain the troops by teaching an economics course. Fortunately I cannot now recall the nature of the bilge I must have spilled over my students' heads. On Christmas Day 1945, the Australian prime minister and federal treasurer, Ben Chifley, suddenly visited us. We formed a large circle round him and he, a man with "the common touch", answered our questions. I was enjoying this when one of my students, a know-all who knew even less than I did, began to query

Chifley about economic policy. I feared that he would soon say "But Mr Hannan taught us that ..." and fled to hide in the officers' lines.

When I returned to Melbourne early in 1946 I decided to take a degree, being eligible for repatriation benefits. I was moved by no noble motives. Examining the university handbook, I first looked at medicine. A knowledge of Latin was, apparently, absolutely essential. Law was the same since, as Dudley Moore once said "you had to have the Latin for the judgin". The university fathers were wiser than they knew, for my lack of the Latin prerequisite saved the city of Melbourne from perhaps considerable damage to its medical and legal systems. The Faculty of Economics and Commerce had minimum prerequisites, so there I went.

The University of Melbourne was very crowded and there were great shortages. For example the library had only one copy of Keynes' *General Theory* [11] published in 1936, and the book could not be bought. Some of the teaching was very bad. It is not uncommon for professors of economics to become so involved in everyday affairs that they lose interest in their subject as it needs to be taught to students. I saw in the university handbook that I could study a course on the theory of statistics, if I first studied Pure Mathematics I, and that the statistics course included such esoteric, and hence exciting, subjects as skewness and kurtosis. I enjoyed Pure Mathematics I and so in my third year also studied Pure Mathematics II. These courses were rather old-fashioned, being concerned with basic real-variable analysis, but they were well taught. The theory of statistics course was quite good for its time.

Though I did well in my examinations I did not remain at the university for a fourth year, and an honours degree, as I wished to marry and needed a job. I returned in early 1949 to the bank, but to the economic adviser's department in the central bank in Sydney. My work was mainly statistical, but rarely theoretical. For example, I was concerned with reconstruction of an index of import prices. I continued to read mathematics in my spare time. I bought Bôcher's *Higher Algebra* and Goursat's *Cours d'Analyse*. I tried to learn some measure theory from Saks' *Theory of the Integral* [13] (in the departmental library) but found that difficult. I saw a review of Cramér's *Mathematical Methods of Statistics* [4], and got the bank to buy it. This great book was very important to me. I tried to work on more interesting topics than index numbers and spent a great deal of time on the seasonal adjustment of data and on reading the developing literature on modelling economic systems through constrained systems of linear equations.

I also tried applying this knowledge to Australian data. However, economics is a difficult subject in which to do elaborate statistical research, especially if you are in a policy-making department where you are faced with the realities both of the world and of the nature of the data. Economic systems are continually evolving as society evolves, and reactions are often highly nonlinear. This is not to say that there is not a great deal of useful work done through econometrics, but the early postwar hopes for the subject in relation to the big issues of macroeconomic policy and control seem to have faded. Of course economists are often harshly and foolishly judged by natural scientists. The system they are studying is incredibly complex, and their understanding is considerable. It is not the kind of study on which a new flood of light will be thrown by some Einstein come from physics.

The economist's position is made worse by the need to predict, which cannot be avoided. I recall the Australian Treasury trying to avoid saying they were predicting by using all kinds of euphemisms. I also recall seeing the Australian meteorologist doing his best to avoid the suggestion that he might use, for prediction, data on the southern oscillation (of equatorial atmospheric pressure between the east and west Pacific) which is certainly correlated with world weather. The relation is, however, very complicated and no doubt the meteorologist knew that if he made long-term predictions on this basis he would now and then be grossly wrong and would lead some farmers to take action (e.g. reducing sheep flocks) that would be costly. I myself tried my hand at predicting the consumer price index while working in the bank, for that was a determinant of wages through an automatic adjustment. The first time I got the figure badly wrong, through failing to take account of some relaxations of rent controls that had taken some time to have effect. However, later predictions were fairly accurate and in particular I successfully predicted the considerable reduction in inflation in Australia in the early 1950s. However, that was all rather easy, as I was predicting a figure that would be published about two months after the date of the actual measurements and I knew the nature of the measurements.

In retrospect I did not really take advantage of the opportunity offered to me at this time. I think an almost ideal way to begin research is first to acquire the basic theoretical knowledge and then to spend some time in an applied field learning the nature of the real problem. The bank was a good place for this. If I had spent more time studying statistics before going to the economic adviser's department, or perhaps also if I had arrived a little later on the scene, after computers had developed, I might have done more; but I feel that the real reason is that I am not the kind of man who does best in economics. I am more the mathematical, scientific type. Economics calls also for elements of judgement that, perhaps, I lack.

In 1949 The Australian National University (ANU) was founded in Canberra, as a pure research institution with special privileges. P. A. P. Moran was the first professor of statistics. The governor of the bank, H. C. Coombs, had played a part in founding the university and he suggested that one young man from the economic adviser's department be sent to Canberra to study for a year. I was chosen, and arrived in Canberra in May 1953. Pat Moran was impressed with the amount of mathematics I had taught myself, and went to see Coombs who agreed to give me two years' leave without pay. Pat had me appointed as a research fellow at about the same pay. I did not return to the bank, and Canberra has now been my permanent home for over 30 years.

The next few years in Canberra were perhaps the happiest in my life. I read a great deal of mathematics and recall enjoying such books as Riesz and Nagy's *Functional Analysis*, Weyl's *The Classical Groups*, Van Der Waerden's *Modern Algebra*, Loomis' *Abstract Harmonic Analysis* and Schwartz's *Théorie des Distributions*. I also recall reading Doob's *Stochastic Processes* [6], but failed to appreciate the importance of martingales. If I had recognized that, I could have been well ahead of my time, in time series. I was greatly influenced by Grenander and Rosenblatt's *Statistical Analysis of Stationary Time Series* [8], which really made me understand Fourier methods for data analysis and which I saw in a prepublication form. I seemed to spend most of my day reading. Of course I was still scientifically immature, since by the end of 1956, when I was nearly 36, I had spent only about the same time studying mathematics and theoretical statistics, as a man would normally have done by the end of his Ph.D. course. I do not regret this way of proceeding. What I do regret is having failed fully to appreciate the importance of computers.

In 1953 and 1954 I did an extensive analysis of some Australian rainfall data (about 36,000 observations) using the computer at the Radiophysics Division of the Commonwealth Scientific and Industrial Research Organization (CSIRO). The analysis would have been impossible without a high-speed computer. If I had been more perceptive I would have oriented my work more towards computer use. This is not to say that I did not recognize the importance of computers in time series analysis, but rather that I thought mainly about theoretical questions and not about algorithms in relation to computers. To be sufficiently cunning to sit back and consider what is the essential aspect of one's subject is important. I often regret having failed to make such connections. For example, a little later I became very involved in generalized Fourier methods, arising from group representations of the classical (semi-simple) Lie groups. I was familiar also with experimental design and recognized the general context which related analysis of variance with the spectral theory of time series. (Of course this had been recognized by others, such as Tukey, though not in an algebraic context.) I was familiar with Yates' technique for calculating all of the interactions in 2^p factorial experiments and with later elaborations such as the Bainbridge technique. I should have seen that this could be used to cheapen the calculation of Fourier coefficients, for I certainly appreciated the importance of those. For that matter I failed to appreciate the part that differential geometry might play in probability and statistics.

Isolation in Australia was also a disadvantage. ANU had no engineering school, and the first I heard of the Kalman filter was in 1964 when I was passing through Stanford. Even then I did not fully appreciate its importance, though contacts in the following year at The Johns Hopkins University (where Geof Watson had invited me) also made me aware of its use. It is not that I did not understand, but rather that the state-space approach did not fully invade my way of thinking, as it should have done. Thus in my book *Multiple Time Series* [9], published in 1970 but written in 1968–9, I devoted only six pages to establishing the properties of the discrete-time Kalman filter and establishing how a system might be represented in state-space form. Over 50 pages is given to the classical prediction theory, and the state-space approach is not used in the chapter on rational transfer function models. Of course at this time I was very harried and was running a large department of 14 academics, teaching econometrics as well as the statistics courses to noneconomists, and was also responsible for computer science. I was the only full professor in the department. A few years later there were three.

2. The Development of Scientific Ideas: An Example

In the first section I have emphasized my failures. I shall try in this section to indicate how adherence to a few good principles in research led me to solve one problem.

When I first went to Canberra in 1953 Pat Moran set me the task of studying data $y(t)$, $t = 1, 2, \ldots, T$ generated by a stationary system

$$y(t) = z(t) + \xi(t), \qquad z(t) = \rho z(t-1) + \eta(t), \qquad |\rho| < 1, \qquad (1)$$

and in particular of testing for $\rho = 0$. Here $(\xi(t), \eta(t))$ is a sequence of independent Gaussian random vectors with zero mean and constant covariance matrix. Thus I was presented with a state-space representation. Though such ideas were unknown in statistics in 1953, nevertheless this also may be counted as a failure in perception on my part, for I failed to see (1) as a special case of a very general situation, namely that of all rational transfer function systems. I got nowhere with the problem because I was obsessed with the calculation of exact distributions for estimators and I could not even see how to calculate the maximum likelihood estimator. Of course I was very inexperienced, and other people in time series then had the same preoccupations. I propose to discuss how, much later, I came to reconsider this problem in a more useful way.

Scalar rational function time series models became important by the late 1960s. I had for some years been concerned with the vector case because of my association with economics and the importance of multiple equation systems in that subject. Thus in one aspect we are concerned with a stationary Gaussian system

$$y(t) = \sum_0^\infty K(j)\varepsilon(t-j), \qquad E\{\varepsilon(s)\varepsilon(t)'\} = \delta_{st}\Sigma, \qquad K(0) = I_s, \qquad (2)$$

where $k(z) = \Sigma k(j)z^{-j}$ is a matrix of rational functions with $\det\{k(z)\} \neq 0$, $|z| > 1$. This is what is usually called a vector ARMA model. Here also the $\varepsilon(t)$ are the linear innovations, i.e. $\varepsilon(t) = y(t) - y(t|t-1)$ where $y(t|t-1)$ is the best predictor of $y(t)$ from the (infinite) past. If $s = 1$ then we may uniquely write $k(z) = b(z)/a(z)$, where $a(z)$ and $b(z)$ are relatively prime and each is monic (i.e. has unity as the coefficient of its highest power). Then, fixing the maximum degrees, n, of $a(z)$, $b(z)$, we may use their remaining coefficients,

$\alpha(1), \ldots, \alpha(n), \beta(1), \ldots, \beta(n)$, say, uniquely to coordinatize the space of all such rational functions of "order" n. The case where $n = 1$ is that presented in another way in (1), if $\rho \neq 0$. The $\varepsilon(t)$ are, of course, functions of the $\xi(t), \eta(t)$ in (1); indeed

$$\varepsilon(t) = \sum (-\beta)^j \{\eta(t-j) + \xi(t-j) + \alpha\xi(t-j-1)\},$$

where I have now written α, β in place of $\alpha(1), \beta(1)$. Also $\alpha = -\rho$.

In this case $s = 1$ there is, these days, no real problem in calculating the maximum likelihood estimator once n is prescribed, though there is a continuing stream of papers about how to do that. However, testing for $\rho = 0$ is not so simple. Of course you can use the first autocorrelation of the $y(t)$, whose distribution for $\rho = 0$ is substantially known. Obtaining this distribution is the kind of problem that formed the basis of the research I referred to (just below (1)) as the mainstream of the early 1950s. However that is not the likelihood ratio statistic, which will in fact be, as an asymptotic approximation, $\lambda_T = -T \log(\hat{\sigma}^2/s^2)$ where s^2 is the sample variance of the $y(t)$ and $\hat{\sigma}^2$ is the estimate of the variance of $\varepsilon(t)$. The parameter space for $s = n = 1$ may be taken to be $-1 < \alpha, \beta < 1, \alpha \neq \beta$. (Actually we may allow $|\beta| = 1$ but for brevity of statement below I neglect that.) This maps the "manifold" $M_1(1)$ of all $k(z)$, for $s = n = 1$, into $\mathbb{R}^2$. However, $k_0(z) \equiv 1$ is a limit point and is evidently not an interior point of $\overline{M_1(1)}$. Equivalently, in coordinates, $k_0(z) \equiv 1$ is the inverse image of the line $\alpha = \beta$.

In my efforts to understand the general case (2) I was aided by knowledge of some algebra and familiarity with a little differential geometry. I do not mean to exaggerate the knowledge needed, merely to point out that knowledge of the appropriate mathematics enables the nature of the situation to be seen. The algebra I had learnt from Macduffee's brief but encyclopaedic treatise *Theory of Matrices* which I had read years earlier, after seeing it referred to in a book by Schwerdtfeger on matrices. Schwerdtfeger had taught me Pure Mathematics II at Melbourne, but no mention of matrices was made in that course. I had recognized that in general we could write $k(z) = a(z)^{-1}b(z)$ where $a(z), b(z)$ were matrices of polynomials with $\det a(z), \det b(z) \neq 0$, $|z| \geq 1$. (Again we may allow $\det b(z)$ to have zeros on $|z| = 1$, but again I neglect that.) Also some uniqueness may be achieved by requiring that $a(z), b(z)$ be left coprime so that if $a(z) = u(z)a_1(z), b(z) = u(z)b_1(z)$, all of these being matrices of polynomials, then necessarily $\det u(z) = $ constant $\neq 0$. Such $u(z)$ are the units in the ring of matrices of polynomials since they are the matrices whose reciprocals are in the ring. Thus uniqueness would be achieved by choosing $u(z)$ so as to reduce, say, $a(z)$ to canonical form (e.g. lower-triangular form).

I was concerned with these ideas in the mid to late 1960s. Again isolation hindered me. At the same time an extensive analysis of the structure of such systems was being effected by systems engineers and mathematicians associated with them. Rosenbrock's book *State Space and Multivariable Theory* helped, but in 1974 light flooded in when I read a paper [3] by J. M. C. Clark of

Imperial College that had been presented at a conference of the International Federation of Automatic Control, held at Purdue University. As I knew, any system (2) may be represented in the form

$$y(t) = Hx(t) + \varepsilon(t), \qquad x(t+1) = Fx(t) + K\varepsilon(t) \qquad (3)$$

and we may assume F to be of minimal dimension, n. In the case (1), H will be unity, $F = \rho$ and $x(t) = y(t|t-1)$ while $K = (\beta - \alpha)$. In general $k(z) = H\{zI_n - F\}^{-1} + I_s$. Clark pointed out that the set $M_s(n)$ of all such $k(z)$ of order n is a smooth surface, a kind of algebraic variety, but that it cannot be globally coordinatized if $s > 1$. Thus the problem is to determine n and, given n, the point on the manifold, $M_s(n)$, which manifold is of dimension $2ns$. If n_0 is the true order, then in estimating it one will certainly have to examine $n > n_0$. For example, for $s = 1$ and $n_0 = 0$ one will have to examine the case $n = 1$ and this is evidently the problem of testing for $\rho = 0$ in (1). Akaike [1] had suggested formulae which, in the present situation, became

$$\text{AIC}(n) = T \log \det \hat{\Sigma}_n + 4sn, \qquad \text{BIC}(n) = T \log \det \hat{\Sigma}_n + 2sn \log T/T. \qquad (4)$$

Here $\hat{\Sigma}_n$ is the estimate of Σ from an nth-order model and $\hat{n}$ is obtained by maximizing $\text{AIC}(n)$ or $\text{BIC}(n)$. The general situation parallels that of (1). for $n_0 < n$ then $k_0(z)$ is an edge point of $M_s(n)$ and any coordinate mapping covering an open set in $M_s(n)$ for which $k_0(z)$ is a limit point will map $k_0(z)$ into an affine subspace of $\mathbb{R}^{2ns}$, of dimension $(n - n_0)s$, along which the likelihood is constant. These questions were discussed in a paper with M. Deistler [5].

Apart from the mathematical knowledge of which I spoke above, I was helped in my research by collaboration with students and especially with M. Deistler who visited Australia from Vienna over this period, and whom I later visited in Vienna. Good research students are an invaluable asset, but they usually leave about the time that they become most valuable as collaborators. For a research worker in an isolated country, such as Australia, the only hope is to travel and to have visitors. I believe that one of the mistakes I made early in my career was to work too much alone, and indeed to some extent this has continued.

As $T \to \infty$ if $k_0 \in M_s(n_0)$, $n_0 < n$, then $k \to k_0$ a.s. (see Dunsmuir and Hannan [7]). Thus in coordinates $\hat{k}$ converges to the affine subspace. We may institute new coordinates (θ, ϕ) where θ varies along the affine subspace and ϕ is orthogonal to that. Thus $\hat{\phi} \to 0$, and at the rate of the law of the iterated logarithm. In the case $s = n = 1$, $n_0 = 0$ and taking $|\beta| < 1 - \delta$, $\delta > 0$, then

$$\theta = \tfrac{1}{2}(\alpha - \beta), \qquad \phi = \tfrac{1}{2}(\alpha + \beta).$$

In this case also λ_T becomes

$$\lambda_T = \max_{-1+\delta < \theta < 1-\delta} \chi_T(\theta)^2,$$

$$\chi_T(\theta) = \frac{1}{(1-\theta^2)^{1/2}} \sum_0^\infty \theta^k T^{1/2} \left\{ \sum_1^T \varepsilon(t)\varepsilon(t-k-1) \Big/ \sum_1^T \varepsilon(t)^2 \right\}$$

(5)

as follows by expanding $\sigma^2(\theta, \phi)$ about $\phi = 0$ and evaluating at θ, $\hat{\phi}$. The details are elementary but tedious, and are omitted here. (The reader may find these details in [10].) The general case is essentially the same, but more complicated. In the case $s = n = 1$, $n_0 = 0$, then as $T \to \infty$, $\chi_T(\theta)$ converges weakly to a Gaussian process on $-1 + \delta \leq \theta \leq 1 - \delta$ and if we put $\tau = \log\{(1 + \theta)/(1 - \theta)\}$ then $\psi_T(\tau) = \psi_T(\theta)$ becomes, in the limit, a stationary Gaussian process on $|\tau| < \log\{(2 - \delta)/\delta\}$ with spectral density cosh $\pi\omega$. Thus the limiting distribution of λ_T is known, as that of the square of the maximum of a Gaussian process on an interval, the process having a very "regular" spectrum. It is also evident from the transformation from θ to τ and the nature of (5) that for δ very small the optimal θ will be near to ± 1, i.e. the maximizing $\hat{\alpha}$, $\hat{\beta}$ will be near to $(-1, -1)$ or $(1, 1)$. Thus an asymptotic solution has been given to the problem of finding the distribution of the likelihood ratio statistic for testing $\rho = 0$ in (1). Of course that answer could hardly be given in 1953 since, for example, the theory of weak convergence did not then exist.

To discuss $\hat{n}$ (see (4)) we need to be able to deal uniformly with the quantities

$$r_\varepsilon(k) = \sum_1^T \varepsilon(t)\varepsilon(t - k) / \sum_1^T \varepsilon(t)^2, \quad k \geq 1,$$

occurring in (5). Because $|\theta| < 1$ it is sufficient to consider these for $k \leq c \log T$, for c large enough. The discussion needs to be accurate to $o\{((\log T)/T)^{1/2}\}$ even for BIC(n). In fact we can show (An et al. [2]) that

$$\max_{1 \leq k \leq H_T} |r_\varepsilon(k)| = O\{((\log \log T)/T)^{1/2}\}, \text{ a.s.}, \quad H_T \leq (\log T)^a, a < \infty \quad (6)$$

so that for BIC(n) then $\hat{n}$ will be strongly consistent. As Akaike and others had shown, AIC(n) does not lead to consistent estimates, but that is not too important as in practice there will be no true n_0 and the virtue of consistency should not be exaggerated.

The example is designed to show:

— the considerable range of probabilistic and mathematical tools needed for the solution of a realistic statistical problem;
— the understanding derived from contact with another subject, in this case systems engineering;
— the virtues of collaboration;
— the way the advance of mathematics opens the way to understanding.

I should like to emphasize the second point. Perhaps I do not need to do so for, in defence of statistics as a subject, I must say that it keeps in very close contact with fields of application and draws a great deal from them.

I wish to conclude my discussion by saying two further things. The relation (6) extends to a wider context and indeed holds in the vector case, for $y(t)$ from (3). Indeed even when $k(z)$ in (3) is not rational but if, say $\Sigma j^{1/2}||K(j)|| < \infty$, then a similar result holds but with $\log \log T$ replaced by $\log T$ and $H_T =$

$o\{(T/\log T)^{1/2}\}$. The techniques of proof are essentially those of the proof of the law of the iterated logarithm for martingales. I was aided again by collaborators, this time in the person of two visitors from China, An Hong-Zhi and Chen Zhao-Guo. (See above (6).) These kinds of results are useful over a wide range. The old paradigm, that statistics is concerned with data generated by a random process that is known in its entirety except for a finite number of parameters, is fading (especially in time series), and model fitting is coming to be regarded more as an approximation process. This broader approach seems to have been especially emphasized by systems engineers; in particular Jorma Rissanen [12] has attempted to produce a complete theory of inference based on the idea of a "minimum description length" for the data, using the model set. In consequent asymptotic theory the number of parameters will increase with T and theories dealing with uniform behaviour of statistics must then become important.

The second point I wish to discuss is the relaxation of the Gaussian requirement implicit in everything above. What is the purpose of this relaxation? Certainly there are situations where such relaxations of conditions are of doubtful value. The model being considered may not represent the truth, as we have said, and the value of the theoretical analysis may then, in any case, be qualitative only. Anyone who acts as an editor or referee for journals will have been troubled by papers that are technically sound but sterile, since they merely relax conditions slightly. On the other hand, an increase in generality makes the range of validity of methods wider. In the present case there is also an aesthetic element involved. The models used are all linear. But what does linear mean? Certainly models of the form of (3) would fit a very wide range of data so far as second moments are concerned. A minimal linearity requirement might be that *linear prediction be optimal*, i.e.

$$\varepsilon(t) = y(t) - y(t|t-1) = y(t) - E\{y(t)|y(t-j), j = 1, 2, 3, \ldots\}.$$

This is evidently equivalent to

$$E\{\varepsilon(t)|\varepsilon(t-1), \varepsilon(t-2), \ldots\} = 0 \qquad (7)$$

and in fact this is the essential requirement that justifies the whole theory. Some additional regularity conditions are

$$E\{\varepsilon(t)^4\} < \infty, \qquad E\{\varepsilon(t)^2|\mathscr{F}_{-\infty}\} = \sigma^2$$

(where $\mathscr{F}_{-\infty}$ is the σ-algebra determined by the indefinitely far past) but these seem fairly inconsequential. It is pleasing that the whole theory can be made to go through on the basis of the very condition, (7), that gives meaning to the model set.

The condition (7) says that the $\varepsilon(t)$ are martingale differences. I said before that I had failed to comprehend the chapter on martingales in Doob's book [6], in 1953–4. However that was too early, because the limit theorems on which the above theory rests were not then constructed. I really could not have got far at that time.

3. My Impressions of Statistics

In the first two sections of this article I have been giving lofty advice about how to do research and about the nature of my subject. I feel a little ashamed, for I was early trained in irreverence. My father, an artist and something of a wild man, was of Irish Catholic parentage, and my mother's father was also a Catholic. However, my parents were not reverent, even about their religion. I was sent to a Catholic school (an excellent Jesuit college in Melbourne), but that may have been partly because my father would have been ashamed if his devout parents, brothers and sisters had seen him do otherwise. My father was very impractical, and apart from his work and some exercise (mainly sea bathing) he seemed to do little other than read. He had a discriminating literary taste. He read a great deal to me and my twin sister, Josie, when we were very young, and as a consequence I have retained an interest in literature.

As I have said, I feel ashamed to be giving advice to the young. Yeats, as a young man, wrote of scholars as "bald heads forgetful of their sins, old learned respectable bald heads". How unbearable we can become as we grow old, stressing the obvious, repeating the same advice and telling the same stories! (God knows how many times I have heard the tale of G. H. Hardy and the point in his lecture that was obvious, after five minutes' thought.) Moreover as we grow old the world gives us more opportunity to so indulge ourselves as "with flattering tongue it speeds the parting guest". This article is an example of that.

Of course irreverence can be overdone and may lead to cynicism. Perhaps my irreverence makes me unreasonably feel, as I do, that statistics is overrated as a subject by statisticians. Of course the subject is not unimportant, as the demand for statisticians shows. I mean rather that it lacks that exciting quality associated with the great parts of natural science or, for example, history. We are often seen by scientists as useless, and with the same fearful eye as that with which a bank manager views an auditor.

If statistics is of such a central and compelling nature, as some of my colleagues tell me it is, why, then, has there been such variation between nations in its development? The French and the Germans are people great in science and technology, yet statistics does not seem to have been so important to them as it has been in England or the USA. But one cannot imagine a history of French and German science without major contributions to physics and chemistry.

Today, however, it is becoming less reasonable to neglect statistics. This seems partly to be due to the importance in relatively wealthy societies of a host of social problems such societies can afford to consider. An example is government-funded, or government-instituted, research in medicine and pharmacology. There is also a close association of statistics with some parts of modern technology, more likely to be important in wealthy societies. An example here is system-theoretic engineering studies.

The subject of statistics has changed markedly in recent years, with a welcome move from the excessively theoretical emphasis of the 1960s to an emphasis on relevance. The mood may be changing again as a new need for theoretical progress is felt. At the present moment a high proportion of those beginning postgraduate research seem to have a (relatively) much smaller mathematical equipment than would have been the case 10 years ago. This may partly be due to the mathematically gifted studying other subjects (e.g. computer science). However, I feel that the present is a good time for a strong mathematician to begin in the subject because the subject is rapidly changing and there is both the scope and the equipment for deep theoretical work.

4. Conclusion

The lessons I have drawn from my academic career are, I am afraid, fairly obvious ones. This kind of advice does not help much, like advice on how to solve problems (consider the converse, try induction and so on). The advice merely tells you what you already know. Research is a very personal thing. However I do feel that there is something very pleasing about theoretical research in a subject that has reasonably important applications, and with which applications the theoretical research reacts closely, as is the case with time series analysis. I have been singularly fortunate to have been given the opportunity to lead a life doing such research.

References

[1] Akaike, H. (1969) Fitting autoregression models for prediction. *Ann. Inst. Statist. Math.* 21, 243–247.

[2] An, Hong-Zhi, Chen, Zhao-Guo and Hannan, E. J. (1982) Autocorrelation, autoregression and autoregressive approximation. *Ann. Statist.* 10, 920–936.

[3] Clark, J. M. C. (1974) The consistent selection of parametrization in systems identification. Proc IFAC Conf., Purdue University, West Layafette, IN.

[4] Cramér. H. (1945) *Mathematical Methods of Statistics*. Almqvist and Wiksell, Uppsala.

[5] Deistler, M. and Hannan, E. J. (1981) Some properties of the parametrization of ARMA with unknown order. *J. Multivariate Anal.* 11, 474–497.

[6] Doob, J. L. (1953) *Stochastic Processes*. Wiley, New York.

[7] Dunsmuir, W. and Hannan, E. J. (1978) Vector linear time series models. *Adv. Appl. Prob.* 8, 339–364.

[8] Grenander, U. and Rosenblatt, M. (1957) *Statistical Analysis of Stationary Time Series*. Wiley, New York.

[9] Hannan, E. J. (1970) *Multiple Time Series*. Wiley, New York.

[10] Hannan, E. J. (1980) The estimation of the order of an ARMA process. *Ann. Statist.* 8, 1071–1081.

[11] Keynes, J. M. (1936) *General Theory of Employment, Interest and Money*. Macmillan, London.

[12] Rissanen, J. (1983) A universal prior for integers and estimation by minimum description length. *Ann. Statist.* 11, 416–431.

[13] Saks, S. (1937) *Theory of the Integral*. Warsaw.

G. S. Watson

Geoffrey Stuart Watson was born in Bendigo, Victoria, Australia on 3 December 1921. He was educated at Bendigo High School and Scotch College, Melbourne and later read mathematics at the University of Melbourne, graduating in December 1942. During and just after the Second World War he did research and teaching in applied mathematics.

In 1947 he went as a graduate student to the Institute of Statistics, North Carolina, for two years and wrote his thesis while in the Department of Applied Economics, Cambridge University. He then returned to teach at Melbourne University and later became a Senior Fellow at the Australian National University.

In 1959 he became an associate professor in the Department of Mathematics, University of Toronto and in 1962, Professor of Statistics at The Johns Hopkins University. He moved to Princeton University as Professor and Chairman of Statistics in 1970. He is a fellow of the International Statistical Institute, the Institute of Mathematical Statistics, and the American Statistical Association.

Professor Watson is the author of one book, *Statistics on Spheres* and numerous articles on applied probability and statistics. Among his other interests are travel, mountains, music and painting. He married Shirley Elwyn Jennings in 1952. They have four children, Michael (a linguist), Catharine (a writer and publisher), Rebecca (an opera singer), and Madeleine (an undergraduate student of the arts).

A Boy from the Bush

1. Beginnings

Probability, expressed as odds, dominated my childhood because my father and his brothers were inveterate (some professional) gamblers. Their father was driven off his farm in the western part of the state of Victoria, Australia, by a drought in the 1890s, and bought a hotel in the coastal town of Portland. So began some 50 years of Watson association with hotels and horses, surely an indication of their Ulster origins. Had they been content to be great horsemen (as indicated by the daring feats described to me, and some witnessed by me, as a child) all might have been well, but they had many racehorses which never seemed to win. My father thought of himself as a Scot (Scots did settle in Ulster) and despised the Irish. When young he won prizes for playing the bagpipes and for Highland dancing and later became quite infatuated with golf. Though he was Australia's champion rifle shot for some years, he would have given away all his shooting medals for a low golfing handicap. He first wanted me to be a doctor, but changed this to dentist because he felt this would give me more assured time on the golf course, which would make up for the slight drop in social status. To ensure this rosy future, I was bribed to be at the top of my class. He hated to pay up when I managed to do this only by my unmanly skills in drawing and painting.

By contrast my mother's family was cultured and respectable. They came from Cornwall and Canada to the gold mining town of Bendigo, Victoria, where I was born on 3 December 1921. My grandmother's house was full of books and music, and reflected her dramatic personality. On a recent visit to Bendigo I was delighted to see that it had been most beautifully restored by a young couple whose little children were thrilled to be told what games we had played, so long ago, in the stables and various trees, as well as where we had dug for gold. When my cousins and I were small my grandmother read aloud to us many books by Dickens, Dumas and other classic writers. She told us of the changes of the North American seasons, the beauty of maples in the autumn, and great stories about Red Indians and about her ancestors who fought them in New England. They eventually left Massachusetts, as Empire Loyalists, for Canada. I often think of them on the way to our country place, Blood Hill, in the Adirondacks when I pass the site of Fort William Henry, where one ancestor was captured in 1757 and taken by Indians in a canoe to Quebec from where he was later ransomed. My grandmother was the major formative influence of my childhood. Labouring with my wife to make Blood Hill more beautiful each year is an important part of my life now, and I wish that she could see it.

She would have approved of my international wanderings, as she never accepted the stuffiness of an Australian country town. I recall its strict social strata—as a publican's son I was not very acceptable. I knew only two people there who had been to Europe, and they had only been "Home", as England was then still referred to by everyone. Neither the high school nor the town had a reference library, though there was no lack of English novels. I found a copy of the proceedings of one annual meeting of the British Association for the Advancement of Science and gazed at it with awe. Fortunately my parents sent me as a boarder to Scotch College, Melbourne. Though this lasted only two terms because I came down for a second time with rheumatic fever, I discovered a wider world in Melbourne. There were, I found with the help of a school friend Peter Roberts, second-hand book stores containing the most extraordinary books including popular expositions of science by such famous scientists as Tyndall, Jeans, Lodge, and Andrade. They had prefaces which had been written in incredible places like Switzerland, where it seemed these learned men went walking with one another and had great thoughts and discussions. It was to Switzerland, as well as to other places, that my watercolour hero, J. M. W. Turner, went to sketch. Here then was an alternative to the provincial life lived by everybody around me, one that I could dream about. At this point I should explain that the title of this article is a family joke, since in Australia a "boy from the bush" is a country or farm boy and this I was not. But I often claim to be so when doing something outlandish!

I was saved from dentistry by lack of money and a permanent hotel guest, Dr E. V. Keogh. While I was at high school Dr Keogh ran the laboratory at the Bendigo hospital, enjoyed a drink, a flutter on the horses, and unbounded respect from my father. Of course no one in Bendigo knew that he was a first-rate scientist. He suggested that I attend Melbourne University in the evenings and work for the Commonwealth Post Office during the day until a post could be found for me as his assistant at the Commonwealth Serum Laboratories where he then worked. I was to study mathematics and chemistry at government expense, and he would teach me biology so that we could do joint research in immunology. By the time the post became available, I was completely carried away by the beauty and power of mathematical physics, and the joys of the cultural and intellectual life at the University. As soon as Keogh realized this, he offered to lend me money to pursue these studies full-time. Furthermore, he was able to persuade my father that this incomprehensible career was a sensible aim for me. Soon after this Keogh became a colonel in the Australian Army Medical Corps where he played a role in Oliver Lancaster's life (see [21]). I still cherish Keogh's copies of Lamb's *Calculus* and Mellor's *Higher Mathematics for Students of Physics and Chemistry* and was glad, many years later, to write several papers on immunology. I have always tried to follow his research advice—always work on a problem for a while before reading the literature on it! Since working is more fun than reading, I often overdo it!

Thus in 1940 I became a full-time student in the honours mathematics program at Melbourne University directed by Professor (later Sir Thomas) Cherry. Having taken Honours Physics I the previous year as an evening student, I had a spare course and, on Keogh's advice, took Theory of Statistics I, given by M. H. Belz. The text was Yule's book with additional reading from Fisher's *Statistical Methods for Research Workers*. Statistics became something I knew that my friends did not. The mathematics program was, by today's standards, all classical applied mathematics and physics with almost no algebra and only projective geometry but a great deal on differential equations, complex variables, calculus of variations, dynamics, continuum mechanics, all with an emphasis on doing hard problems. All its graduates went into physics. However Cherry, K. E. Bullen, E. R. Love, M. L. Urquhart (from whom I got my passion for the mountains in summer and winter), M. H. Belz and F. Behrend (who brought to Melbourne a taste of more modern mathematics as well as of Europe and its music) were all excellent teachers, and I am grateful to them. I recall giving an undergraduate talk on integrals and derivatives of fractional order, at the end of which I solved a differential equation of order $\frac{1}{2}$. Professor Cherry immediately caught me out with—"Er, Watson, what happened to the half a constant?" It was unsettling to be a student during the Second World War, but manpower controls and two childhood bouts of rheumatic fever ensured that I should be one.

When I graduated at the end of 1942 with a thesis on "The Effects of Non-normality", first-class honours and the Dixon Research Fellowship in Mathematics, I joined G. K. Batchelor's CSIRO research group in aerodynamics in preference to taking another year to complete my physics degree. After one year of largely experimental turbulence work, I returned to the university to help with the teaching of analysis, mechanics, the theory of elasticity, and statistics. In the evenings Belz and I ran some classes in quality control for the munitions people. I worked briefly with Cherry on the use of complex variable methods for the biharmonic equation, but I have little to show for these years. They came to an end when the department decided that I should go abroad to study statistics so it could be taught better. In August 1947 I went to the newly-formed Institute of Statistics in North Carolina. This choice was made because W. G. Cochran, then at North Carolina State, and my teacher in applied mathematics K. E. (later Sir Keith) Bullen had been together in Cambridge. I had rather lost momentum and had, at that time, no clear vocation.

2. Apprenticeship

In my two years as a graduate student in North Carolina, I was influenced most by Cochran and E. J. G. Pitman who was visiting from the University of Tasmania, Australia. Superb courses from R. C. Bose on multivariate analysis and the design and analysis of experiments were technically very helpful in improving my algebra, as was reading notes taken in lectures given by P. L. Hsu before I arrived. As Cochran's assistant, I learnt many tricks and got much insight into Fisherian methods—Cochran was my model for years. Pitman's ability to go to the heart of any question took my breath away. He is justly famous for his contributions to statistical inference, and these two courses were naturally extraordinary. But the only lecture notes I still read were those from his course called "Applied Probability". It is a tragedy that they were not typed and circulated, like his lecture notes on "Nonparametrics", for they contained much novel material, and many results that were later proved and published by others, usually by much more cumbersome and less insightful methods. I have in mind, in particular, his results on the gaps between n independent uniform random variables on the unit interval. There were no courses or books on stochastic processes or time series analysis available then and, tired of design and analysis of variance, these were what I wanted to study. In particular I wanted to follow up some work of T. W. Anderson and R. L. Anderson on testing the residuals from least-squares regression for serial correlation. I had found a vocation.

A visit by J. M. Wishart of Cambridge University led to an appointment in the Department of Applied Economics (DAE), Cambridge University, where I intended to work on time series regression problems. I also wished to enjoy life in Cambridge where many of my Australian friends were students or members of faculty. James Durbin was already at the DAE and, almost immediately, we started our very fruitful collaboration [13], [14], [15], [16] which continued after (indeed long after) he left for the London School of Economics.

In considering least squares on the model $y = X\beta + \delta$, where δ is a vector of errors with zero mean and a covariance matrix Σ, the key is the relationship of the eigenvectors of Σ to the column space of X. Our test of $\Sigma = \sigma^2 I_n$ versus (an approximation to) the covariance matrix of a first-order autoregressive process was an instant hit in econometrics but was ignored in statistics. In my opinion it has been very badly used in practice, whereas I think the technical side of the paper deserved a warmer reception. Our version of Aitken's 1935 treatment of least squares [1] has now become standard.

My two years in Cambridge with occasional trips to Europe were very happy and busy. I never became interested in economics or econometrics, despite the brilliance of my colleagues at the DAE. I could not see the sense of probabilistic models that could not be even roughly checked for fit. To this day the social sciences interest me very little (though I can see the elegance of mathematical economics), and the natural sciences very much. But I was kindly adopted by the Statistical Laboratory whose members were then F. J. Anscombe, H. E. Daniels, D. V. Lindley and J. M. Wishart. I talked a great deal with Henry Daniels who had the endearing habit of declaring anything I suggested to be obvious, perhaps after some days of thought! As my first serious research was done in the Cambridge atmosphere, it is not surprising that my style has remained British though I have made my career in the United States.

I recall giving a talk in the Statistical Laboratory on my work with Durbin which turned upon inequalities between the eigenvalues of a symmetric matrix A and those of MA where M was a symmetric idempotent. While Wishart made a joke about transforming MA to Ph.D., Sir Harold Jeffreys, from whom I expected a Bayesian blast, murmured some kind words and said that the inequalities reminded him of something in Courant and Hilbert. Years later when I read Courant and Hilbert in translation, I saw that the inequalities were indeed trivial consequences of the Courant–Fischer theorem and realized how kind Jeffreys had been to me!

In the fall of 1951, my Ph.D. thesis was examined in Raleigh and I returned to Melbourne University as a Senior Lecturer in Statistics, and as acting head of the department while Professor M. H. Belz went on leave for 15 months. My three years in Melbourne were very educational. I reorganized courses, studied Feller's Volume I, worked with F. E. Binet on a number of topics including genetics which he taught me [4], [5] and, most importantly, married Shirley Elwyn Jennings and had a son, the first of our four children.

During this period we looked after R. A. Fisher when he visited Melbourne. On his return to Cambridge, he sent me a reprint of his 1953 paper "Dispersion on a sphere" [19]. This paper started me on one of my lifelong interests. In it Fisher began the development of statistical methods (using his fiducial ideas over which we later fell out) for the distribution on the unit sphere in three dimensions with density proportional to $\exp k\mu^t x$ where $k \geq 0$ and x and μ are vectors of unit length or "directions". μ is the mode of this distribution but is most often called the mean direction. μ is parallel to Ex. I subsequently called this the Fisher distribution, although it had been used by Langevin in a statistical mechanical model for paramagnetism in 1905.

I also wrote a short note pointing out that the asymptotic distribution of the largest value in n successive observations from an m-dependent process is the same as in the case of independence. G. N. Alexander interested me in hydrology and I tried to formulate a theory of dams. To get explicit results I tried diffusion approximations (unlike Moran who wisely used Markov-chain methods) and, to make a bad pun, got into deep waters where I drowned. However, P. A. P. Moran heard me talk about this and invited me to The Australian National University where my research career began in earnest, if somewhat late, in January 1954.

3. Canberra

The department then consisted of Professor P. A. P. Moran, myself as a senior fellow, E. J. Hannan as a research fellow and J. M. Gani as one of two research students. Initially I wrote up the work I had done at the DAE on serial correlation in regression analysis [25], [26], [27], [61] and enjoyed collaborating with Hannan who was beginning his distinguished career in time series analysis. It was also a pleasure to see Moran's ingenuity and modelling skills in action.

The Australian National University (ANU) was created after the Second World War to encourage research and graduate training in Australia and to reverse the "brain drain". The faculty (now members of the Institute of Advanced Study of the ANU) had no duties apart from their research and the training of graduate students. They were favoured with better salaries, working conditions and leave arrangements than members of the other universities in Australia. Certainly when I was there, most people were conscious of these privileges and worked very hard to justify them. It was an exciting place to be.

E. J. Irving was working there on palaeomagnetism. The basic observation was the original direction of magnetization of a rock specimen and it was with this data in mind that Fisher had written his 1952 paper. I was able to find out from Irving what statistical methods needed to be developed to help in the analysis of this sort of directional data [28], [29], [62], [68]. In the palaeomagnetic applications of Fisher's distribution, the small dispersion of the observations leads to values of k so high that I was able to develop approximate but adequate statistical methods in terms of the analysis of variance or dispersion. Exact results being rather unattractive, this approach became very popular as it was so easy for anyone to understand—it made simple geometric sense. This has since become almost an axiom of mine—namely, that one should only expect users to apply methods that make simple common (usually geometric) sense. Users do not care about the last drop of optimality! In thinking about how the Fisher distribution actually fitted my data sets I was also led to work on goodness-of-fit problems, then chi-squared methods [30], [32], [33] and later other methods. Irving interested me in what might be deduced from palaeomagnetic data, for example, continental drift. Thus I have had the pleasure of watching the progress of plate tectonics theory from the beginning.

Early on I became friendly with all the members of the department of microbiology, especially H. J. F. Cairns and S. Fazekas de St. Groth. Their work was mainly in the fields of virology and immunology and opened my eyes to the thrills of modern biology; this became my major interest for many years. The yield in terms of papers was not much at the time [7], [17], [31], or later, but I would not have missed it for anything. The last two papers were on immunology, at that time fascinating but rather arcane, and the former arose when I asked Fazekas to show me the most mathematical paper that had been written in this subject! On reading his selection, Goldberg's theory of the precipitin reaction, I was able to make a simple conditional branching chain argument that destroyed the theory. This exposure to basic biology led me to take laboratory courses in bacterial genetics, tissue culture and bacteriophages at Cold Spring Harbor in the summer of 1958, and enabled me to watch the development of molecular biology with informed eyes. Thus my time in Canberra put me to touch with two of the great scientific revolutions of this century!

At Christmas 1957, I left Canberra for study leave at Cambridge University, Cold Spring Harbor, North Carolina and Princeton. In Cambridge I spent a lot time with Fisher talking happily about many scientific topics on which he was incredibly lucid. But I ended up being severely chastised by him for my views on statistical inference. I have always said that my decision to stay in North America was not unrelated to this incident: I felt a poor letter from Fisher might prevent me from getting a chair in Australia. Now I think it was just a trigger: I wanted to live abroad. Parenthetically, though I am now legally an American citizen, I still think of myself as an Australian and am very proud of Australian achievements in the arts and sciences.

At Cold Spring Harbor, I started some non-Mendelian genetical work (I had a hunch that this might bear on the speciation problem) with Ernst Caspari which lead to several papers [9], [35], [36], [60]. In North Carolina I finished off the last of a series of papers on chi-square [30], [32], [33]. In Princeton, where I spent the academic year sharing John Tukey's office, I got to know S. S. Wilks, W. Feller and some of the other mathematicians. Apart from collaborating with Caspari, I did more work on directions [34], largely tests for the coplanarity of the mean directions of several Fisher distributions.

After my work on chi-square and the effects of estimating population parameters it seemed natural to take the same approach to the Cramér–von Mises statistic W_n^2 which is the integral of the square of $F_n(x) - x$. To get this into the form of a sum of squares it was natural to find the Fourier expansion of $F_n(x) - x$,

$$n^{1/2}\{F_n(x) - x\} = n^{1/2}(a - 1/2) + \sum_{-\infty}^{\infty} \left[\frac{-\sum \exp 2\pi i m x_j}{2\pi i m n}\right] \exp(-2\pi i m x), \quad (1)$$

where a is the arithmetic mean of the sample. The coefficients in the series are orthogonal to one another, but not to the $(a - 1/2)$ term, and I found the distribution of the sum of squares of their moduli. Anderson and Darling [2] had tabulated the distribution of W_n^2, so my result did not, at that moment, seem of much use. Other things came up and I lost interest and did not pursue the effects of parameter estimation, which needed some further (and decidedly nontrivial) ideas. This was done quite independently by Durbin [12] who also pursued the decomposition of the Cramér–von Mises statistic.

In Princeton I also got involved with G. E. P. Box's activities at the "Gauss House". This was a very lively place where I first met E. Beale, C. Mallows, Henry Scheffé and many others. One of the products of this interaction was a joint paper [6] on the robustness of least squares. My insight here came from my work mentioned above on least squares: it seemed that some X or design matrices must yield analyses more robust against outliers, than others. Similarly, if one can choose the X-variates in a regression situation but with stationary errors (imagine altering slightly some control variables in an industrial process) then some X-choices will make the estimation of the regression coefficients more efficient than others.

4. Toronto and the Research Triangle Institute

When I took up my appointment at the University of Toronto, in the fall of 1959, my biological work came to a halt. Indeed, the need to earn extra money to establish our family (now with three children) in North America meant that consulting had to come before research for several years. I spent one day a week at the Ontario Research Foundation. Gertrude Cox had just established the Research Triangle Institute (RTI) in North Carolina and wanted me to direct theoretical statistics there. This I did for several years by visiting and by mail. Industrial problems suggested a paper on group screening [38] which was a transposition of Dorfman's (1943) idea on blood tests to factorial designs with many factors, and one on reliability [67]. The latter explored the gain in useful life that may come if items are run in the factory for a short period before shipping. Efforts to formulate reliability problems in terms of the maxima of processes led to Ross Leadbetter's coming to North Carolina and our joint work on the estimation of density and hazard functions [63], [64], [65] and so indirectly, to his book with Cramér [10].

In my first summer at RTI, Klaus Schmidt-Koenig came over from Duke University, to ask me about tests for uniformity for the vanishing bearings of his homing pigeons. Kuiper [20] had adapted the Kolmogorov test to the circle which has no natural origin. For a circumference unity, his statistic is

$$V_n = \sup_x \{F_n(x) - x\} - \inf_x \{F_n(x) - x\}.$$

Using the series representation (1) it is clear that the $(a - 1/2)$ terms cancel and that the statistic does not depend on the arbitrary choice of the origin. Indeed it suggests some alternative statistics such as

$$\sup_x n^{1/2}(F_n(x) - (x - 1/2) + a),$$

which I justified in a 1974 paper [47] and whose asymptotic distribution has just been found by Darling [11]. At this time I realized that the statistic I had studied the year before, namely

$$U_n^2 = n \int_0^1 [F_n(x) - (x - 1/2) + a]^2 \, dx, \tag{2}$$

was a variant of the Cramér–von Mises statistic which, as a consequence of (1), was unaltered by the arbitrary choice of an origin on the circle. The article [38] develops the distribution theory differently as I did not know how to make a rigorous proof using Fourier series! In Toronto M. A. Stephens wrote a thesis under my supervision on directional statistics which was the beginning of his many contributions to this field.

5. Baltimore

In the fall of 1962 I went to The Johns Hopkins University to create the Department of Mathematical Statistics on its Homewood campus to supplement the Department of Biostatistics in the School of Hygiene on the East Baltimore campus, both departments being chaired by A. W. Kimball. While I was able to spend more time again with biologists in my eight years at Hopkins, the 20-odd papers written there are spread across my earlier areas of interest, with geology as a major motivation. The high points of this period were a trip to Antarctica to develop a method for counting penguins (which has formed the basis of innumerable elementary public lectures and was a marvellous adventure), some good graduate students (R. Beran, G. Chase, E. Rothman, Woolcott Smith), the outstanding academic visitors we had, and a 15-month sabbatical in Europe.

My work with the molecular biologist C. A. Thomas and W. Smith [43] on models for the annealing of DNA strands led to the following elegant problem. Suppose that a point is dropped on a triangle with sides 1, 1 and $2^{1/2}$, uniformly at random and then the point moves so each of its coordinates is an independent Wiener process. What is the chance that it exits from the triangle via the hypotenuse? Surprisingly the fool's answer, $2^{1/2}/(2 + 2^{1/2})$, is quite close to the exact answer, given in Smith and Watson [66]. Beran [3] in his thesis developed a beautiful and very general theory of locally most powerful and invariant tests of uniformity on compact homogeneous spaces. The test statistic is always the integral of the square of a function of the sample values. Beran showed that my U_n^2 was in this class, although for a strange class of alternatives.

To have a successful graduate program with such a small faculty, it was necessary to have visitors of distinction to provide stimulation and lectures on a wider variety of topics than we could cover. R. G. Miller and E. J. G. Pitman were followed by C. R. Rao, J. Durbin, D. G. Kendall E. J. Hannan, J. M. Gani and V. P. Godambe. Elliot Montroll and George Weiss both gave special courses on stochastic processes. Miller wrote much of his book on simultaneous inference, and Rao was writing his book on linear statistical inference, when visiting Hopkins. Hannan's book *Multiple Time Series* arose out of the lectures he gave at Hopkins. The program thus had little structure, since most courses were on what people were currently doing, and important things might never be covered, but it seemed to be enjoyed by the students and justified by their subsequent successes.

I wasted a lot of time on an unfinished book on mathematics for biologists, begun in Toronto where I taught such a course. I spent time on various federal advisory committees. I began a book on orientation statistics in the earth sciences which was also abandoned after much study of geology. Throughout my life I have had trouble in finding the right balance between developing statistics for a particular field of science and acquiring the amount of knowledge of the field that should lie behind it. Too little means that one's efforts might be irrelevant. But too much study does not leave enough time to do the statistical job and one just becomes an amateur scientist. It is essential to see the problem clearly and without the clutter of too many details. I think this is the greatest difficulty for the applied mathematician to resolve.

I had my first taste of being a chairman at Hopkins when Kimball became a dean and my apparent success led to so many offers that I actually believed I was good at it—for a while!

After a summer in Rome and a dramatic time in Prague (when the Warsaw Pact troops invaded Czechoslovakia), our sabbatical leave continued at the Istituto di Genetica in Pavia for nine months, after which we spent the summer at the Mathematical Institute of the University of Aarhus with Barndorff-Nielsen. In Pavia I had my second-last unsuccessful shot at finding some ways to contribute to molecular biology. In Aarhus I found the statistical atmosphere very much to my taste and have, since then, returned there as often as I have been able. But this period of living in and crisscrossing Europe was not solely for research—I was fulfilling my boyhood dream of extensive European travel.

6. Princeton

In 1970 I come to Princeton to be chairman of the Department of Statistics which was created in 1966. Before then S. S. Wilks, J. W. Tukey, and later F. J. Anscombe and J. Hartigan (who both left later for Yale), had been members of the Department of Mathematics. With the stimulus of D. R. McNeil, the skills of P. Bloomfield, and the fund-raising abilities of J. W. Tukey, we became one of the first departments to have its own computing equipment. We have since remained in the vanguard with hardware.

I have, as a consequence, had several large-scale data-analysis contracts; one with the Energy Information Administration to examine the oil and gas resources of the United States, which ran for several years, and a much longer-term study of stratospheric ozone and atmospheric temperatures. The latter began when I was a member of NRC committees charged with assessing the possible effects of the release of fluorocarbons (e.g. from spray cans) on stratospheric ozone. A brief summary of some of this work is given in [54]. It was continued under the sponsorship of the Chemical Manufacturers Association and led us to study temperatures as well. Just recently, by using the vast meteorological global database, we seem to have verified that atmospheric temperatures are changing in the way the modellers have suggested the increase in carbon dioxide in the atmosphere will cause them to. P. Bloomfield, S. Zeger and G. Oehlert have been successively involved in this work.

These activities partly arose out of my advisory work with the Federal Evironmental Protection Agency; these came to an abrupt halt when my name was put on the Reagan "hit list" as a "smooth but extreme environmentalist". We suspect that this came about because my wife donates to every organization concerned with nature and, for humans, only to Planned Parenthood!

As the department at Princeton has always been small, I have tried to have many visitors to leaven the loaf; the list is too long to give here. During the "Robustness Year", they outnumbered the staff. As at Hopkins, Australian visitors have been frequent, and the visits of Eugene Seneta, Terry Speed and Alan James all led to theses which they inspired and supervised, even after returning home!

With the exception of my brief experience with economic time series, the atmospheric work just described and the mining field to be discussed later, my work has mostly been concerned with "small-scale" science where experiments can be planned to investigate a particular point. This is what most theoretical statistics now has in mind—Fisher set the stage for it in his Rothamsted years. In the environmental work just mentioned, and in another project on mandating the use of helmets for motor cyclists and its effect on fatalities [69], [70], things are very different! This always seems to be the case in questions of public moment, especially environmental problems. It is most unusual to be able to use data collected to answer the specific question posed. One has to make do with what is available and seems relevant; sometimes its quality is dubious and one is not sure of the definitions used in collecting it, whether it is truly representative or not, since "random samples" are rarely available. The instruments used may vary from place to place or over time. They are maintained by very different people, and the data tabulated and preprocessed by others. It is very tedious to obtain datasets, involving lots of letters and phone calls. People are naturally very possessive about their data. Thus "outliers" are very common. It is not too hard nowadays to do robust analyses that effectively ignore them, but sometimes they are said by experts to be quite possible values. "Experts" tend to try to "blind statisticians with science". Thus, instead of spending one's time practising one's "trade", it seems that most of one's effort goes on value-assessments of matters that are not really statistical at all! An alternative view is that the "trade" of a statistician is not confined to what we teach or read in statistical textbooks!

My activities as an advisory editor for the Probability and Statistics Series of John Wiley and Sons have greatly increased since I have been in Princeton. Few weeks go by without a call or letter from Beatrice Shube, the Wiley editor.

In the mid 1970s the work of Georges Matheron and Jean Serra of the Center for Mathematical Morphology at the Paris School of Mines, attracted my attention. They seemed to be breathing new life into the application of statistics to geology and mining. As a result, I spent a lot of time persuading English-speaking geologists and statisticians that this was so, while trying to persuade the Fontainebleau School to integrate their writings with that of the anglophones! Our geologists receive very little mathematical training (indeed they seem less mathematically inclined than any scientific group I know) so French-style "geostatistics" was just too much for them. The basic idea is to regard the concentration of a mineral as a random function f. If parts of the region have been mined, then one knows the integral of f over these regions. With this data one wishes to predict the integral of f over unmined regions. Their methods are based on the variogram function, $E\{f(x+h) - f(x)\}^2$, which hopefully depends only on h. The computation this leads to is called kriging, and this technique is now well known; my comments on it are to be found in [46], [53], [58]. Their other contribution, mathematical morphology, was, to my mind, more novel. It is a mathematical formulation, suitably generalized, of how to handle the information from a digitized picture. An introduction to these ideas is given in [48], [49], and a complete account is now available [22]. This subject is closely related to stochastic geometry. While I never persuaded Matheron to adopt the "Anglo-Saxon optique", I enjoyed his hospitality on many occasions, thereby sampling French family life (at its best, I suspect) which one can never know as a tourist.

For the last few years John Aitchison and I have worked hard on a quixotic task—collecting material for a biography of A. G. McKendrick who did such original and unrecognized work in medical statistics, especially the development of stochastic models. This has taken me to a remote town in South Africa where his eldest brother died of tuberculosis in 1892, to the India Office in London to find the original papers documenting his appointments, leaves and promotions, to the London School of Hygiene and Tropical Medicine where some of his correspondence with Sir Ronald Ross is preserved, to India where he served for 20 years in the Indian Medical Service and to Edinburgh where he ended his career. It has given us some insight into the historian's trade but, so far, no book.

It is hard to describe my more recent work with any objectivity. Only current research comes to mind, perhaps because I spent too much time for a while on committees. In the last three years or so I have realized that this sort of activity is not satisfying to me and I have gone back to statistical problems in a geometric setting. Much of this work is contained in my monograph *Statistics on Spheres* [55]. While the motivation for this theory lies in the analysis of data coming from animal behaviour, geology, and geophysics, I feel it has an intrinsic mathematical interest. Many basic quantities of common statistics (like means and variances) no longer make sense when the observations are unit vectors. Of course the primitive notions of central tendency and dispersion continue to make sense for any kind of data and the trick is to adapt them to "curved spaces", in this case a sphere. Again, most real problems deal with two and three dimensions but it is both fun and instructive to theorize in a space of any number of dimensions. Earlier in my life I would have thought it pretentious to go beyond what seems scientifically necessary. It is certainly dangerous because one has to depend on one's taste and not on the obvious utility argument. But at my age I can aim only to please myself (and perhaps the NSF)! A recent enthusiasm following a long-term interest in the more practical problems of geometrical probability or, as it is now often called, stochastic geometry, has been the study of the shapes of sequences of random triangles, or more generally, figures or sets of points in any number of dimensions under random cyclic transformations. (See [56], [57], [60].)

7. Epilogue

The field of statistics and probability has changed completely from what it was in 1947 when I became a graduate student. Then one had only to master a small amount of literature and could rapidly feel a bit of a pioneer. The common level of knowledge of probability and stochastic process theory was negligible by today's standards. Statistics was dominated by the brilliant advances of Fisher in small sample inference and distribution theory, by the Neyman–Pearson theory and later by Wald's decision theory. All of these led to so much developmental research that only a few people were able to stand back and ask whether these formulations captured all the aspects of "statistics".

The applications of statistics and the technological advances in computing may have been more influential than the philosophical arguments (such as Bayes versus non-Bayes) in changing the subject. In any event, while there is a rather dauntingly large literature now, the cutting edge of the subject, (but not most of its courses or textbooks) seems so much more realistic (for which John Tukey can claim fame for his emphasis on informal exploratory techniques and robust methods), nondoctrinaire, mathematically elegant and powerful. Fisher, Neyman and Cramér gave statistics its mathematical and theoretical form and, after the Second World War the centre of gravity of theoretical statistics moved to the United States. Most Americans still probably feel that Berkeley and Stanford are its U.S. homes. There must be more statisticians in the United States than in the rest of the world combined, so this might be numerically true. However, despite the smaller number of people involved there, I would argue that the centre of mathematical statistics has now moved back to Europe: Scandinavia (Denmark this time, due primarily to the work of Barndorff-Nielsen) and England (with David Cox and others). Once again (of course this is all wildly simplified) the intuitive ideas come from England and the heavy mathematical tools from Denmark. The applied side of the subject is more active in the United States. This, and the practical and intellectual influence of computers which is much greater here, are the forces driving the very considerable theoretical developments in the United States, rather than mathematical structure as in Europe.

Of course the computer allows us to display data in order to *see* (literally) possible regularities, as well as to carry out calculations that could hardly be envisioned in 1947. It is an object lesson to realize what a long time elapsed before the computer was used in statistics for anything beyond the calculations needed by the pre-computer age.

A result of all this is (or should be) that much of the literature of the past is now irrelevant, as I suppose is common in all subjects that advance rapidly. Many of the innovations in statistical thinking in the past have come from people working outside academe in specific areas (Fisher in agriculture, Box in the chemical industry, to give but two examples). This will presumably continue, and it behooves academic statisticians to watch for these new inputs. Much the same general remarks can be made about applied probability, which does not, however, have any philosophical problems. Applied probability also seems to me to be more interesting, at the moment, outside the United States: Australia (thanks to Moran) and Europe (D. G. Kendall and others). Driven by the increasing quantification of all aspects of the social and natural sciences, of government and business activities, and by the astonishing progress in the recording and handling of data in computers, as well as by the influx of a younger generation with more powerful mathematical tools, we may expect probabilistic modelling and statistics to advance even more rapidly than it has in the past. However, I would predict that probability will play a less central role in statistics than it has to date, and computing and data display will become more important. Certainly much more attention will be paid to nonexperimental data, which constitutes most of the data a statistician is presented with.

The editor urged me to make comments, where I could, that might be helpful to young people. Let me make one more. The world is such a wonderfully fascinating place with its different cultures, arts and sciences, all offering great satisfactions and insights, that it is a tragedy to see only one little part of it, no matter how well you do it! If I were entering graduate study now I *might* (from among the bewilderingly large number of attractive alternative careers) still chose statistics and probability. It has certainly led me into all sorts of interesting corners of science and of the world, and left me plenty of time to pursue my many other interests, and surely it will do the same for others in future.

The editor did not urge me to write for my contemporaries, though they may find this more relevant to their own lives. Though I began this account with misgivings, it has been another new experience. It has cleared my vision of my past. I have always loved reading autobiographies, but this restricted attempt has shown me more clearly what makes some so much better than others. So often the writer says he or she did nothing for years when young, or just had odd bursts of activity. Most of my friends who still work put in much, much, longer hours than they did when young; I certainly do. This may reflect the difference between hard work and hard thinking. Though we complain about professional responsibilities, these are much less demanding than caring for young wives and little children. We thus have more time for work. I suspect that the youthful periods that now seem so empty were filled to the brim with emotional experiences whose memory one has suppressed. The great writers can fill these in.

Acknowledgements

The writing of this article was partially supported by Grant MCS 821831 from the National Science Foundation to Princeton University. A number of friends, notably J. M. Gani and R. J. W. Beran, have been most helpful in criticizing drafts.

Publications and References

[1] Aitken, A. C. (1935) On least squares and linear combinations of observations. *Proc. R. Soc. Edinburgh* 55, 42–48.

[2] Anderson, T. W. And Darling, D. A. (1952) Asymptotic theory of certain 'goodness of fit' criteria based on stochastic processes. *Ann. Math. Statist.* 23, 193–212.

[3] Beran, R. J. (1969) Testing for uniformity on compact homogeneous spaces. *J. Appl. Prob.* 5, 177–195.

[4] Binet, F. E. and Watson, G. S. (1956) Algebraic theory of the computing routine for tests of significance on the dimensionality of normal multivariate systems. *J. R. Statist. Soc.* B 18, 70–78.

[5] Binet, F. E., Sawers, R. J. and Watson, G. S. (1958) Heredity counselling for sex-linked recessive deficiency diseases. *J. Hum. Genet.* 22, 144–152.

[6] Box, G. E. P. and Watson, G. S. (1962) Robustness to non-normality of

regression tests. *Biometrika* 49, 93–106.

[7] Cairns, H. J. F. and Watson, G. S. (1956) Multiplicity reactivation of bacteriophage. *Nature* 177, 131–132.

[8] Carlson, F. D., Sobel, E. and Watson, G. S. (1966) Linear relationships between variables affected by errors. *Biometrics* 22, 252–267.

[9] Caspari, E., Smith, W. K. and Watson, G. S. (1966) The influence of cytoplasmic pollen sterility on gene exchange between populations. *Genetics* 53, 741–746.

[10] Cramér, H. and Leadbetter, M. R. (1967) *Stationary and Related Stochastic Processes*. Wiley, New York.

[11] Darling, D. A. (1984) On the asymptotic distribution of Watson's statistic. *Ann. Statist.* 11, 1263–1266.

[12] Durbin, J. (1973) *Distribution Theory for Tests Based on the Sample Distribution Function*. SIAM, Philadelphia.

[13] Durbin, J. and Watson, G. S. (1950) Testing for serial correlation in least squares regression—I. *Biometrika* 37, 409–428.

[14] Durbin, J. and Watson, G. S. (1951) Testing for serial correlation in least squares regression—II. *Biometrika* 38, 159–178.

[15] Durbin, J. and Watson, G. S. (1951) Exact tests of serial correlation using non-circular statistics. *Ann. Math. Statist.* 22, 446–451.

[16] Durbin, J. and Watson, G. S. (1971) Testing for serial correlation in least squares regression—III. *Biometrika* 58, 1–19.

[17] Fazekas, S., Reid, A. and Watson, G. S. (1958) The neutralization of animal viruses I—Theory. *J. Immunol.* 80, 215–224.

[18] Feller, W. (1957) *An Introduction to Probability Theory and its Applications*. Wiley, New York.

[19] Fisher, R. A. (1953) Dispersion on a sphere. *Proc. R. Soc. London* A217, 295–305.

[20] Kuiper, N. H. (1960) Tests concerning random points on a circle. *Ned. Akad. Wet. Proc.* A63, 38–47.

[21] Lancaster, H. O. (1982) From medicine, through medical to mathematical statistics: some autobiographical notes. In *The Making of Statisticians*, ed. J. Gani, Springer-Verlag, New York.

[22] Serra, J. (1982) *Image Analysis and Mathematical Morphology*. Academic Press, New York.

[23] Veitch, J. and Watson, G. S. (1986) Random cyclic transformations of n points. In *Essays in Time Series and Related Processes*, ed. J. Gani and M. B. Priestley, Applied Probability Trust, Sheffield, 369–382.

[24] Watson, G. S. (1954) Extreme values in samples from m-dependent stationary stochastic processes. *Ann. Math. Statist.* 25, 798–800.

[25] Watson, G. S. (1955) Serial correlation in regression analysis—I. *Biometrika* 42, 327–341.

[26] Watson, G. S. (1956) On the joint distribution of the circular serial correlation coefficients. *Biometrika* 43, 161–168.

[27] Watson, G. S. (1956) A note on the circular multivariate distribution. *Biometrika* 43, 467.

[28] Watson, G. S. (1956) Analysis of dispersion on a sphere. *Monthly Notices R. Astronom. Soc. Geophys. Suppl.* 7, 153–159.

[29] Watson, G. S. (1956) A test for randomness of directions. *Monthly Notices R. Astronom. Soc.* 7, 160–161.

[30] Watson, G. S. (1957) The χ^2 goodness-of-fit test for normal distributions. *Biometrika* 44, 336–348.

[31] Watson, G. S. (1958) On Goldberg's theory of the precipitin reaction. *J. Immunol.* 80, 182–185.

[32] Watson, G. S. (1958) Some recent results in χ^2 goodness-of-fit tests. *Biometrics* 15, 440–468.

[33] Watson, G. S. (1958) On χ^2 goodness-of-fit tests for continuous distributions. *J. R. Statist. Soc.* B 20, 44–72.

[34] Watson, G. S. (1960) More significance tests on the sphere. *Biometrika* 47, 87–91.

[35] Watson, G. S. (1960) The behavior of cytoplasmic pollen sterility in populations. *Evolution* 14, 56–63.

[36] Watson, G. S. (1960) Cytoplasmic 'sex-ratio' condition in *Drosophila*. *Evolution* 14, 256–265.

[37] Watson, G. S. (1961) Goodness-of-fit tests on a circle. *Biometrika* 48, 109–114.

[38] Watson, G. S. (1961) A study of group screening methods. *Technometrics* 3, 371–388.

[39] Watson, G. S. (1962) Goodness-of-fit tests on a circle—II. *Biometrika* 49, 57–63.

[40] Watson, G. S. (1964) Smooth regression analysis. *Sankhyā* A 26, 359–372.

[41] Watson, G. S. (1965) Distributions of organisms. *Biometrics* 21, 543–550.

[42] Watson, G. S. (1965) Equatorial distributions on a sphere. *Biometrika* 52, 193–201.

[43] Watson, G. S. (1966) Recombination of a pool of DNA fragments with complementary single-chain ends. In *Progress in Nucleic Acid Research* 5, ed. W. E. Cohen and H. N. Davidson, Academic Press, New York.

[44] Watson, G. S. (1967) Another test for the uniformity of a circular distribution. *Biometrika* 54, 675–677.

[45] Watson, G. S. (1971) Estimating functionals of particle size distributions. *Biometrika* 58, 483–490.

[46] Watson, G. S. (1972) Trend surface analysis and spatial correlation. In *Quantitative Geology*, Special Paper 146, Geological Society of America, 39–46.

[47] Watson, G. S. (1974) Optimal invariant tests for uniformity. In *Studies in Probability and Statistics*, ed. E. J. Williams, Jerusalem Academic Press, Jerusalem, 121–128.

[48] Watson, G. S. (1975) Mathematical morphology. In *Proc. Internat. Symp. Statistical and Linear Models*, ed. J. N. Srivastava, North-Holland, Amsterdam, 547–553.

[49] Watson, G. S. (1975) Texture analysis. In *Memoir* 142, Geological Society of America, 367–391.

[50] Watson, G. S. (1976) Counting penguins. *Vinculum* (The Mathematical Association of Victoria: M. H. Belz Memorial Issue), 16–23.

[51] Watson, G. S. (1976) The analysis of mortality data. *Proc. 4th Symp. Statistics and the Environment*, 114–118.

[52] Watson, G. S. (1977) Age incidence curves for cancer. *Proc. Nat. Acad. Sci. USA* 74, 1341–1342.

[53] Watson, G. S. (1977) Review of *Advanced Geostatics in the Mining Industry*. *J. Amer. Statist. Assoc.* 72, 687–689.

[54] Watson, G. S. (1982) Methods of analysis of stratospheric ozone data. In *Interpretation of Climate and Photochemical Models, Ozone and Temperature Measurements*, Conference Proceedings 82, American Institute of Physics, 201–222.

[55] Watson, G. S. (1983) *Statistics on Spheres*. Wiley, New York.

[56] Watson, G. S. (1983) Random triangles. In *Memoirs* 6, Institute of Mathematics, University of Aarhus, 183–199.

[57] Watson, G. S. (1983) Random points on hyperspheres. In *Stochastic Geometry, Geometric Statistics, Stereology*, Teubner Texte zur Mathematik 65, 225–232.

[58] Watson, G. S. (1984) Smoothing and interpolation by kriging and with splines. *J. Math. Geol.* 16, 601–615.

[59] Watson, G. S. (1986) Shapes of random triangles and a limit theorem. *Adv. Appl. Prob.* 18, 156–169.

[60] Watson, G. S. and Caspari, E. (1959) On the evolutionary importance of cytoplasmic sterility in mosquitoes. *Evolution* 13, 568–570.

[61] Watson, G. S. and Hannan, E. J. (1956) Serial correlation in regression analysis—II. *Biometrika* 43, 436–448.

[62] Watson, G. S. and Irving, E. (1957) Statistical methods in rock magnetism. *Monthly Notices R. Astronom. Soc.* 7, 289–300.

[63] Watson, G. S. and Leadbetter, M. R. (1963) On the estimation of the probability density—I. *Ann. Math. Statist.* 34, 480–491.

[64] Watson, G. S. and Leadbetter, M. R. (1964) Hazard analysis I. *Biometrika* 41, 175–184.

[65] Watson, G. S. and Leadbetter, M. R. (1964) Hazard analysis II. *Sankhyā A* 26, 101–116.

[66] Watson, G. S. and Smith, W. (1967) Diffusion out of a triangle. *J. Appl. Prob.* 4, 479–488.

[67] Watson, G. S. and Wells, W. T. (1961) On the possibility of improving the mean useful life of items by eliminating those with short lives. *Technometrics* 3, 281–298.

[68] Watson, G. S. and Williams, E. J. (1956) On the construction of significance tests on the circle and the sphere. *Biometrika* 43, 343–352.

[69] Watson, G. S., Zador, P. L. and Wilks, A. (1980) The repeal of helmet use laws and increased motorcyclist mortality in the United States 1975–1978. *Amer. J. Public Health* 70, 579–585.

[70] Watson, G. S., Zador, P. L. and Wilks, A. (1981) Helmet use, helmet use laws and motorcyclist fatalities. *Amer. J. Public Health* 71, 297–300.

2
THE CRAFT ORGANIZED

Norman T. J. Bailey

Norman T. J. Bailey was born in London on 27 May 1923. He was educated at Christ's Hospital, and in 1941 went on to read mathematics at Cambridge University. During the latter part of the Second World War he worked as a Scientific Officer at the Admiralty in London; he then returned to Cambridge to take the Maths Tripos III and the Diploma in Mathematical Statistics. He later obtained his D.Sc. from Oxford University.

In 1948 he was appointed statistician to the Medical School, Cambridge; from 1953–4 he worked as statistician to the Division for Architectural Studies, Nuffield Foundation. In 1955 he was appointed Reader in the Design and Analysis of Scientific Experiment (later Biometry) at Oxford University.

Norman Bailey travelled to the USA for the first time as a visiting professor during 1961–2, visiting both the Department of Biostatistics, The Johns Hopkins University, and the Department of Statistics, Stanford University.

He returned to the USA in 1966 as Professor of Biomathematics and Chairman of the Biomathematics Division, jointly sponsored by the Cornell University Medical School and the Sloan–Kettering Institute for Cancer Research, in New York. In 1967 he joined the World Health Organization (WHO) in Geneva as the senior statistician in the new Division of Research in Epidemiology and Communication Science. He later transferred to the Health Statistical Methodology Unit, which he headed from 1974 until statutory retirement from WHO in 1983.

In 1981 he was elected to the post of Professor of Medical Informatics in the University Cantonal Hospital, Geneva, where he now works on a full-time basis.

Professor Bailey has been closely involved in the application of mathematics, statistics and operational research to a wide variety of medical and biological problems, especially in the area of clinical, epidemiological and public health research. He is the author of eight books and numerous papers on applied mathematical and statistical topics. He is a fellow of the Royal Statistical Society; he is also a member of the International Statistical Institute, the American Statistical Association, the International Biometric Society, and the International Epidemiological Association. He was for many years vice-president of the WHO Medical Society.

Norman Bailey was married to Jean Mackenzie in 1959, and they have two sons. His recreations are literature, music, walking and cross-country skiing.

An Improbable Path

1. Childhood and School: Elementary Mathematics

I have always been intrigued by numbers, symbols, formulae, and foreign alphabets. Perhaps because they seem to hold some promise of cabbalistic knowledge or insight not otherwise readily obtainable? Anyway, I took to elementary arithmetic early, and needed little encouragement to fall in with my mother's idea that I should look rather foolish if, on going to school at the age of five, I did not know all the multiplication tables up to 12×12! After that, school seemed rather dull, though I do remember being fascinated by a boy whom I can still visualize and whose distinguishing characteristics were that his ears were always very green (due to lack of washing) and his "sum cards" involved division (which I could not do) with weird "$\div$" signs.

Fortunately, I was afflicted with septic tonsils which meant that, in the winter at least, I spent most of the time at home nursing a series of upper respiratory infections. This naturally accelerated rather than retarded my education. By the time I was seven or so a tonsillectomy became essential. The only thing I really needed in hospital was the little "table book", a small cheap compendium of weights and measures, natural constants, chemical symbols, and elementary formulae, which I was still clutching in my hand while being dragged off for the ritual cleansing after admittance to a ward.

A big factor in my life was that my parents moved house just before I was five from Worcester Park in south London to an address in Norbury, the significance of the latter being that it was just inside the London County Council area, as it then was—the boundary was at the bottom of the garden. This meant that I would become eligible in due course to sit for a certain kind of scholarship, although of course no one had even heard of this at the time.

It was usual, round about the age of 10, for children to sit for examinations that would give them a chance of entering various local secondary schools. Those living in the London County Council area could also try for scholarships to Christ's Hospital, a "public school" originally founded by Edward VI in London in 1553 but transferred in 1902 to a large site in the country several miles from the nearest town, Horsham. My septic tonsils gave me the opportunity of having a trial shot at this examination while still only $8\frac{1}{2}$, which was a valuable experience. Naturally, I achieved nothing: my health was none too good, I misread the examination papers, and generally failed to do my best.

My lasting detailed memory of this examination is one of the mathematics questions which required the candidate to write down *all* the prime numbers between 100 and 200. I was somewhat annoyed by having no idea of how to proceed—nor had the headmaster, as it turned out. In fact it was only several years later that I succeeded in devising an algorithm that was both simple and rapid, being based on a skeleton grid of numbers up to 200 in which whole rows and columns of nonprimes could be rapidly crossed out. A bright child *could* do this in five minutes, but it was asking rather a lot! Nothing daunted, however, I sat the examination the next year, got onto the short list, somehow survived a searching interview by a panel of the school's masters and governors, and was packed off to school in a special train from Victoria Station in September 1933.

Christ's Hospital was a good school in healthy surroundings where, as a town boy, I developed a lasting love of the countryside. Discipline was strict, games were compulsory, academic standards were high. One-third of the entrants were scholarship boys, mostly but not entirely from London, the rest being hand-picked by the school governors. Most came from working-class or lower-middle-class homes, and a means test operated to exclude the more affluent. The school uniform was a great social leveller—Elizabethan dress, worn most of the time and consisting of the well-known long blue gown, knee breeches, neck bands, yellow stockings and black shoes.

Although I am not now in favour of this kind of narrow, forced, non-coeducational, geographically and socially isolated education, it was at the time the only available way ahead. I was always glad to get home and away from school at the end of term, but always glad to get back to school and away from home at the end of the holidays.

I was greatly encouraged in my last years at school by the head mathematics master, William Armistead, who prepared me for the Cambridge entrance scholarship examination which resulted in my obtaining a Minor Scholarship to Trinity in December 1940. Armistead also guided me through the Higher Certificate examination which, at my second attempt in June 1941, resulted in a State Scholarship. These were rather scarce at the time, and the financial award involved was crucial in enabling me to stay at Cambridge and support both myself and my widowed mother after my father died early in 1942.

2. Cambridge University: Maths Tripos II

Wartime Cambridge was very restricted for obvious reasons, such as the black-out, food rationing, fire-watching, home guard, and senior training corps, though compared with school it provided a first taste of personal freedom. I certainly found it very stimulating, both intellectually and socially. My friends and acquaintances were drawn from many different countries, races and creeds. At school we had been nearly all white, English, Anglo-Saxon, Protestants. There had been only an occasional Catholic or Jew. Foreigners had been almost entirely absent, and even the Welsh and Irish had been suspect! Cambridge provided a culturally diversified setting which I greatly appreciated.

When I went up to Trinity in October 1941 I found a strong preponderance of scholarship entrants because of stringent wartime regulations on entrance standards. This considerably increased the intellectual pressure, and in mathematics it meant that most of us were expected to embark on a two-year course leading to Part II of the Maths Tripos. The latter was the normal degree-level examination, while Part I was considered equivalent to the last year at school.

My introduction to probability and statistics was a somewhat curious one, indeed largely a comedy of errors, as it turned out. When I first went to Cambridge I knew virtually nothing about either of these two subjects. And when I finished the two-year wartime course in June 1943, taking a disappointing second class in Part II of the Tripos, I still knew next to nothing. We had been encouraged to attend certain special courses that might provide technical knowledge that would be useful later in support of the war effort. I attended lectures on statistics given by Oscar Irwin who recommended A. C. Aitken's *Statistical Mathematics*, but I failed to discern any practical relevance in either the lectures or the textbook and gave up after a few weeks. A parallel course on electronics was more promising, consisting largely of being let loose in an electronics laboratory full of mysterious apparatus, meters and oscilloscopes, but with a minimum amount of direct instruction.

This greatly intrigued me, and I had already had some experience at school in successfully putting together large multivalve radio receivers. So at least I emerged from Cambridge with a second-class honours degree (my strength, such as it was, being primarily in applied, rather than pure, mathematics), plus an apparent acquaintance with the elements of electronic engineering.

3. War Work in London: Operational Research

On the strength of expertise supposedly acquired from the electronics course, I was drafted into Section 9 of the Signal Division, at the Admiralty in Whitehall, to work as a junior scientist on technical problems associated with locating German U-boats as accurately as possible from an examination of sets of direction-finding (D/F) bearings obtained from short-wave radio transmissions. Of course the cipher experts at Bletchley Park were already reading German Naval Enigma, and many U-boat signals provided highly accurate estimates of position. But no one knew how long this would last, and in any case good independent estimates of probable locations were essential.

It soon became clear that little could be done to reduce the often considerable inaccuracy of D/F bearings. Electronic engineers had already developed efficient aerial arrays for the usual goniometric devices, and the direction of arrival of a moderately strong signal could be accurately measured. Unfortunately, the signals often arrived from misleading directions because of the way they had been deflected by perturbations in the ionosphere, over which no control was possible. In short, we were compelled to accept a large degree of basic statistical variation in D/F bearings as measured. The problem was then to consider how we might extract information from a set of bearings on a given signal as to the most probable location of the source of the signal, together with an indication of the likely accuracy of the estimated position.

So here I was, confronted with the necessity of developing a probabilistic model of the whole D/F process, as well as devising a practical procedure for estimating the position of the radio transmitter sending any given signal. Fortunately, Maurice Kendall's *Advanced Theory of Statistics* [13] had just appeared, and I read this avidly on my journeys to and from work on the London Underground. Later I attended the excellent lectures of Egon Pearson at University College London, together with the stimulating but more difficult courses by J. E. Walsh. Motivation was clearly a major determining factor for me. First and foremost, there was the situation of wartime urgency, plus the fact that all the scientific workers in the group to which I belonged cooperated directly with operational staff and had open access to the adjacent operations plotting room for the North Atlantic control. In addition, the D/F problem itself was a practical, technical challenge.

Under these conditions it is not surprising that it soon became possible to represent the reigning First World War technology of plotting bearings with lengths of string on gnomonic projection charts (which preserve angles) in terms of a mathematical and probabilistic model. We assumed that the differences between true and observed bearings at any given D/F station were normally distributed, the mean and variance appropriate to any subjective reliability classification (e.g. A, B, C, etc.) being estimated from past records on sufficiently accurate determinations of transmitter locations.

It is easy to see that, for any given position (X, Y) of a transmitter, where X and Y are the parameters to be estimated, the likelihood of the observed set of bearings is of bivariate normal form in X and Y. Lines of equal likelihood are given by concentric ellipses, with the best estimate of position $(\hat{X}, \hat{Y})$ being at the centre. In practice $(\hat{X}, \hat{Y})$ was specified as the "most probable point", and a convenient rectangle (rather than an ellipse) was also supplied corresponding to an operationally appropriate level of probability such as 80%, 90% or 95%.

For computational purposes, reference tables were calculated for all possible bearings for each D/F station. It was then a simple matter, for any given set of bearings, to read off the relevant constants from the tables and calculate on an electric desk calculator the coefficients determining the equation for the concentric ellipses, and hence the centre and appropriate rectangles to be searched (if sufficiently small) by Coastal Command or used as a basis for deciding avoidance strategies. The computations would take around 30 minutes for a set of 40 bearings. This was quite good in the circumstances, but a bit slow for urgent situations. The opportunities for electronic computation are obvious, but I had left the Admiralty before this took place.

4. Return to Cambridge: Maths Tripos III, Diploma in Statistics, University Medical School

In 1946, after the war was over, I applied for admission to the Scientific Civil Service, but was rejected outright, probably (I like to think) because I was not allowed to talk about my successful D/F work during the interview (the only test applied) as it was still classified as Top Secret!

As a result, I returned to Cambridge in October 1946 to read for Part III of the Mathematics Tripos. I rejected the advice of both my supervisor and Director of Studies, who wanted me to "offer" *at least* eight special courses for the examination, and decided to concentrate on four only, namely Waves and Tides, Dynamical Meteorology, General Mathematical Statistics and Stochastic Processes. (So far as the latter were concerned, I am ashamed to have to record that I failed to understand the time-series part of Maurice Bartlett's lectures, but was very stimulated by the theory of evolutionary processes, which became of great importance to my subsequent work.) This strategy paid off handsomely and I achieved a distinction, more through the motivation of doing what I enjoyed than through special ability: several of my friends failed to reach distinction level although they were quite evidently much better mathematicians than I was.

During this year I had attended some of R. A. Fisher's lectures on mathematical genetics and found this an exciting subject. As a result I signed up in October 1947 for the new one-year course leading to the postgraduate Diploma in Mathematical Statistics, with genetics as a special field of application. About this time I was also offered the post of statistician to the Medical School in Cambridge, to be taken up after completing the diploma in mid-1948. This was a challenging opportunity, as I had always been very interested in medical subjects but did not really have the kind of emotional and intellectual make-up that is needed for a successful career in medicine. And in any case I did not have either the time or money required to survive the long-drawn-out apprenticeship.

I was still having some difficulty with statistics as such, though I was certainly up to pass level in the diploma. In fact, I emerged this time with an unexpected and ill-deserved distinction, a marvel that I believe was correlated with the amount of time I had spent in assiduous devotion to the breeding of mice in Fisher's laboratory at Whittingham Lodge.

I gradually came to know Fisher quite well, having discovered early on that he either liked or hated people. Fortunately I was on the right side, but one still had to be very careful. Fisher was certainly a great source of inspiration in regard to the application of mathematics and statistics to biology, and was always ready to discuss and help with awkward problems. It was in Fisher's laboratory that I first met people like Victor Rothschild and Luigi Cavalli-Sforza, the latter in particular arousing my interest in the genetics of bacteria and viruses.

At the Cambridge Medical School I was very much concerned with providing a statistical consulting service to all the medical departments, and with undertaking applied research on all kinds of biological and medical subjects: genetic linkage, capture–recapture techniques, medical biostatistics, socio-medical surveys and operational research in medicine. The latter arose through the consulting work which I started in 1950 as a member of the

Investigation Team for the Nuffield Provincial Hospitals Trust project on the Function and Design of Hospitals. I was soon involved in queueing problems in outpatient departments, and was able to demonstrate in 1952 the trade-off effects between consultants' waiting-periods and patients' waiting-times by using a simple simulation model for the finite queueing processes involved, the relevant calculations all being performed with paper and pencil! I have always liked this particular piece of work [1] because the operational consequences were so easily obtained and understood, without recourse to heavy analytical theory or even sophisticated electronic computation, which was only just becoming available.

The main project was inspiringly directed by the architect Richard Llewelyn Davies, and pioneer work on hospital design was achieved. I, as statistician, worked closely with Brigadier Digby Welch, the field work organizer who had had wartime experience of operational research in south-east Asia, and also with Jack Nightingale, a research assistant for time study who had worked on a variety of technical problems in business and industry. Neither of these men was academically inclined, but it was a great pleasure to work with them since their practical skills complemented my more theoretical approach. It was also around this time that I first met Jean Mackenzie, who worked for Digby Welch as a research assistant and who later became my wife.

At the Cambridge Medical School I came in close contact with Dr Alan Macfarlan, the Reader in Human Ecology. He confronted me with various problems in the epidemiology of infectious diseases. My first theoretical paper in this area, on the simple stochastic epidemic, dates from 1950. Later I became involved in developing chain-binomial models for the spread of infectious disease within families, as previously introduced by Major Greenwood in the context of measles. Greenwood had already suggested that inadequacies in such a model to represent family micro-epidemics might be due to variations between families in the average chance of infection. My contribution (1953) was to postulate a beta-distribution to account for this possible variation, and to estimate the two parameters involved using maximum likelihood. Excellent fits of theory to data were then obtained.

5. Nuffield Foundation: Statistics in Architecture

I left Cambridge early in 1953 to take up the post of statistician to the Division of Architectural Studies at the Nuffield Foundation in London. This involved a variety of practical operational research investigations, including the design of agricultural research laboratories. In particular, I designed a modified "snap-round" survey of scientific work carried out in some 16 Agricultural Research Service stations throughout the United Kingdom. Over 300 individuals, covering 3 grades and 11 disciplines, were investigated. Some 40 "observations" were made on each individual in the course of 12 months. Each "observation" consisted of a long, detailed checklist of items on type of work in progress, and the use made of bench length, sinks and draining boards, electric services, cold and hot water supplies, gas supplies, drainage, and fume cupboards.

All data were recorded in the field at time of observation on Hollerith mark-sensing cards, subsequently to be punched automatically, and then analysed mechanically on a Hollerith Sorter Counter. What has continued to impress me about the use of such old-fashioned methods 30 years ago is that a detailed analysis of the very extensive survey material was completed within three months of the end of the year-long survey in 1955. Only one or two clerks working under my general direction were required at this stage (I had by then moved to Oxford and was acting only on a consultant basis). If only we could achieve a comparable efficiency nowadays with all our computers and text processors!

There is no space here to describe the results in detail. Suffice it to say that the statistical distributions of the use made of various laboratory facilities could be combined by simple convolutions to predict usage patterns for teams of research workers in different proposed architectural designs, thus providing what was at the time a new approach to laboratory planning.

While at the Nuffield Foundation I continued with the application of statistics and operational research to medical problems, particularly those involving queueing processes. I also had the opportunity to extend my previous stochastic investigations of micro-epidemics in families by using the first-class material collected by R. E. Hope-Simpson, of the Cirencester Public Health Service. These data, based on very careful field work, showed actual dates of onset of measles in families of various sizes. In 264 families with two susceptible children, 45 had one case only, and 219 had two cases. The time interval between the latter cases had a bimodal distribution and showed great variability. Standard chain-binomial theory was no longer adequate to describe such observations. Various hypotheses were tried and found unsatisfactory. A fairly promising model, however, was to assume that infection was followed by a variable latent period (normally distributed, say, as a first approximation, with mean μ and variance σ^2), after which there was an extended infectious period (of fixed length α) during which infection of other susceptibles might take place as a Poisson process with parameter λ. I first described this model in 1954, and later carried out maximum-likelihood estimation of the four parameters for families with two and three individuals, obtaining satisfactory goodness-of-fits. A very significant finding was that the average latent period of measles came out at about $8\frac{1}{2}$ days, with an *extended* infectious period of $6\frac{1}{2}$ days, as compared with the previous chain-binomial assumption of infection being closely concentrated about a single point of time. The overall incubation period was about 15 days, with a case-to-case interval of 11 days.

The application of mathematical modelling to infectious disease dynamics was thus beginning to develop as a major interest.

6. Oxford and the USA: Biometry

Towards the end of 1954 I had the good fortune to be elected to the new readership in the Design and Analysis of Scientific Experiment (as the unit was then called), in succession to David Finney who had vacated the previous lectureship to become Reader in Statistics at Aberdeen. I took up my post in Oxford from 1 January 1955, and the year started well when my first substantial review paper on the modelling of epidemics was read to the Research Section of the Royal Statistical Society on 5 January.

The name of the department was changed after two years to the more suitable and euphonious "Unit of Biometry", following a sensible proposal previously put forward by David Finney. My duties involved providing a statistical consulting service to other departments in the Faculty of Biology, to which the unit belonged, and also, by simple extension of subject matter, to departments within the Faculty of Medicine. I was ably supported by two lecturers: John Scott, who was an excellent teacher with very practical interests, and A. Morris Walker on whom we all relied for support in the intricacies of mathematical and statistical theory. We provided courses of lectures for the postgraduate Certificate in Statistics (one year) and the Diploma in Statistics (usually two years). I learned a lot about the design of experiments by lecturing to those who knew less than I did. Undergraduate-level lectures on elementary applied statistics were also given on an increasing scale for individual biological departments.

I was involved at the same time in a wide range of applications of biostatistics, stochastic modelling and operational research to problems in biology, medicine and public health. The work on infectious disease dynamics continued, and I was able to find time to expand my Royal Statistical Society paper into a monograph entitled *The Mathematical Theory of Epidemics*, [2] which was published by Griffin in 1957 and was (I think) the first general survey of the field to be published in book form.

Although I was pleased and grateful to have secured an important and virtually tenured academic post in a prestigious university, there were some depressing undercurrents that became more explicit as time went by. There was, somehow, a certain local lack of enthusiasm for statistics and biometry at Oxford, in the university in general and in the Faculty of Biology in particular. The same was also true for the Ministry of Health in London, with which I had a number of distinctly unfruitful contacts.

Perhaps it was not without some degree of psychosomatic aetiology that I fell ill with a spontaneous pneumothorax while on holiday in San Remo, Italy, in September 1957. I elected, for various reasons, to stay in the local Ospedale Civile in a public ward. Since the treatment there was very conservative I was confined to bed for five weeks. This turned out to be a highly interesting and valuable experience, and for the second time in my life I derived considerable benefit from illness. In particular I was able to plan the contents of a book, requested by the then English Universities Press (later Hodder and Stoughton) and published by them in 1959 under the title *Statistical Methods in Biology*. [3] This has sold well over the years, having been continually reprinted and appearing in a second edition in 1981. I was also able to plan, while in hospital, *The Mathematical Theory of Genetic Linkage* [4]. I greatly enjoyed writing this, though it was never so popular as the other books.

Looking back I see my time in Oxford very much as a period of consolidation, though not without its excitements. In 1959 I married Jean Mackenzie, whom I had first met many years before at the Nuffield Provincial Hospitals Trust. In 1961 we travelled together on sabbatical leave to the USA, where I had appointments to two visiting professorships: the first in the Department of Biostatistics, School of Hygiene and Public Health at The Johns Hopkins University, Baltimore (fall and winter quarters, 1961–2); and the second in the Department of Statistics, Stanford University, California (spring and summer quarters, 1962). These arrangements were initiated by Charles Flagle and Lincoln Moses, respectively. We took our own car with us and did a lot of travelling within the USA, visiting other universities and research institutes.

This visit was a tremendously stimulating experience, socially, culturally and intellectually. I found myself giving lectures at Johns Hopkins on elementary applied stochastic processes, which of course I based on those parts of Maurice Bartlett's Cambridge lectures that I could understand. In no time at all Wiley and McGraw-Hill were competing for options on a possible future book. (*The Elements of Stochastic Processes* [5] was eventually published by Wiley in 1964 and did very well for several years.)

What impressed me most right from the start in university circles was the way in which both students and staff interacted and communicated. One was certainly not listened to as an oracle! But one *was* listened to, and then expected to take part in lively debate, Moreover there was not only respect in medical circles for mathematics and statistics but actual competence amongst many of the staff and students. Several of the latter would take a year off from their medical studies to obtain qualifications in biostatistics. I also developed a number of close personal and professional ties with biostatisticians at the National Institutes of Health (to which I made regular weekly visits), including Jerry Cornfield, Sam Greenhouse, Ed Gehan, and many others.

We were away for just over a year, and after returning to Oxford in October 1962 everything seemed distinctly dull and low key. Before I left for the USA, the Biology Board had been giving serious consideration to possible future developments and expansion in biometry. When I returned, all this had gone. Moreover, many plans in other departments had also been postponed. There was a general mood of disillusion, exacerbated by an increasing shortage of funds.

Before my year away I had already established a small research team, supported by the Nuffield Foundation, to work on operational research problems in medicine. Spurred on by what I had seen in the United States, I again approached the foundation for a grant to buy a small computer for the teaching, service and research work of my unit. I had in mind an Elliott 803, and £30000 was soon made available. All our computing had previously been done centrally in the University Computing Laboratory, but the problems of their administration were such that computations requiring only milliseconds of CPU time often had an overall turn-around time of a week!

I was naively surprised when the Computing Laboratory tried to block my proposal. Fortunately, I was a member of the committee which dealt with these matters and was able to hear at first-hand the director's view that I and my colleagues were quite incompetent in matters of numerical analysis and would, in particular, lead undergraduates astray. We argued for a whole term, after which I sent in my own carefully considered case to the university's General Board. They decided that we could have our own independent computer *provided* we kept it to ourselves and did nothing for anyone else! The Elliott was duly installed, and our whole handling of biostatistical methodology was rapidly transformed and modernized.

By mid-1964 a succession of further minor discouragements brought me to the point where I felt the need for a change and some more positive action. It only needed a few telephone calls to the United States to get my name on a short list for the chairmanship of a new Division of Biomathematics, jointly sponsored by the Cornell University Medical School and the Sloan–Kettering Institute for Research in Cancer, both in New York City. I was soon on my way for an interview, and was delighted to learn in due course that I had been selected for the post.

It naturally took a little while to extricate myself from Oxford. In the meantime my first son James was born in August 1964 and I was very preoccupied with my family. I did find time, however, for carrying out a lot of computer work, and developed an efficient ALGOL program for studying simulation models of the spatial spread of epidemic disease from an initial focus. This resulted in a substantial paper that was eventually presented to the Fifth Berkeley Symposium in 1966.

We all finally set sail for the USA as immigrants in December 1965, taking a long holiday on Sanibel Island off the west coast of Florida before going to New York. I spent many happy hours on the sunny beach completing the writing of *The Mathematical Approach to Biology and Medicine* [6]. This was a kind of semiphilosophical, personal credo (eventually published by Wiley in 1967) that I found of great benefit in systematizing my own ideas and attitudes with regard to the applications of mathematics, statistics, computing, and operational research.

7. Cornell and Sloan–Kettering Institute (SKI): Biomathematics

The attraction of the Cornell–SKI post was that it involved a more or less autonomous biomathematics facility within a diverse but highly stimulating medical environment—the kind of thing I had often dreamed about. I had been assured that administration would be minimal and that fund-raising had already been taken care of. The Director of Sloan–Kettering, Frank Horsfall, was as good as his word. But the Dean of the Medical School, for his part, had already embarked on a course of action that was doomed to failure from the start. As a consequence, I did have more administration to begin with than I had expected, but things gradually got sorted out. Originally, the family plan was to stay four or five years in New York and then perhaps return to Europe (United Kingdom or continent). But there was a complication.

For a number of years, the World Health Organization (WHO) had been playing with the idea of some kind of health research institute to be established in Europe along the lines of the National Institutes of Health in America. I was on the committee concerned with technical planning while still at Oxford and would have been glad to join WHO from Britain if the occasion had arisen. However, negotiations dragged on and I took up the New York appointment instead. Six months later an opportunity did arise quite unexpectedly, and I had a crucial decision to make. After much thought and discussion I decided to take up the offer to join what was now to be established as a regular division of WHO headquarters in Geneva, entitled Research in Epidemiology and Communication Science (RECS). Such a chance might never be repeated.

Again, it took time to disengage myself from Cornell–SKI, and I felt badly about wanting to leave so soon after arrival. However, they were very kind and understanding and there were no hard feelings or unpleasantness. My second son, Michael, was born in New York in April 1966, but it was not until June 1967 that we all recrossed the Atlantic and drove across France to Geneva.

Although my stay in New York was relatively short I did some work on spatially distributed birth, death and migration processes, and also developed a perturbation approximation to the simple stochastic epidemic. More valuable, however, was the experience of helping to develop and organize a biomathematical facility explicitly geared to the practical support of a wide range of medical and health activities.

8. The World Health Organization: Biostatistics, Modelling and Disease

It would be impossible to summarize adequately my varied experiences in WHO in anything less than a whole book. However, I will try to pick out the points of greatest relevance to the present subject.

I joined the new research division in July 1967 with the prospect of a five-year programme, and was initially classified as a "theoretical epidemiologist". Later I was in charge of the biostatistical, operational research and computer science components of the group. There was a galaxy of talent in the division including different kinds of epidemiologists, public health experts, ecologists, sociologists and behavioural scientists.

The director was given a free hand to develop the division's research programme, and this proceeded apace until there was an average of one project per professional staff member (about 35 in number). Although a lot of good applied health research was carried out, it all became very diffused and was increasingly difficult to justify as time went by.

One of the best projects was a large multidisciplinary study of malaria control in Nigeria, excellently described by Molineaux and Gramiccia in *The Garki Project* [14]. This involved members of RECS, as well as a large number of other WHO staff, both in Geneva and in the field. Klaus Dietz, who was one of my own group in RECS, played a primary role in the development of a realistic mathematical model, ably assisted by Louis Molineaux as a mathematically competent epidemiologist. This work is generally recognized as a major breakthrough in the quantitative understanding of the population dynamics of malaria, but most of the real practical implications and potentialities for management and control on a world-wide basis have never been realized.

Generally speaking, the WHO *milieu* made me more vividly aware of a fascinating and challenging range of medical, epidemiological and public health problems, largely of course in the Third World where the needs are most urgent. There were considerable opportunities for direct contact with field efforts to control infectious and parasitic diseases on a vast scale—affecting hundreds of millions of people.

In June 1971 a scientific group was invited to make a critical analysis of the scientific work of RECS, especially "in the development of models for disease and health systems". David Finney was elected chairman, and Joe Gani acted as a special advisor. I and my mathematical and statistical colleagues found the discussions at the meeting and the ensuing constructive criticisms extremely helpful in support of expected future developments. Some, however, read all this as personal criticism, and the group's report was effectively suppressed. A few months later I put forward a carefully reasoned proposal for a new, broad communication science facility within WHO to provide practical mathematical, statistical, computing and operational research support to all technical WHO programmes. This proposal met exactly the same fate as the scientific group's report, and for the same reasons. It was no surprise therefore when I was told early in 1972 that my contract would not be extended beyond the end of the year, and I received an official confirmatory letter from the head of personnel thanking me for my services.

I protested strongly against this treatment, and made it clear that I would not lightly accept such a summary termination of my involvement in the development of important ongoing and successful biostatistical activities. It was not long before an Assistant Director General, who later became the present Director General, undertook to transfer me and some of my colleagues, including Klaus Dietz, to the Health Statistical Methodology unit

(HSM) which provided the main statistical consulting service to the whole organization. This was a perfectly satisfactory solution for me; I joined that unit in November 1972, and subsequently became its head in August 1974.

In the meantime I was able to consolidate my acquaintance with the rapidly expanding field of infectious disease modelling, and prepared a greatly enlarged second edition of my first book, now to be called *The Mathematical Theory of Infectious Diseases* [7].

With the exception of my last year in WHO, my work as chief of HSM was about 40–50% administration, the remaining time being allotted to collaborative work with my statistical colleagues, in headquarters in Geneva, in various regional offices of WHO around the world, and in a number of individual countries. On the research side I became increasingly concerned with infectious and parasitic disease modelling (*The Biomathematics of Malaria* [8] was published by Griffin in 1982), and with the development of health-care systems models, entailing of course a major emphasis on Third World requirements.

My background in operational research, biostatistics and stochastic modelling was always very prominent in my approach to operational problems arising in field contexts. Although there has been a growing interest in these approaches at the technical level, especially in the Third World itself, the top WHO administration has over the years become increasingly opposed to such work.

One would have thought, as global problems became visibly and admittedly more multidisciplinary and multisectoral in character, that there would have been a growing perception of the need for the operational modelling of large interactive systems, using appropriate combinations of deterministic and probabilistic components.

I am not suggesting that such developments are easy—far from it—only that they appear to me to be urgently needed, and special efforts have to be made if theory and practice are to be integrated in operational contexts where real-life decision-making goes on. There is little doubt that patients with diseases want to be cured, while those who are threatened prefer the risks to be reduced. Again, technicians, whether they be medical workers or scientists, find it professionally challenging to meet these demands. So at this level there is no conflict of interest. But administrators see their established methods of intuitive judgement questioned by those who count, measure, assess probabilities, check logic, test predictions, reveal operating mechanisms, and criticize implicit trade-offs; thus, they tend to regard these activities with disfavour. However, I see probabilistic modelling, in the context of health systems development, as a very important contribution to the fight against disease and for better health.

When I first took over HSM in 1974 there was a total of some eight or nine professional statisticians, who were not only personally responsible for the statistical aspects of a large number of field projects but also constituted amongst themselves a lively group for interactive applied research. In spite of constant efforts by myself and my colleagues to establish a firmer foundation for our work, deliberate attrition by higher policy over the years reduced the numbers to five by the time I had to retire in 1983, well below the critical mass required for really effective and productive activities. Two years later there are only four of them left, spending a high proportion of their time travelling and having no appointed leader to promote, coordinate and integrate the ever-decreasing biostatistical component.

I should, however, emphasize that I am very much in favour of WHO as *the* international World Health Organization. There is a general consensus that it has done much excellent work and has a lot of potential for future achievements. At the same time, the moral and physical pressures on many of its staff, at both the professional and general service levels, are often quite excessive (I got off pretty lightly as it turned out), and this considerably degrades the quality and effectiveness of its enterprises.

This is quite apart from the huge amount of bureaucratic activity that one finds in any large organization. During my final year in WHO, the time spent on internal comments, memos, reports, committees, etc., rose to nearly 100%. In my view most of this was totally self-serving, and quite irrelevant to our proper job of providing statistical support to field-based operations in preventive medicine and health service development. I was glad to leave at the end of May 1983: it seemed like the end of an era.

9. University of Geneva: Medical Informatics

Although our two boys, James and Michael, had been brought up at various schools in Geneva and were fluent in French and English, both wanted to go to a boarding school in England, which they duly did—being one year out of phase with each other. After three years at school they disliked both the school and England, and insisted on returning to Geneva. They had obtained quite good O levels and were able to enter the second year of one of the Geneva *collèges*, roughly equivalent to an English sixth form, emerging after three years with a Swiss *maturité* in science. This includes not only the usual science subjects and mathematics, but also three languages, history, geography, and philosophy. The elder boy is now reading biology at the University of Geneva, and the younger is studying economics.

Since my wife and I had decided to stay in Geneva after my departure from WHO, it was necessary for me to look for a job. I had several local university contacts, and as early as September 1981 I had managed to obtain an appointment to a professorial post in the Division of Medical Informatics, headed by Professor J.-R. Scherrer, in the University of Geneva. While I was still in WHO this was largely honorary, but I gave occasional lectures and gradually became acquainted with the division's activities. After leaving WHO the post was put on a more formal basis.

My current duties entail the usual mixture of teaching, consulting and research, with strong emphasis on the latter and virtually no administration. A major concern at present is the development of improved methods of computing the solutions of sets of differential equations describing general compartmental models that can be applied to a wide range of nonlinear biological and medical problems. Since the law of mass action, and modifications thereof, is very prominent in chemical kinetics and many physiological processes, as well as in infectious disease dynamics, there is a lot of common ground with my own previous special interests. In fact if my current research, pursued jointly with the Department of Mathematics, is successful I expect there to be far-reaching implications for more realistic modelling and prediction in complex systems representing the spread of infectious and parasitic disease through spatially heterogeneous communities.

I am of course glad to be back once more in an academic environment and free of the all too frequently self-serving activities of bureaucratic in-fighting. Nevertheless, my stay in WHO has had a very strong impact on my thinking about medical and health problems under two main aspects. First, contact with real disease situations affecting hundreds of millions of people in the Third World acts as a constant reminder that biomathematical theory is not just a private amusement, but is to be used for practical application in solving horrendous real-life problems. Second, dealing with such problems involves, not only technical medical and health action in the field, but, inevitably, a whole range of managerial and administrative actions. If the latter are to be facilitated by biomathematical theory the implications of this theory must be made available in a readily understandable and quickly usable form. This implies an appropriate integration of biomathematics, biostatistics, computer science and operational research, plus practical demonstrations of effective action.

10. Biomathematical Modelling: Practical Problems and Pitfalls

I suppose, looking back, my primary interest has always been in the use of some kind of applied mathematics for the clarification, understanding and solution of various practical problems in biology and medicine. This has covered genetics, ecology, epidemiology, clinical trials, hospital design and public health control. The major biological aspects of all such particular topics guarantee a considerable degree of natural variability, and this of course makes the inclusion of probabilistic and stochastic elements inevitable, even if we decide on occasion that numbers are sufficiently large for deterministic approximations to be good enough.

Any serious work in the areas mentioned entails crucial assumptions in quantitative terms, from which mathematical development must start—whether in the form of analytical derivations or computer simulations. These assumptions, suitably formulated, constitute what is nowadays usually recognized explicity as the "model". These may be supplied largely by the experts working in the field of study, or may be generated by interested mathematicians, but should arise preferably from a synthesis of these two different but complementary approaches. I have always found this modelling stage the most difficult, and first learned to handle it in the radio direction-finding context already mentioned. When one was sufficiently motivated by involvement in practical problems, it usually seemed that an effective working solution could be found sooner or later. But it appeared to be as slow and difficult as it was fascinating. I remember being staggered at the Admiralty by colleagues, with mainly engineering backgrounds, who seemed to be very fluent and rapid in generating models of real-life situations, e.g. of the structure and dynamics of convoys under threat of air or sea attack.

Nothing in my mathematical education prepared me for this, not even when I went back to Cambridge after the war to take Part III of the Maths Tripos. I should not of course imply that I found the mathematics easy either! But if one's main purpose is to solve a practical problem then there should be no psychological objection to consulting mathematical experts for the solution of mathematical difficulties that one cannot handle oneself. Alternatively, one looks for numerical approximations to analytically intractable problems—not that this does not require careful handling as well. And when this fails, one falls back on computer simulation, which with careful consideration is not quite so restrictive nowadays as used to be thought. It seems to me that in real-life contexts, practically useful solutions can *always* be obtained if only one can find the right formulation of the problem.

This latter point is of course crucial. It may be an article of philosophical faith, or may simply derive from the notion that, of the infinite number of models available to portray a given situation, some at least are very likely to be both approximately realistic and approximately soluble—and that is good enough.

Consider, for example, the epidemiological problem of finding an adequate description of the highly variable, bimodal distribution of the time interval between two successive cases of measles in a family with two susceptible children, as already briefly outlined at the end of Section 5. Because of the substantial variability, a strict chain-binomial model, with *fixed* incubation period and a very short interval of high infectiousness, is clearly inapplicable.

There is moreover a question of the epidemiological interpretation of the bimodal data. Of the 219 families with two cases, 190 provided a fairly orderly unimodal curve with mode at about 10 days—the approximate serial interval for case-to-case transmission. The remaining 29 cases constituted a fairly sharp peak close to the origin, and were considered to represent the consequences of both susceptibles having been simultaneously infected by contact with the same external source of infection, i.e. a "double primary" situation.

The first modification to the model worth trying was to envisage a variable period with variance v, say, but still retaining the very short infectious period. On this assumption it is easy to see that the second moment about the origin of the 29 simultaneously infected pairs should be $2v$. But the observed value V was only about $\frac{1}{2}v$, and significantly less than the hypothesis required.

The next obvious extension was to suppose the latent period to be followed by an *extended* period of infectiousness. Even with the latter being kept fixed, with length α say, this introduces a new element of variation. It was assumed that the infection of the second susceptible occurred as a Poisson process with parameter λ. To facilitate calculation the latent period was originally assumed to be normally distributed with mean μ and variance σ^2, though this restriction could easily be removed. There were thus four parameters, λ, α, μ and σ^2, to be estimated from the data. Although the use of maximum-likelihood estimation was quite straightforward, the analysis turned out to be fairly intricate and the original computations were very time-consuming.

To summarize briefly, there are three sources of information. First, let there be A cases of simultaneous infection, for which the time interval ω is distributed as

$$f_1(\omega) = \sigma^{-1}\pi^{-1/2} \exp\{-\omega^2/(4\sigma^2)\}, \qquad 0 \leq \omega < \infty. \tag{1}$$

Second, suppose that there are B families with case-to-case transmission, and C families with only one case. The probability of observing C, given $B + C$, is

$$f_2(C|B+C) = \binom{B+C}{C}\{\exp(-C\lambda\alpha)\}\{(1-\exp(-\lambda\alpha))\}^B. \tag{2}$$

Third, the B families involve a variable $\zeta = \xi + \tau$, where ξ has the latent-period distribution

$$f(\xi) = (2\pi\sigma^2)^{-1/2} \exp\{-(\xi-\mu)^2/(2\sigma^2)\}, \qquad -\infty < \xi < \infty,$$

and τ is the time-interval from the beginning of the infectious period of the first case to the point at which infection of the second case actually occurs, having distribution

$$f(\tau) = \lambda\{\exp(-\lambda\tau)\}\{1-\exp(-\lambda\alpha)\}^{-1}, \qquad 0 \leq \tau \leq \alpha.$$

The distribution of ζ then comes out to be

$$f_3(\zeta) = \frac{\lambda\exp\{-\lambda(\zeta-\mu-\frac{1}{2}\lambda\sigma^2)\}}{1-\exp(-\lambda\alpha)} \int_{b'}^{b} \frac{\exp(-\frac{1}{2}t^2)}{(2\pi)^{1/2}} dt, \tag{3}$$

where $b = \sigma^{-1}\{\zeta - (\mu + \lambda\sigma^2)\}$, $b' = b - \alpha\sigma^{-1}$. The actual data involved the numbers A, B and C, and the sets of observed values ω_i and ζ_j for the variables ω and ζ.

The original calculations in 1955 were based on R. A. Fisher's method of maximizing the likelihood by the well-known iterative scoring procedure, involving repeated calculation of four scores and a 4×4 information matrix. Initial trial values were obtained by setting the observed V, C, $\bar{\zeta}$ and v, equal to their expectations, and then solving to give simple formulae for λ_0, α_0, μ_0 and σ_0. Several person-weeks on a desk calculator were required to obtain satisfactory results. The actual figures obtained were

$$\hat{\lambda} = 0.256 \pm 0.032,$$
$$\hat{\alpha} = 6.57 \pm 0.76 \text{ days},$$
$$\hat{\mu} = 8.58 \pm 0.32 \text{ days},$$
$$\hat{\sigma} = 1.77 \pm 0.13 \text{ days}.$$

The average incubation period was $\hat{\alpha} + \hat{\mu} = 15.2$ days, and the average serial interval $\bar{\zeta}$ came out as 11.0 days.

It was of course easy to carry out a χ^2 goodness-of-fit test, and a nonsignificant value of 12.9 was obtained on eight degrees of freedom, after apparent excesses of observations at 7 and 14 days had been smoothed out.

The lengthy infectious period required by the model was viewed with suspicion by some epidemiologists, but it seems to be an inevitable deduction from a close visual investigation of Hope-Simpson's original data, with detailed modelling and analysis bringing this out quite clearly.

The above account only summarizes the beginning of the work—it was in fact extended from families of two to families of three, with a considerable increase in the complexity of both data and analysis. The estimated values of the main parameters were, however, very similar, and not significantly different from the figures found for families of two.

This kind of modelling could have important epidemiological implications, especially if based on data obtained both before and after some kind of public health intervention such as an immunization campaign. We should then have specific measures of the biological consequences of the intervention undertaken, in terms of changed infection-rates on lengths of latent or infectious periods. Admittedly, this has never been done, partly because of the cost involved, but largely no doubt because the methods of analysis are beyond normal epidemiological competence. Personally, I believe that the modelling is *not* particularly difficult to comprehend—it is the mathematical analysis that is the primary cause of bafflement.

Later, in 1970, a modern computer approach was used by Cynthia Steinberger and myself [10] to streamline the whole procedure. In essence, this involved no more than writing down the relevant log likelihood function, L, namely

$$L = \sum_{i=1}^{A} \log f_1(\omega_i) + \log f_2(C|B+C) + \sum_{j=1}^{B} \log f_3(\zeta_j), \qquad (4)$$

where f_1, f_2 and f_3, were given by (1), (2) and (3), and a suitable optimization programme was then used to maximize L. We used the CERN program MINROS, which then automatically calculated the maximum-likelihood estimates, their standard errors, and the relevant goodness-of-fit χ^2. This reduced computing time from several weeks down to a few minutes, and provided rapid, error-free calculation. A few additional complications, such as uncertainty of chain classification (e.g. double primary or case-to-case transmission) or variability between families in chance of infection (especially important for families of three or more), could also easily be incorporated.

We were able to correct some minor errors in the earlier analysis, and easily carry out another new investigation, requested by the epidemiologist Dr K. Petersen, on the spread of infectious hepatitis in families with three children, drawn from hospital material collected in the Hamburg area.

More work is needed on this kind of model, investigating for example the incorporation of a gamma-distributed latent period for greater realism and flexibility. Also, more attention should be paid to the larger-scale community phenomena, where stochastic aspects are less important though not negligible. Much insight is available, but completely satisfactory validation has yet to be achieved. The main difficulty seems to be that, while the use of deterministic compartmental models may entail computable sets of differential equations linked to optimization procedures for estimation, there are serious problems in handling spatial, demographic and geographical heterogeneity. Some progress has, however, been made in the field of influenza modelling and prediction, particularly in the USSR.

I believe that the difficulties at both micro-level in families and macro-level in communities will only be resolved by the advent of general operational efforts involving real-life decision-making.

I should therefore like to emphasize my considered opinion that not nearly enough importance is attached to developing theoretical work in the context of, and as a support for, practical operations. While the seriousness of this charge varies from field to field, it is certainly justified in the area of infectious and parasitic disease control, in which I have long been interested. Although one could hardly impugn any given theoretical paper (which might have just the insight required to solve a major practical difficulty), I have a strong quasi-statistical impression that at least 90% of the mathematical literature in this area is either irrelevant or has a very low degree of relevance. Since many hundreds of millions of people are affected by unpleasant or repulsive infectious and parasitic diseases, I believe that a far greater contribution could be made to the control of these scourges if the mathematical brainpower available were more effectively utilized in the context of public health interventions.

11. Disease Control: Adaptive Environmental Management

I have already mentioned in Section 8 the highly significant work of Dietz, Molineaux, and others in developing a useful model to assist the top world priority of malaria control, and pointed out the sad fact that "most of the real practical implications and potentialities for management and control... have never been realized". A somewhat similar situation exists in regard to the second global priority, namely schistosomiasis (bilharziasis), where the depth and intensity of available modelling is even greater but the actual applications in the field are practically negligible. An important recent review by Ingemar Nåsell [15] lists nearly 70 relevant technical references.

As a part of my recent work in Geneva, I have set out at some length what I believe would be a possible development strategy in regard to the more effective use of mathematical modelling in the control of schistosomiasis [9]. The purpose here is to place a new emphasis on the direct development of quantitative support in field contexts, rather than simply multiplying purely academic discussions at an ever-increasing rate.

The disease currently affects some 200 million people, with perhaps three times that number being threatened. My proposal is that a special, operationally-oriented, project should be mounted in some country, like Sudan, where there are serious schistosomiasis problems, but where there is also a good chance of demonstrating the beneficial consequences of the right kind of applied mathematical modelling. The work would consist of two major efforts. The first would consist of direct support for ongoing work at the decision-making level in the country itself, involving biostatistics, applied mathematics, modelling, operational research, and systems approaches, all to be carried out as required in relatively simple down-to-earth ways. The second effort would entail activity based in a university or research institute (perhaps employing several different people in different countries) designed to develop the kind of schistosomiasis modelling that would be of relevance to the practical activities at country level.

These two efforts would be integrated, over a two-year period say, by a series of three or four workshops that were especially designed to facilitate the flexible flow of information between theoreticians and actual decision-makers in such a way as to promote and strengthen public health control of the disease. Work of this kind has already been carried out on various serious ecological problems by C. S. Holling and his co-workers, and has been reported under the title of *Adaptive Environmental Assessment and Management* [12]. There are two reasons for drawing special attention to this work in the present autobiographical context. First, I believe that this whole approach, or some suitable variant of it, is highly relevant to the development of more effective quantitative, operational research and systems support, not only for schistosomiasis control but also for the reduction or eradication of many other infectious or parasitic diseases. Second, probabilistic modelling has a major role to play in developing the underlying epidemiological and demographic population dynamics models.

It should perhaps be emphasized that although we may settle at some point for an approximate deterministic model of such processes, the underlying mechanisms are nearly always probabilistic. The infection process itself, the occurrence and duration of latent and infectious periods, the removal from circulation, the development and loss of immunity, the phenomenon of drug resistance, all entail major stochastic elements. Of course, if numbers are sufficiently large it may be valid to work with deterministic formulations regarded as equivalent to stochastic means. But if we study small populations such as family groups, school classrooms, or small villages, probabilistic modelling is inescapable. And we must not forget that in his studies of fade-out effects for measles epidemics, Maurice Bartlett found that stochastic phenomena could be important even for towns with up to 250,000 inhabitants. Again, in schistosomiasis it is probably legitimate to use deterministic approximations for many large-scale population phenomena, but when it is important to distinguish between male and female parasites in human hosts a probabilistic formulation at this level appears to be indispensable.

I have concentrated attention upon the importance of probabilistic modelling for facilitating the control of infectious and parasitic diseases, because these happen to be of special interest to me. But similar considerations, *mutatis mutandis*, are equally applicable to the modelling of chronic diseases, in which there is a great deal of current interest and development. In fact the whole area of clinical diagnosis, treatment, surveillance, prognosis, drug development, clinical trials, epidemiology, preventive medicine, public health control, decision-making, strategy choice, and policy setting, contains major probabilistic components at all levels. These components can be handled satisfactorily only by giving due recognition to the craft of probabilistic modelling which is partly science and partly art. There is no royal road to avoiding errors and wrong directions, but I believe that these can be kept to a minimum, if we accept the view that theory and practice must develop hand in hand, each supporting, criticizing and collaborating with the other.

12. Retrospect and Prospect

As I have already expressed my personal views on a variety of topics associated with probabilistic modelling, there is no need to repeat these remarks here. However, I would like to conclude with some general observations and speculations on historical trends and future directions in my own area of special interest.

Probabilistic modelling has been with us, in one form or another, implicitly or explicitly, since the birth of probability science in the seventeenth century. In infectious disease epidemiology the first deterministic model appeared in 1760, in the work of the Swiss physician and mathematician, Daniel Bernoulli, on smallpox inoculation [11].

And the first probabilistic model in this field was published in 1889 by En'ko, an Estonian physician, who used an essentially chain-binomial approach to study the spread of infections in population groups of given size. A more deliberate use of such methods had to wait until around 1930, although mathematically inclined epidemiologists, like the Nobel Laureate Ronald Ross, were well aware of the probabilistic basis of infectious and parasitic disease transmission from the beginning of the twentieth century.

The real impetus for substantial progress came only after 1945, when standard statistical theory on estimation, significance testing and goodness-of-fit, was greatly enhanced by new advances in the theory of stochastic processes, soon to be potentiated by the rise of electronic computing. When applied to epidemiology this led to a wide range of new developments. And, indeed, a similar expansion took place over the whole area of biological and medical science.

At the same time, it must be admitted that, so far as practical applications are concerned to clinical and preventive medicine, infectious and chronic disease epidemiology, and public health planning and control, there are still enormous gaps between the potentialities exhibited in the literature and the actual benefits provided to the public.

I believe that this is in large measure due to the existence of big interactive systems of dynamic processes in all the areas mentioned, that are only now beginning to be understood, modelled, handled and controlled. These systems, moreover, entail not only the natural systems occurring in the physical and biological world—and this can already involve great complexity—but also the man-made systems inherent in all socio-economic activity which bring in an additional dimension of stochastic variability.

While the handling of large systems is certainly facilitated by high-speed computing, we are also in need of a greatly improved conceptual basis. Big systems are not easy to think about, and since system behaviour may exhibit new properties not deducible from the component parts it is not sufficient merely to supply a computer with an astronomical number of detailed items.

In my view this is possibly the major problem facing mankind, though there are of course other ways of formulating it in terms drawn from other disciplines. Probabilistic modelling has a major role to play in helping to construct concepts and models that can lead to the solution of practical problems. But there are many other interlocking conceptual tools that are equally vital, including a wide range of applied mathematics, statistics, modelling, computing, systems analysis and operational research. It is also essential to achieve an effective integration of theory and practice, so that science and technology can be mutually supportive and more effectively used for the benefit of mankind in furthering socio-economic development.

The doctrine of the "union of theory and practice", that I have strongly and repeatedly advocated, was first put forward in clear and insistent terms by Charles Babbage in England in 1832, and was later developed and given a wider interpretation by Marx and Engels. In the event, there arose in England a regrettable and continuing separation between pure science and applied technology, with subsequent disastrous consequences in Britain for industry in particular and society in general. The stimulus of the Second World War did promote a temporary flowering of operational research, but the malaise still continues.

Many other countries have done better, including Germany, the United States, France, and Japan. But none has done so consistently, and all have serious unsolved problems in using the potentialities of science for the betterment of their own populations. This constitutes, of course, a global tragedy, especially when one reflects that it is possible to walk on the moon or recover damaged satellites from space; it seems possible to maintain sophisticated multi-billion-dollar "defense" programmes, but often impossible to distribute food supplies that are abundant in one country to millions of people who are starving in another.

Looking back, I am very conscious that my entry into the field of biological and medical applications of mathematics and statistics has provided me with over 35 years of challenging and satisfying work. Progress has in some ways been slow, especially in regard to the provision of real benefits to mankind. However, the variety of possible applications has expanded enormously in recent years. At least we can now see more clearly just what are the potential applications of mathematics, statistics, probability theory, mathematical modelling, operational research, and systems approaches to facilitate the solution of some of mankind's major disease problems, especially the infectious and parasitic diseases that affect as many as a billion people. It could be said that we know how to proceed with finding out reasonably effective practical methodologies for controlling such diseases, using the available medical, epidemiological and scientific resources. The real practical difficulties are: (a) finding situations in which we can demonstrate the power of these methodological approaches within the inevitable social, economic and political constraints; and (b) overcoming administrative and bureaucratic resistance to actually undertaking the detailed applied research that is necessary.

Indeed, these constraints and difficulties are partly effects and partly causes. This is why I believe that in most cases effective results will be achieved only by adaptive managerial approaches, in which the requisite scientific methods are pursued in a fully conscious and deliberate interaction with the socio-economic scene.

Finally, I cannot help but be grateful that my somewhat improbable entry into a field containing probabilistic modelling as a vital element has led me in so many fascinating, provocative, if unpredictable directions. I can only hope that more effective personal, social, administrative and political efforts will be made in the future by those with the vision to see what must be done if science and mathematics are to make their own special contribution towards ameliorating the lot of mankind.

References

[1] Bailey, N. T. J. (1952) Study of queues and appointment systems in hospital outpatient departments, with special reference to waiting times. *J. R. Statist. Soc.* B 14, 185–199.

[2] Bailey, N. T. J. (1957) *The Mathematical Theory of Epidemics*. Griffin, London.

[3] Bailey, N. T. J. (1959) *Statistical Methods in Biology*. English Universities Press, London.

[4] Bailey, N. T. J. (1961) *Mathematical Theory of Genetic Linkage*. Clarendon Press, Oxford.

[5] Bailey, N. T. J. (1964) *The Elements of Stochastic Processes.* Wiley, New York.

[6] Bailey, N. T. J. (1967) *The Mathematical Approach to Biology and Medicine.* Wiley, New York.

[7] Bailey, N. T. J. (1975) *The Mathematical Theory of Infectious Diseases.* Griffin, London.

[8] Bailey, N. T. J. (1982) *The Biomathematics of Malaria.* Griffin, London.

[9] Bailey, N. T. J. (1984) *The Mathematical Modelling of Schistosomiasis: Implications for Public Health Control.* (Report to the Edna McConnell Clark Foundation, 31 May 1984, Grant No. 284–0040).

[10] Bailey, N. T. J. and Alff-Steinberger, C. (1970) Improvements in the estimation of latent and infectious periods of a contagious disease. *Biometrika* 57, 141–153.

[11] Bernoulli, D. (1760) Essai d'une nouvelle analyse de la mortalité causée par la petite vérole, et les avantages de l'inoculation pour la prévenir. *Hist. Acad. Roy. Sci., Mém. Math. Phys.*, 1760, 1–45.

[12] Holling, C. S. (ed.) (1978) *Adaptive Environmental Assessment and Management.* Wiley, New York.

[13] Kendall, M. G. (1943) *The Advanced Theory of Statistics.* Griffin, London.

[14] Molineaux, L. and Gramiccia, G. (1980) *The Garki Project: Research on the Epidemiology and Control of Malaria in the Sudan Savanna of West Africa.* WHO, Geneva.

[15] Nåsell, I. (1984) *Mathematical Models of Schistosomiasis.* Unpublished.

J. W. Cohen

Professor Jacob Willem Cohen was born on 27 August 1923 in Leeuwarden, the Netherlands, the first child of Benjamin Cohen and Aaltje Klein. He studied mechanical engineering at the Technological University of Delft, where he obtained his master's degree in 1949; in 1955 he was awarded a doctorate in technical sciences for his thesis on "Stress Calculations in Helicoidal Shells and Propeller Blades". From 1948 to 1950 he held the position of assistant in applied mechanics at the Technological University of Delft, and in 1950 became an engineer with Philips Telecommunication Industries. Here, he worked on telephone engineering and teletraffic theory and became the leader of the laboratories' mathematical group. In 1957 he was appointed to a chair of mathematics and mechanics at the Technological University of Delft. In 1973 he moved to the chair of operational analysis in the Mathematical Institute of the State University of Utrecht.

He has travelled widely as visiting professor and research fellow, and spent shorter or longer periods at Purdue University, the Technion, Haifa, the University of Maryland, the University of Clermont-Ferrand, The Negev University, the IBM Research Center at Yorktown Heights and the Bell Laboratories at Holmdel and Murray Hill. He is one of the founders of the journal *Stochastic Processes and their Applications*, and is also editor of various journals in applied probability and operations research. He was one of the founders of the Conferences on Stochastic Processes and their Applications, and was from the very beginning an active member in the organization of the International Teletraffic Congresses; he is a permanent member of its International Advisory Council. He was elected to membership of the International Statistical Institute in 1977. He has written three books, and contributed to several others; he has also published about 85 papers.

Jacob Willem Cohen is the proud father of three married daughters, Channa, Alice and Thirza, who have presented him with seven grandchildren. He married Annette Betty Waterman in 1964, and they have a younger son, Arjon. He enjoys chess, history, travel and above all good friends.

Some Samples of Modelling

1. Engineering

I was born in Leeuwarden, a provincial city located in a cattle-breeding and agricultural area in the north of the Netherlands. This is a rural part of the country with its own culture and language, a tough-minded rather stubborn and independent population, often well versed in the scriptures. Here my family had lived for about three centuries, making a living as small salesmen, their culture rooted in Jewish tradition. My father was a businessman, who had to solve week in, week out, the problem of earning a living. The records of his weekly activities were among my first experiences of problem-solving. At his workshop he had several mechanical devices, like jackscrews and tackles; when accompanying him on his business visits to machine mills, I noticed the driving wheels and axles which powered the machines. Such mechanical devices fascinated me, and with my extensive Meccano set I tried to imitate these mysterious constructions.

My schooling followed the usual lines. I was not considered a bright pupil; mathematics held my attention, but outside school, activities like soccer and chess were of greater interest to me. As a young boy, perhaps six years old, I once saw people playing chess. I still do not know why, from then on, I wanted to understand the game; we had no chess set and my parents did not play it. When I was 12, I got hold of a set and a little instruction book, and taught myself the game. I soon became an active player and won prizes. Chess has given me much pleasure, and my desire was to play a perfect, logically consistent game. Many years later I played two or three games which approached this ideal, but I also learned that chess had more to offer. I once defeated a famous grandmaster in a simultaneous game with a pure *schwindle*; both of us enjoyed it. I later recognized a similar desire for perfection among my engineer colleagues who constructed telephone systems. A telephone system is a rather abstract engineering structure; we dreamt of designing the perfect system, pure logic realized in structured hardware. The older engineers quickly talked us out of our dream; it was too expensive, and no telephone administration would pay for it. Apparently reality was an important component of modelling.

At high school we were thoroughly trained in mathematics. In those days, Dutch schools had a solid curriculum and well-trained teachers. I am still benefiting from their learning. That the square root of -1 did not "exist" puzzled me then, and I could not believe that mathematics was so poor that it did not have a solution to this problem. This reflected my rather optimistic feeling about mathematics, a feeling which I still have and which has helped me in solving many a problem. One of the schoolteachers (not the mathematics teacher) told us about the existence of other geometries (noneuclidean), of different types of logic, of another system of mechanics (relativity); they were all mysteries.

This was the situation when I reached the age of 17. My father had handled the problems connected with the economic crisis of the 1930s well. I was a rather quiet youth who had become interested in science. Then the war broke out: one morning it was there, totally unexpected although it could have been foreseen. In 1940 the Netherlands were occupied by Germany.

The Nazis acted gradually in their persecution of Dutch Jews. They followed policies different from those practised in eastern Europe, where, as we learned later, it was their habit to execute large numbers of Jews on the day after their arrival. In 1941 I graduated from high school; at that time the Nazis were rounding up boys of my age. I had to hide, and lived for some time with a farmer and his family. I learned to milk cows (main problem: preventing the cow's leg from entering the milk-pail) and to assist in the birth of calves and lambs. I became acquainted with the life of farmers, with their routine and culture. I could not stay long on this farm because too many people knew of my presence. In the summer of 1942 the Nazi persecution became serious; it posed a direct threat to the life of every one of us, my parents and their six children, five boys and a daughter.

Jewish families had to go to camps in Holland from where they were transported to unknown destinations. We made preparations to go, but we also discussed plans for hiding the whole family. My mother was the fighter; my father, who had a narrow escape from a Nazi round-up, solved the problem. He had many business relations with Dutchmen of orthodox Christian faith, who after two years of German occupation summed up the situations in terms of their conviction that "God's people is at stake". They were prepared to help and acted courageously, risking their lives and those of their families. On 11 November 1942 our family left home and went into hiding. We learned to live with danger and had to keep the lowest possible profile, since there were traitors. Fortunately, the whole family survived safely through the war, which was a great exception.

This time my hiding place was with a family with two children, who lived in a rather modest house. An elderly couple was also hidden with this family. My host was a chemical engineer, who possessed a large collection of books on mathematics and philosophy, among them the nearly complete Springer yellow series of that time. My days were passed in the study, my nights in a secret cabinet. Extreme silence was a must; the children were never to know that three people were hidden in the house. I studied mathematics, reading one book after another. It was an unguided tour, but I became acquainted with the field and the right books during the 12 months of my stay. Since I could afford to buy books, the nucleus of my present library was formed then.

There was much to read in the engineer's study, but when the cleaning woman came in twice a week, I had to hide under the floor in the dark for more than five hours. It was mainly during those hours that I played with the tetrahedron, and obtained many relations between its edges, angles and sides. I was happy to find that the ratio of the product of the lengths of two opposing edges and the product of the sines of the angles at those edges is an invariant. Later I discovered this property in a book; I wonder whether I can still derive it without paper and pencil, as I did then.

In the fall of 1943 I went to live with my parents at their hiding place. I was able to bring my books and continue my studies; I was particularly interested in classical mechanics, and I experimented with a self-made double pendulum. At the position of the loads, I attached burning shoe-laces, and when the pendulum swung in the dark I saw beautiful curves; but my measurements of gravity were never very accurate. For several months I was taught about Kant's categories by a clergyman, who also had to hide. These were my first and only oral lessons in philosophy, a subject in which I browsed quite a bit during my first year after high school. Its problems attracted me, but its discussions seemed too subjective. Mathematics appeared to me to be more certain; at that time I had not yet heard of Gödel's theorem on "undecidability". The clergyman was a nice man, who liked tobacco, but cigarette paper and writing paper were in short supply. The only paper we had consisted of very thin sheets, rather like blotting paper. It hardly served both purposes, but I still have a number of those sheets of blotting paper covered with studies on the line of Wallace.

My father always encouraged my study of mathematics; "it keeps the brain supple", he used to say, but he did not want me to become a mathematician. His argument was partly a practical one: the possibilities for making a living as a mathematician were limited; a proven argument in the 1930s. He wanted me to become an engineer, and I had few objections, since I was interested in technical mechanics.

During our hiding period we had several critical, dangerous moments. When I think of that period after 40 years, I am amazed at the intensity with which I could then study; maybe studying shut out other thoughts, maybe we became used to the dangers of our situation.

In later years I benefited much from my studies during the war; I had more or less mastered the subjects of a university curriculum. I could understand and handle the techniques, but my feeling for the deeper problems remained vague. Let me elaborate. I studied a good book on relativity theory and learned to handle the involved tensor calculus. However, I did not understand why the model underlying the theory was built the way it was. Here I had missed a teacher; later on I realized that such a gap could also be filled by studying the historical development of the subject. The craft of modelling can be learned by experience, but another valuable approach, easier to incorporate into a university program, is confrontation with history (if the model has a history).

On 15 April 1945 we were liberated by the Canadian Army, and soon afterwards my family was reunited. About two months later I enrolled at the Technological University of Delft to study mechanical engineering. Thanks to my wartime study of mathematics I was able to finish the five-year curriculum in less than four years, and in 1949 I was awarded my master's degree with honours.

The curriculum for mechanical engineering at that time was a mixture of mathematics, physics, technology, and performance of engines and structures; a selection of the science and technology which had developed since the industrial revolution. A subject like systems engineering hardly existed at the end of the war. Engineering was taught by starting from basic models, such as the Carnot cycle for heat engines and the simple spring system (a spring with a heavy load attached to it) as the basic model for vibrations of structures. Basic models such as these form the starting point in the design of every man-made construction.

I specialized in elasticity theory, a branch of continuum mechanics, at that time an established field to whose developments many great scientists had already contributed, particularly in the nineteenth century. As a science it is a part of classical physics, as a discipline part of engineering. Its tools are mathematics and experimentation, its goal the development of techniques for stress, strain and strength analysis of structural elements. I was very much attracted by this field, which was taught in Delft by renowned scientists. It contains beautiful mathematics and has a direct contact with reality, although reality can be very complicated. Quite often model analysis results in long and intricate formulas. This is a fact to be accepted, since one cannot run away from it, and one therefore has to develop the appropriate numerical analysis. The adage that "a problem in applied mathematics is not solved before it has been shown explicitly that the analytical solution can be numerically evaluated with sufficient accuracy" was taught to us from the very start (see also Von Mises [21]). Elasticity theory, fluid mechanics and aerodynamics were at that time typical examples of applied mathematics, a field of activities so well characterized by Von Mises in his motivation for establishing the *Zeitschrift für angewandte Mathematik und Mechanik* (cf. [20]). In prewar applied mechanics, geometry played an interesting role; in Galileo's time geometry was the language of mathematics and mechanics. In engineering it played this role much longer, and actually served as the main tool for numerical evaluation. Descriptive geometry, shaped by Monge, the first rector of the École Polytechnique in Paris, and the large variety of graphical techniques (where beautiful applications of projective geometry occur) served as the main tools for solving large systems of linear equations directly or by iteration. The appearance of the modern computer has made much of this beautiful geometrical craft disappear; maybe computer graphics will lead to a comeback in due course.

I did some research in elasticity theory, assisted with teaching and wrote some reports. One was for a prize question on Shanley's theory of nonelastic buckling which got an award, and several others considered Southwell's relaxation method, an iterative technique for solving systems of linear equations and partial differential equations. Southwell's technique was then a revolution in numerical analysis. My main research concerned the stress analysis in rather wide propeller blades for ships. This topic formed the content of my doctoral thesis [3]; in 1955 the doctoral degree was awarded to me with honours.

My research on propeller blades is a nice example of modelling; I shall therefore digress a little on it. The blade is actually a shell, i.e. a body of which one dimension, the thickness, is small compared to its other two dimensions. The middle surface of the blade can be thought of as a sector of a right helicoidal surface. The finite width of the blade (in the direction of the helices) causes extra mathematical difficulties of a kind already known in the elasticity theory of flat plates. To explore the influence of the helicoidal shape I therefore started with a helicoidal shell of finite radial length but unbounded in the direction of the helices. The existing theory of shells, developed by Love, used the lines of principal curvature of the shell's middle surface as the reference system. For helicoidal shells the generators and helices, i.e. the asymptotic lines, seemed to be a more appropriate coordinate system. I therefore derived a theory of elastic shells for a generic set of parameter curves, and then applied it specifically to the helicoidal shell. The stress analysis of an elastic shell is a first-order approximation to the three-dimensional stress analysis in generic bodies, an extremely complicated problem. The approximation is (almost everywhere) based on the fact that the square of the ratio ρ of the thickness of the shell, and the torsion or a principal curvature radius, is small with respect to 1, and so can be neglected. By using differential geometry and the first principles of elasticity theory I derived the partial differential equations for the determination of the stresses and displacements. Then there came several surprises.

A right helicoidal infinite strip can be parametrized in Cartesian coordinates by

$$x = r\cos\beta, \quad y = r\sin\beta, \quad z = a\beta; \quad r_1 \leq r \leq r_2, \quad -\infty < \beta < \infty.$$

The resulting differential equation becomes somewhat more pleasant if the parameter r is replaced by $r = a\tan\alpha$, $0 \leq \alpha_1 \leq \alpha \leq \alpha_2 \leq \pi/2$. Then the basic differential equation for the shell deflection w to be solved is

$$\frac{d}{d\alpha}\left[D\frac{\cos^3\alpha}{a^2}\{w_{\alpha\alpha} - (2-v)w_\alpha\tan\alpha + (3-v)w\}\right]$$
$$- D\tan\alpha\frac{\cos^2\alpha}{a^2}\{w_{\alpha\alpha} - w_\alpha\tan\alpha + 2w\}(1+v) = f(\alpha),$$

where $w_{\alpha\alpha} = d^2w/d\alpha^2$, $w_\alpha = dw/d\alpha$, D and v are constants, and $f(\alpha)$ is a known function.

The first surprise was that this third-order differential equation could be solved explicitly, viz. $\sin\alpha$, $\cos^{-2}\alpha$ and $\sin\alpha\log\tan(\frac{1}{4}\pi - \frac{1}{2}\alpha)$ are independent solutions of the homogeneous equation. This was good luck, which happens perhaps once in a lifetime. The second surprise was that the results obtained from the solution of the above equation appeared not to be consistent with the equilibrium conditions; the difference, which should have been of order ρ^2, could in fact be much larger. The physical assumptions on which Love's modelling of shell theory were based had to be revised. I refined my model; a fully consistent theory was later developed by my thesis supervisor Professor Koiter. The revised differential equation, again of the third order, had to be integrated numerically.

The twisted character of the helicoidal shell suggested that the stresses in it could be approximated by those occurring in a beam if the loading of the infinite helicoidal strip was rotationally symmetric. This conjecture appeared to be justified by my theoretical results. The influence of the finite width of the propeller blade again required a numerical approach; unfortunately, modern computer power was not yet available. The final result of my research was a strength analysis sufficiently accurate for engineering purposes, scientifically motivated and in agreement with available experimental results. The combining of mathematical techniques, physical assumptions, conjectures based on mechanical insight and numerical analysis was exciting; the approach and the results obtained still give me satisfaction.

In 1950 I left the university. The general atmosphere in Holland at that time was very positive. People enjoyed their freedom and cooperated to build a better society. Industrial activities were increasing rapidly, and young engineers were much in demand. I was offered several positions, and for a time considered a research appointment, but finally decided to work in engineering development, a field intermediate between design and research. I became a member of the electro-mechanics group of Philips Telecommunication Industries; the group needed a worker with interests in the theoretical engineering sciences. At that time the company was entering the field of telephone engineering. The mechanical selector and the relay were the main elements of a telephone system. My knowledge of Southwell's relaxation technique brought me a quick success. With only paper and pencil I was able to determine the leakage flux between the armature and core of a relay [9]; it required the solution of the Poisson partial differential equation for a domain of connectivity 2. The engineers were impressed, but did not really trust results obtained by such simple means. Half a year later a competing company published experimental results which showed complete agreement with my model. The analytic approach, however, was quicker, more flexible and less costly; at that time, this was a nice example of the application of mathematics.

My colleagues and superiors were all electrical engineers concerned with construction and production. I had to learn to speak their language and to understand the structure of telephone and telegraph systems. This was the time when Wiener published his book on cybernetics, the time of Wiener–Kolmogorov prediction theory, and of Shannon's switching algebra and information theory, all important topics in telecommunications. It was part of my task to investigate the direct relevance of these new ideas to engineering. In particular, this was the case with switching algebra. A telephone system is a highly organized structure, a huge database system in present-day computer terminology. At that time its logical structure was realized by intricate configurations of relay contacts, which operated in the binary mode. Shannon showed that the models of symbolic logic could be used to describe the behaviour of relay circuits. I studied books on propositional and predicate calculus, could soon understand Shannon's theory and was able to instruct my colleagues who were dealing with circuitry design in its use. The results were rewarding, even experienced designers could sharpen their techniques; the main advantage was, however, in the teaching of apprentices. The design of logical circuitry could be systematized, and several trial-and-error methods could be avoided.

In setting up calls, a telephone system acts as a data-handling system; in its performance it provides channels for speech transmission. The models by which these activities are scheduled differ essentially from those in continuum mechanics, and so do the mathematical techniques required for their analysis. Information theory modelled the reliability of encoded signals, group theory and combinatorics became useful tools in the analysis of coding and decoding. A telephone system should be economically feasible; hardware, such as telephone cables, selectors and relays, is costly. The relations between hardware, traffic handling capacity and performance are of paramount importance. That system performance, in so far as it concerns the handling of traffic, should be measured in terms of congestion probabilities was a brilliant idea, presumably stemming from M. C. Rorty in 1898, then an engineer with American Bell Telephone (see [17] where an interesting account of the introduction of probability theory in teletraffic engineering is given). Wilkinson's contribution to [17] is a fine account of early modelling in operations research.

Soon after I had entered the Philips company I became involved in teletraffic theory, then called congestion theory. In 1950 teletraffic theory had already built up a tradition, and several theoretical and experimental techniques had been developed, among them the method of "artificial traffic" at present known as "simulation". The available theoretical studies were nearly all of a pioneering type, using seemingly *ad hoc* techniques. The state of the art was rather chaotic. Fry's book [11] was the only available text which tried to present a connected account of available models and to explain what are currently referred to as birth-and-death processes. It was, in those early days, difficult to develop the intuition needed in creating the right models; i.e. models which were, on the one hand, sufficiently close to reality, and on the other hand were accessible to mathematical analysis. An intuition for modelling, and a feeling for the potential powers of mathematical techniques, do not appear out of the blue, at least not in my case. It requires hard work leading to many dead ends, particularly if systematic theoretical expositions on the subject are not available. And that was the situation with the theory of stochastic processes in 1950. There were of course the books by Paul Lévy, Fréchet and Romanovsky; there were the fundamental papers by Feller and by Kolmogorov, but their results did not seem very relevant to the direct needs of teletraffic theory.

The publication of Feller's famous book was a landmark in the development of stochastic modelling; the book by Blanc-Lapierre and Fortet [2] has also been influential, although to a lesser extent, partly because it was written in French. Feller's book has eliminated much existing vagueness in the basic definitions of probabilities and random variables; it presented a lucid text on fundamental probabilistic models and processes. Blanc-Lapierre and Fortet's book united the developments in mathematical probability theory with the experience in probabilistic modelling of physical processes, e.g. Brownian motion and noise. There were other books published in the 1950s aiming at similar goals; they came at the right moment. The possibilities of operations research became apparent, and the need for probabilistic modelling made itself felt in many fields. Such modelling was no longer hampered by vague definitions of probabilities and stochastic variables, nor by the barriers encountered in solving large systems of linear equations. The emergence of fast computing machinery moved that barrier considerably in the direction of infinity. In the late 1950s probabilistic modelling became a tool in industrial engineering and management, and randomness was accepted as a phenomenon which could be modelled.

2. Queueing Theory

In 1898 Blood (cf. [17]) observed that the distribution of the number of busy calls in a group of telephone subscribers agreed well with the terms of a binomial expansion $(p + q)^n$. Presumably this was the first probabilistic model in teletraffic engineering. Dissatisfaction with empirical knowledge probably urged him to think the way he did. Rorty, a colleague of Blood's, realized that probability theory could provide the appropriate concepts and tools to describe the fluctuations in the number of busy subscribers. Rorty wrote reports, produced engineering graphs and lectured on his ideas; history records that some of his listeners were sceptical about the possibility of describing human activities, like the duration of subscriber conversations, by any theory of random events. An important conclusion of Rorty's research was that with the same grade of congestion, larger trunk groups could handle relatively more traffic than smaller groups, which is of course an example of variance reduction. For further details about Rorty, see [17].

It was soon realized that the congestion probability in setting up a call should be very small (about 0.02), if an acceptable service was to be provided. With the introduction of automatic telephone exchanges this idea led to the concept of the "lost call", i.e. a call which gets the busy signal and is not repeated. In his most important paper, Erlang [10] modelled this concept by considering a group of N servers (telephone lines) with a Poisson arrival process of customers (calls); an arriving customer finding all servers busy disappears and never returns (lost call). Obviously, this is a fine example of modelling if the blocking probability is small. The service times (durations of calls) were assumed to be of constant length. In present-day terminology Erlang considered the $M/D/N$ loss model: M for the Markovian character of the arrival process, D for the deterministic service time, and N for the number of servers. Erlang discusses in only a few lines the concept of *statistical equilibrium*, obviously stemming from kinetic gas theory. With this he motivates his statement that the distribution of the number of busy servers is the normalized Poisson distribution truncated at N. He remarks that constant service times model the duration of long-distance calls very well, whereas local calls can be better represented by varying service times which follow a negative exponential distribution, as measurements had taught him.

Erlang then argued that it was "easy to see" that the distribution of the number of busy servers was the same in both cases, i.e. the $M/D/N$ and $M/M/N$ loss models have the same stationary distribution of busy servers. Parenthetically he warns the reader that both types of service distribution "do not lead to the same results in all problems". Erlang's contemporaries had great difficulties with his arguments. The "insensitivity" of the resulting busy server distribution for both types of service time has puzzled many investigators; it has had important effects on modelling in queueing theory. It is interesting to note that this insensitivity in Erlang's model has been repeatedly rediscovered in relatively recent management science and computing literature. Before sketching these effects I shall try to elucidate "insensitivity" and "statistical equilibrium"; this exposition, published here for the first time, uses an idea of Van Dantzig.

Consider a stationary Poisson point process with intensity λ on the horizontal line. The points are marked as type 1 (·) or type 2 (×) points, independently of each other with probabilities p_1 and p_2, where $p_1 + p_2 = 1$. To a type 1 point, a line segment of length τ_1 is assigned, and to a type 2 point, one of length τ_2; these line segments are plotted vertically (see Figure 1) and their endpoints constitute a second point process in the plane.

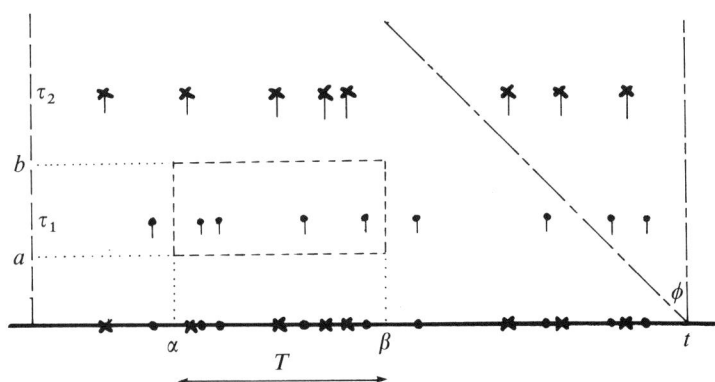

Figure 1   A Poisson process on the horizontal axis generating a Poisson process in the wedge

The number of points contained in any rectangle $(\alpha, \beta) \times (a, b)$ is readily seen to have a Poisson distribution with intensity depending on $\beta - \alpha$ and on the position of a and b with respect to τ_1 and τ_2; moreover, the numbers of points in disjoint rectangles are independent. Application of Carathéodory's extension theorem shows that the distribution of the number of points in any measurable set in the plane is Poisson, so the point process in the plane is a nonhomogeneous Poisson process. In particular it follows that the number of points in the infinite wedge of angle $\phi = 45°$ to the vertical axis and with its top at the point t has a Poisson distribution, with intensity $\lambda(p_1\tau_1 + p_2\tau_2)$, as a simple calculation shows.

Suppose we rotate every point of the point process in the plane clockwise by $90°$ around its projection on the horizontal line. The point process on the line so obtained is the sum of two independent Poisson processes with intensities λp_1 and λp_2, and as such is again a Poisson process with intensity λ. Obviously, the latter process may be regarded as the departure process of a service facility with an infinite number of servers and Poisson arrivals. The line segments with endpoints in the wedge fall on its top t after the rotation and represent the service times of the customers being served at time t. Hence their number has a Poisson distribution. The service-time distribution is obviously given by $p_1 U(\tau - \tau_1) + p_2 U(\tau - \tau_2)$; $U(\tau - \sigma), \tau \in (-\infty, \infty)$ being the probability distribution degenerated at σ. The arguments used apply equally to a service time distribution $\sum_{i=1}^{m} p_i U(\tau - \tau_i)$ with $0 < \tau_1 < \cdots < \tau_m$; $\sum_{i=1}^{m} p_i = 1$. Their extension to a general service time distribution with support in $[0, \infty)$ is a matter of technique (weak convergence). Note that if the line segments are sampled from a distribution with support in $[0, \infty)$, and successively assigned vertically to the points of the Poisson process on the horizontal line, then their endpoints are easily seen to form a Poisson process in the plane; in particular the number of points in the wedge is again Poisson distributed.

Hence we reach the conclusion that in an $M/G/\infty$ system (G for an arbitrary service time distribution) the stationary distribution of busy servers is Poisson and the departure process is a Poisson process, independently of the type of service time distribution. This latter independence is the so-called *insensitivity property* of the model.

Another evident consequence of the argument above is the reversibility of the process, i.e. the stochastic structure of the process formed by the number of busy servers at time t is identical for t increasing or decreasing (if the process is stationary).

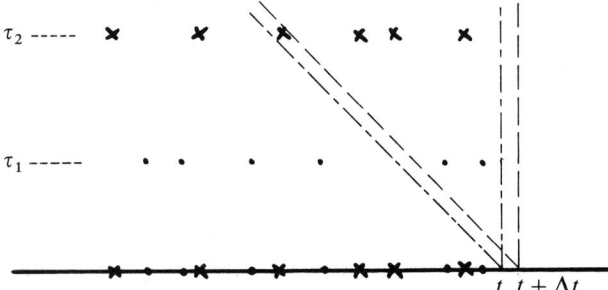

Figure 2

Next consider two wedges (as in Figure 2) a distance Δt apart, again with two levels τ_1 and τ_2. Obviously, the probability that the vertical strip formed by the vertical boundaries of the wedges contains just one point is $\lambda \Delta t + o(\Delta t)$. To calculate the probability that the sloping strip between the sides of the wedges contains one point, note that the probability of having k points in the wedge at t with n points at level τ_2 is equal to

$$[\{\lambda p_2 \tau_2\}^n / n!] [\{\lambda p_1 \tau_1\}^{k-n} / (k-n)!] \exp\{-\lambda(p_1 \tau_1 + p_2 \tau_2)\}.$$

Given that at level τ_2 there are n points, then they are all uniformly distributed on $[0, \tau_2]$, and similarly for level τ_1. Hence the probability of having k points in the wedge at t and no points in the sloping strip is equal to

$$\sum_{n=0}^{k} \left(\frac{\tau_2 - \Delta t}{\tau_2}\right)^n \left(\frac{\tau_1 - \Delta t}{\tau_1}\right)^{k-n} \frac{(\lambda p_2 \tau_2)^n}{n!} \frac{(\lambda p_1 \tau_1)^{k-n}}{(k-n)!} \exp(-\lambda(p_1 \tau_1 + p_2 \tau_2))$$

$$= \left\{1 - \frac{\Delta t}{p_1 \tau_1 + p_2 \tau_2}\right\}^k \frac{\{\lambda(p_1 \tau_1 + p_2 \tau_2)\}^k}{k!} \exp(-\lambda(p_1 \tau_1 + p_2 \tau_2))$$

$$= \left\{1 - \frac{k \Delta t}{p_1 \tau_1 + p_2 \tau_2} + o(\Delta t)\right\} \Pr\{\mathbf{x}_t = k\} \quad \text{for } \Delta t \downarrow 0,$$

when $\mathbf{x}_t$ denotes the number of points in the wedge at t. It follows that for $\Delta t \downarrow 0$

$$\frac{k}{p_1 \tau_1 + p_2 \tau_2} \Delta t + o(\Delta t)$$

is the conditional probability that the sloping strip contains just one point when the wedge at t contains k; that of containing two points is of order $(\Delta t)^2$.

Let u and $w > u$ be two time points with $\mathbf{x}_u = 0$, $\mathbf{x}_w = 0$ and such that $\mathbf{x}_v > 0$ for some $v \in (u, w)$. Then the number of upward jumps from $\mathbf{x}_t = k$ to $\mathbf{x}_t = k+1$ for t lying between u and w is equal to that of the downward jumps from $k+1$ to k, with probability 1; note that jumps larger than 1 have probabilities of $o(\Delta t)$. This is the principle of *statistical equilibrium* as applied by Erlang. It holds if the points u and w exist with probability 1. Application of this principle shows that the intensity of upward jumps from k to $k+1$ is equal to that of downward jumps from $k+1$ to k (a formal proof can be obtained by applying a stochastic mean-value theorem, cf. [5]). Using the probabilities derived above, the equality of these intensities is expressed by

$$\lambda \Pr\{\mathbf{x}_t = k\} = \frac{k+1}{p_1 \tau_1 \times p_2 \tau_2} \Pr\{\mathbf{x}_t = k+1\}, \quad k = 0, 1, \ldots.$$

This statistical equilibrium relation also results from the reversibility of the process.

The relation for $\Pr\{\mathbf{x}_t = k\}$ is obviously true because $\mathbf{x}_t$ has a Poisson distribution with parameter $\lambda(p_1 \tau_1 \times p_2 \tau_2)$, as we have seen above. Conversely, if this relation is assumed to hold then it follows that $\mathbf{x}_t$ has this distribution.

The principle of statistical equilibrium, which in present-day terminology is called "partial balance", obviously applies to the sample functions of the number of busy servers in the $M/G/N$ loss model, which includes the $M/D/N$ and $M/M/N$ case. Here, it is readily seen that the upward jump intensity at $\mathbf{x}_t = k$ is equal to $\lambda \Pr\{\mathbf{x}_t = k\}$; but it is rather hard to prove that the downward jump intensity at $\mathbf{x}_t = k + 1$ is equal to $\{(k + 1)/(p_1\tau_1 + p_2\tau_2)\} \Pr\{\mathbf{x}_t = k + 1\}$. The finiteness of N, the number of servers, complicates the stochastic structure of the process, and the argument used above for $N = \infty$ does not apply. Only if the service times are distributed as negative exponentials (the case $M/M/N$) is a simple deduction possible because of the memoryless properties of this distribution. Erlang's argument for the $M/D/N$ loss model is not acceptable. The principle of statistical equilibrium was not questioned, but its application fails owing to the difficulty of transforming it into a mathematical equation. It took nearly 60 years before the basic idea of transforming global sample function properties into quantitative relations between probabilities was applied again in queueing analysis (cf. [5] and [7]).

The first acceptable discussion of the $M/M/N$ loss model available in the literature seems to be in Fry's book [11]. Here the process is modelled as a birth- and -death process, phrasing it in modern terminology. It is not clear by whom and where this type of local analysis was initiated; from Erlang's work I cannot conclude that he applied the birth- and -death technique. It might well be that Fry, and also Molina, should be credited for it; Fry's discussion of kinetic gas theory in his book points in this direction. Molina, who was self-taught, also made essential contributions in shaping teletraffic theory, (cf. [17]); he is best known for his analysis of the $M/M/N$ waiting time model [15] by means of the birth- and -death process technique.

There was doubt about the validity of Erlang's result concerning the (finite) $M/D/N$ loss model and a proof was needed. Kosten [14] provided the ultimate proof. He used a technique at present known as that of the "supplementary variable", which he developed when preparing his doctoral thesis in the late 1930s. He used it for the analysis of the $M/G/1$ waiting time model (for details cf. [6]): this is the queueing model with Poisson arrival process, general service time distribution, one server, and congested customers who wait for service. The state description of the process was characterized by the number of customers in the system and the elapsed service time of the customer being served.

In the search for mathematical techniques capable of analyzing queueing models with non-specified service or interarrival time distributions, Pollaczek's work holds a special position. In his research the actual waiting time of a customer was introduced as the state variable, and the time-dependent behaviour of the process involved, in particular its asymptotic analysis as $t \to \infty$, received due attention. Pollaczek's approach has led to a powerful technique, analytically rather difficult to handle; singular integral equations play a dominant role in it. For the $G/G/1$ model, Pollaczek's solution is elegant and closely related to the Wiener–Hopf technique: its probabilistic interpretation became understood with the development of "fluctuation" theory (cf. [6] for details and references). Another contribution of great methodological interest came from Kendall [12], who showed that the embedded queue length process at successive departure moments in an $M/G/1$ model can be modelled as a discrete-time-parameter Markov chain. These departure epochs constitute regeneration points of the process; their significance as a starting point for analysis had already been appreciated in Palm's work. Takács has introduced the workload of the server, i.e. the virtual waiting time, as the state variable, and showed that its distribution function satisfies a renewal type integral equation. In many cases it seems that a problem formulation in terms of virtual waiting times leads to a less complicated analysis than that required when the queue length process is used as the starting point.

In the late 1950s and early 1960s research workers in queueing theory had a fairly good set of basic models and appropriate mathematical techniques at their disposal. At that stage, professional mathematicians became interested in the subject; workers in management science and industrial engineering soon recognized that the models of queueing theory could serve their needs. The models became more refined, their time-dependent behaviour was investigated, and other types of service such as priority disciplines were considered. The mathematical techniques became more flexible and their possibilities were better understood.

Of course, there were still fairly basic models which defied analysis, the mathematical problems involved being just too hard. These were mainly models which required a two-dimensional state space for their state description. As an example consider a single server facility with two Poisson arrival streams, say, men and women, who are served alternately, each having their own service time distribution $B_1(\cdot)$ and $B_2(\cdot)$, respectively. The analysis of this model requires the construction of the bivariate generating function $\Phi(p_1, p_2)$ of the joint distribution of numbers of waiting men and women; this involves a linear relation in $\Phi(p_1, p_2)$, $\Phi(p_1, 0)$ and $\Phi(0, p_2)$ whose coefficients depend on $B_1(\cdot)$ and $B_2(\cdot)$. I had been confronted several times with problems of this type, and so must others. The only problem of this type which I could handle in the 1950s was that of two service facilities in series, each with negative exponential service time distribution, and the first one having a Poisson arrival stream. I found the solution by inspection: it made me conjecture that the departure process of an $M/M/N$ queue was a Poisson process. I was able to prove the conjecture and wrote a report on it. When meeting Ryszard Syski in Paris in 1956, where we were charged by the International Advisory Council for Teletraffic Congresses with the preparation of a complete bibliography on teletraffic studies, I informed him of my findings; we were climbing the steps of the Sacré Coeur. Ryszard immediately recognized the meaning of my discovery for the analysis of queueing networks, which were not at that time of such engineering importance as they are at present. I decided to publish my report; however, on returning to Holland, I found that the next issue of *Operations Research* contained Burke's theorem and its fine proof on the output process of the $M/M/N$ queue. I gave up the plan of publishing my report. It was the second time in my life that I was a few months too late; the first was with my results on priority service when Cobham's result also preceded my own by a few months. Nevertheless, I still enjoy and find satisfaction in the research I did at the time.

In 1955 I was confronted with the phenomenon of insensitivity in a traffic network; the model concerned the imbalance in congestion due to variations in the traffic characteristics of the subscribers. Within a group of subscribers the calling rate may differ considerably from one subscriber to the other. There were empirical indications that subscribers with lower calling rates encountered higher congestion probabilities. I therefore investigated the loss model with a finite number N of servers and a finite number M of sources (i.e. subscribers) $M > N$, every source having its individual characteristics with respect to calling rate and service time distributions, these being unspecified. Using Kosten's technique of the supplementary variable, the explicit expression for the stationary distribution of specified idle and busy sources could be found (cf. [4]). It was a surprise to discover that this distribution possessed the insensitivity property, since only the first moments of the service time distributions were involved.

Moreover, it had the *product form*, i.e. the distribution was the product of its marginal distributions apart from a factor. Actually, the model is a two-node closed network. Closed networks currently play a prime role in modelling the performance of computer networks. My results concerning the phenomenon of insensitivity became a central point of research in the East German school of queueing theory, where insensitivity was thoroughly investigated. This school had started with research on point processes which had been initiated in the work of Khinchine [13], who himself based his research on Palm's thesis [16]. In it, Palm investigates the departure process of an $M/M/N$ loss model, which is itself often the arrival process for a subsequent selector stage in a telephone network.

3. Teaching and Research

In 1957, I was asked whether I would be interested in a chair of mathematics or of applied mechanics. These were the times when universities were expanding rapidly. I preferred the chair in mathematics and began to teach courses in linear algebra, Laplace transforms and stochastic processes in the general science department of the Delft Technological University. The department had just initiated graduate education in mathematical engineering. The curriculum consisted of basic courses in mathematics and physics, introductory courses in engineering and a wide range of specialized courses in continuum mechanics, numerical analysis, statistics and operations research. It was a well-balanced program: the students obtained a good training in scientific methodology and were broadly grounded in the theoretical engineering sciences. It was hoped that with such a program, graduates could cope with new developments in technology during their future careers. That expectation turned out to be correct, although the immense and rapid developments in the 1960s and 1970s were not, and could not have been, foreseen. In the late 1960s the student revolts and the changing views on scientific education heavily influenced the balance in the curriculum. Whether students are now being equipped with a training which will protect them from the rapid ageing of their knowledge, and which is sufficiently flexible to meet the requirements of fast-changing social and economic structures, should be a point of prime importance in current discussions on university education.

The transition from an industrial environment to a teaching position required quite some effort. I liked teaching and still do, but no longer having a boss was something to get used to: I missed him! However, during my 17 years' sojourn at the Delft Technological University, I was able to maintain good contact with industry, and in particular with developments in telecommunications and computers. Around the mid 1960s, I felt that queueing theory had reached a level that required a systematic exposition of those analytical techniques which had proved their applicability and power. Syski's eminent book [19] had been oriented mainly towards a discussion of teletraffic models. I therefore wrote and published my book *The Single Server Queue*, and dedicated it to Annette Waterman, my second wife. Her efforts in taking care of my three daughters by my first marriage, and in creating the atmosphere needed for writing and rewriting the book made her a silent co-author, whom I listed in the author index. (I was married for the first time in 1946, but rather unhappy developments led to a separation in 1964.)

There are several reasons for a scientist to write a book. One of my own main motives was that I realized the future importance of queueing theory in the design and performance analysis of computers and computer systems. This was due to the fact that the early engineering developments of computers were rather similar to those in telecommunications in the first decades of this century. Although queueing theory is certainly of value in management science, its applicability in management is often hampered by the difficulties of obtaining accurate data on the model constants; in engineering systems, which by their nature are man-made, such difficulties are easier to overcome.

In early computer engineering hardly any of the deeper results of queueing theory were used, since capacity constraints were hardly felt. The marketing of computers was easy, and the availability of a large variety of new electronic components resulted in an exponential growth of computational speed. If capacity constraints occurred they could often be avoided by flexible programming. When communication between computers was introduced, trunk capacity (which is costly) became relevant, and studies on queueing networks were initiated; their analysis then developed rapidly.

A queueing network consists of nodes connected by links. At the nodes are located the service facilities (processors); a message after having been fully processed at a node is routed via a link to another node; the routing process is frequently modelled as a Markov chain, with the set of nodes as the state space, and the routing matrix as the matrix of transition probabilities. Closed and open networks are distinguished. In the latter type, messages can enter and leave the network, while in closed systems the number of messages is constant. Closed networks turned out to be an appropriate framework within which program processing, input–output activities and the behaviour of terminals could be modelled, in so far as it concerns the internal traffic transport in a computer. Network analysis became a new and important topic in queueing theory. The existing theory already had some results available on the simpler models, but the complexity of the new models posed many questions. Remarkably, the new research encountered the old problem of "insensitivity". It turned out that a large variety of queueing networks possess stationary distributions which are independent of the service-time distributions at the various nodes; in its simplest form Erlang [10] had been confronted with this property (see also above). For the special case of networks with $M/G/\infty$ service facilities at their nodes, the "insensitivity" property could have been expected on the basis of the properties of the $M/G/\infty$ model discussed above. That the "insensitivity" property turned out to be so robust was a surprise, which still stimulates much research. An important open problem concerns networks which are known not to be "insensitive"; the problem is to determine how sensitive they are, in other words to find the effects of variations in the service time distributions on the performance of the network.

Another important topic in the analysis of a network is the "relaxation time". This is a measure of the speed with which the stationary distribution of the traffic process is reached. The speed should be sufficiently large to make this distribution acceptable as a basis for the performance prediction in the design phase of the system. For single-node systems, some information concerning relaxation times is available; for networks the research has just started (cf. [1]).

The need for queueing theory, or more generally for applied probability, in computer system design, is a recognized fact. The large computer companies have strong groups in this field, and the proceedings of the many conferences on performance modelling display the large variety of stochastic models used in the analysis of performance capacity, algorithmic procedures, storage scheduling and database management. Improvement of system performance by using faster components and signalling is likely to reach its limits in the near future, and can presumably be achieved only by sophisticated scheduling and structuring; vector processing is already an example. Mathematical modelling will be an indispensable tool in these developments. A new field of applied mathematics is emerging; its future task in the design of computer systems can be guessed from the role played earlier by classical applied mathematics in the design of structures, ships and aeroplanes. The birth and growth of queueing theory from the needs of telephone engineering suggest that exciting problems will continue to appear.

The writing of my book, which took me four years of long working days, was completed in 1968. I then became head of my department; it was a large department and its management activities contrasted sharply with those necessary for composing a book. But trying to encourage the cooperation of a large group of people is not an uninteresting challenge. Unfortunately, I soon had to cope with the effects of the students' revolt, and after two years I was entirely exhausted. The students' revolt affected Dutch universities considerably; the hierarchically-structured decision system was abolished. Every decision had to be made by councils and committees composed on a representative basis from the groups of full professors, of teaching staff, of students, and of technical and administrative staff. The widely differing varieties of background and expertise of these representatives resulted in endless discussions, in which factual arguments were replaced by political viewpoints. The atmosphere in the department became tense and unpleasant, and it was hardly possible to concentrate on research. In 1972, I realized that I could not shape my future any longer in the Delft department, and decided to leave. In the 1960s I had been approached several times and encouraged to apply for a vacant chair in other departments in the Netherlands as well as abroad; I had always refused. The Delft school, my *alma mater*, was dear to me; it had a number of fine scientists and I had many good friends among colleagues and assistants. The decision to leave was not an easy one. It took me a few weeks to find new possibilities: I could opt for a position in industry or apply for two chairs, one in applied mechanics, the other in operational analysis. All three were attractive. I preferred the chair in operational analysis in the mathematics department of the University of Utrecht. The department had an excellent reputation in pure mathematics, it also had strong workers in the applied field, and above all it had an excellent atmosphere, mathematics being the department's first and last concern. I left Delft in 1973, and have never regretted my decision. In Utrecht I teach more or less in the same field as I did in Delft, but I had to change the emphasis of my lectures. This was not due to an essential difference in the mathematical knowledge of my students in Utrecht and Delft, but rather to the fact that those in Delft had a better training in intuitive thinking; the Delft students knew more physics, whereas those in Utrecht have a better feeling for mathematical formalism and can learn new mathematical techniques more quickly.

O. J. Boxma was a student of mine at the time I left Delft; he soon graduated and became my assistant. His quick understanding and fine devotion have been a great help to me in bridging the gap in my research activities caused by the transition from Delft to Utrecht. In the ten years during which the two of us have been in Utrecht, we have developed a close cooperation; this and his warm personal friendship are among the greatest of the many satisfactions which my work in Utrecht brings me.

The main intent of the research I planned to carry out in Utrecht was to develop queueing theory as a discipline in applied mathematics. Boxma, my students J. P. C. Blanc and G. Hooghiemstra and I have together obtained several results in this direction, in particular for queueing models with a two-dimensional state space. I have already outlined above the character of such problems, which are of basic importance in computer performance modelling. They had intrigued me for many years, and in 1977 I decided to initiate some fundamental research on them. The central idea was to develop a technique which did not require specification of the inherent service time distributions of the model; experience with the analysis of the $M/G/N$ loss, and $M/G/1$ queueing models had taught us that such an approach could lead to powerful techniques, as I have indicated earlier. For models with negative exponential distributions the uniformization technique can be used; this is a method for constructing a parameter representation for the zeros of an algebraic polynomial in two variables. Fayolle and Iasnogorodski were able to avoid the use of this technique in their analysis of several models which, however, had negative exponential distributions. They transformed the problem into a boundary-value problem, which in its simplest form requires the determination of a function which is analytic inside and outside a simple closed contour, with prescribed discontinuities on that contour. The results of Fayolle and Iasnogorodski encouraged us to have a closer look at boundary-value problems as discussed in continuum mechanics.

After some hard work, we have succeeded in constructing techniques which can analyse a fairly large class of two-dimensional state space models with *arbitrary* service-time distributions, and which allow for a numerical evaluation of the analytical results. In our monograph [8] we have given an exposition of the approach; the method developed may be considered to be part of the continuing search for robust analytical techniques in queueing theory, a search which started with Erlang.

4. Epilogue

In this account, I have tried to sketch some of my experiences in the building and analysis of models. I have given special attention to the search for robust analytical techniques in queueing theory; this robustness is not so often encountered in other applied mathematical disciplines, and as far as I know has never received adequate attention in general discussions on aspects of modelling. Such discussions are of direct relevance, following the recent rapid developments in mathematics and computing power and the problems which have arisen in the modern teaching of high-school mathematics. The recent book *Mathematics Tomorrow* [18] and its review by Von Wehausen [22] are of interest in this regard. This book review comes very close to my own views on the essential aspects of modelling, as does also Von Mises' paper: "On the tasks and aims of applied mathematics" [20].[1] Although Von Mises wrote in

---

[1] German title translated by the present author

the early 1920s, his standpoint has not lost its actuality, notwithstanding the revolutionary impact of modern computing power on modelling techniques.

Modelling is an activity which tries to capture reality in concepts and symbols, in logical and numerical relations. I have always found it amazing that human beings can construct models, whether they are directed at the structure of the universe or of man-made systems and organizations; and thus they experience a little of the mystery that is called reality. That I have been fortunate enough to share in such experiences I owe to the courage and actions of my father's friends during the Second World War.

References

[1] Blanc. J. P. C. and van Doorn, E. A. (1986) Relaxation times for queueing systems. *Proc. CWI Symp. Math. Comp. Sci.*, ed. J. W. de Bakker *et al.* North-Holland, Amsterdam.

[2] Blanc-Lapierre, A. and Fortet, R. (1953) *Théorie des fonctions aléatoires.* Masson, Paris.

[3] Cohen, J. W. (1955) *On Stress Calculations in Helicoidal Shells and Propeller Blades.* Waltman, Delft. (See also Netherlands Research Centre for Shipbuilding and Navigation, Report 21S.)

[4] Cohen, J. W. (1957) The generalized Engset formula. *Philips Telecomm. Rev.* 18, 158–170.

[5] Cohen, J. W. (1976) *On Regenerative Processes in Queueing Theory.* Lecture Notes in Economics and Mathematics, Springer-Verlag, Berlin.

[6] Cohen, J. W. (1969) *The Single Server Queue.* North-Holland, Amsterdam. (Revised edition 1982.)

[7] Cohen, J. W. (1977) On up- and downcrossings. *J. Appl. Prob.* 14, 405–410.

[8] Cohen, J. W. and Boxma, O. J. (1983) *Boundary Value Problems in Queueing System Analysis.* North-Holland, Amsterdam.

[9] Cohen, J. W. and Grosser, H. (1952) Calculation of the leakage flux of the pendant-armature relay. *Communication News* 12, 125–131.

[10] Erlang, A. K. (1917) Solution of some problems in the theory of probability of significance in automatic telephone exchanges. *Elektroteknikeren* 13, 5. (See also *The Life and Works of A. K. Erlang*, ed. E. Brockmeyer *et al.*, *Trans. Acad. Techn. Sci.*, Copenhagen, 1948.)

[11] Fry, T. C. (1928) *Probability and its Engineering Uses.* Van Nostrand, New York.

[12] Kendall, D. G. (1953) Stochastic processes occurring in the theory of queues and their analysis by the method of the embedded Markov chain. *Ann. Math. Statist.* 24, 338–354.

[13] Khinchine, A. Y. (1960) *Mathematical Methods in the Theory of Queues.* Griffin, London.

[14] Kosten, L. (1948) On the validity of the Erlang and Engset loss formula. *Het P. T. T. Bedrijf* 2, 42–45.

[15] Molina, E. C. (1927) Application of the theory of probability to telephone trunking problems. *Bell System Tech. J.* 6, 461–494.

[16] Palm, C. (1943) Intensitätsschwankungen im Fernsprechverkehr. *Ericsson Technics* 44, 1–89. (English version to appear.)

[17] Proceedings of the 1st International Congress on the Theory of Probability in Telephone Engineering and Administration (1957) *Teleteknik* 1, 1–130 (English edition).

[18] Steen, L. A. (ed.) (1981) *Mathematics Tomorrow*. Springer-Verlag, New York.
[19] Syski, R. (1960) *Introduction to Congestion Theory in Telephone Systems*. Oliver and Boyd, London.
[20] Von Mises, R. (1921) *Z. angew. Math. Mech.* 1, 1–15. (See also *Selected Papers* II, American Mathematical Society, Providence, RI, 1964).
[21] Von Mises, R. (1931) *Wahrscheinlichkeitsrechnung*. Franz Deutiche, Leipzig.
[22] Von Wehausen, J. V. (1982) Review of [18]. *Math. Intelligencer* 4, 157.

Ryszard Syski

Ryszard Syski was born on 8 April 1924 in Płock, Poland, the first child of Zygmunt and Halina (née Wisniewska) Syski. He studied in Poland, Italy and England, earning his first degree in telecommunications engineering, and his later Ph. D. degree (1960) in mathematics from the University of London, England. From 1952 to 1960 he worked as a mathematician with the Automatic Telephone and Electric Company Ltd., based in London. He joined the University of Maryland in 1960, becoming full professor in 1966, and is currently Professor of Mathematics in the statistics branch of the Department of Mathematics. He served as chairman of the statistics branch from 1962 to 1970.

Professor Syski has written two books and contributed to several others. He has also published about a hundred papers in the area of Markov processes and their applications, including queueing theory and potential theory. He was one of the founders of the journal *Stochastic Processes and their Applications.* He was also a founder and committee member (1972–9) of the Conferences on Stochastic Processes and their Applications, organizing the fifth such conference in Maryland in 1975. He has been involved since their beginning in 1955 with the International Teletraffic Congresses and was a member of the organizing committee of the fifth congress held in New York in 1967.

He was elected a fellow of the Institute of Mathematical Statistics in 1973, and is a member of the Polish Institute of Arts and Sciences of America, as well as of other societies. He has been an invited speaker at several international conferences, and has visited various universities in the USA and in Europe. During 1976–7 he spent a year at the Technion, Haifa, as Paul Naor Visiting Professor, and in the summer of 1982 he visited Warsaw, Poland, under the exchange program of the U. S. National Academy of Sciences.

Ryszard Syski is the father of six children, two of whom have presented him with six grandchildren, including twins. He enjoys good books (especially on history), good conversation and good friends.

Markovian Models—An Essay

1. Modeling

Mathematical modeling, like painting or photography, is an art, requiring proper balance between composition and the ability to convey a message. A good mathematical model, aiming to present an idealistic image of a real-life situation, should be accurate as well as selective in its description, and should use mathematical tools worthy of the problem.

These requirements are the source of a well-known conflict. A complex, or even a simple, but unrealistic model is useless for the practitioner but may be of interest to a theoretician; regrettably, many theoretical papers purporting to discuss "applications" commit this sin. A practitioner interested in realistic, and hence rather complex models, would prefer a simple, or even elementary, mathematical formulation, which is usually of no interest to the theoretician. The situation reflects a lack of communication between theoreticians and practitioners. The users of mathematical models, even if they possess adequate mathematical training, are more interested in concrete numerical results than in the niceties of a mathematical structure. A mathematician, on the other hand, pays more attention to the proper formulation of a problem, examines the existence and uniqueness of its solution, and studies its behavior under varying conditions.

In my scientific experience I have observed many manifestations of this conflict—sometimes severe, sometimes mild, but always painful. Personally, I lean to the theoretical side. I have always been interested in mathematical aspects, in the formulation of elegant and useful theoretical models exhibiting interesting structures. I would much prefer a model which presents a unifying picture embracing a variety of situations, a general rather than a specific model applicable to a particular problem at hand. I have tried to convince users of models that a mathematical structure has a beauty of its own; and, if at the moment the model does not satisfy their needs, it should not be discarded like a worn-out pair of shoes.

In my work I have tried to use general models, with special emphasis on their structure. That is why I find Markovian models fascinating. Their theory has reached a very high degree of development and has enormous power and beauty; their applications are numerous. The use of potential theory terminology provides an additional link between old classical problems and the most modern developments. It has been my aim to show that even the most abstract concepts, when properly explained and interpreted, can prove useful in the most concrete applications.

During my term as editor of a technical journal in London in the late 1950s, I expressed similar sentiments in my editorial column, "On the role of mathematics", and again much later during the discussion on applied probability organized by N. U. Prabhu (see [16], [23]). In my column, I wrote that in order to improve communication and understanding between theoreticians and practitioners, the users of models should familiarize themselves with the language of mathematics. A mathematician can always speak "many languages" by simply interpreting his results. In my contribution to the discussion on applied probability, I listed three highly subjective and rather vague criteria which a good model should satisfy. It should contain good mathematics, deal with a real practical problem, and present useful results. These comments have even greater validity today when applications of mathematical (especially probabilistic) models to the engineering, biological, and social sciences are so widespread.

2. Early Influences

2.1. As the lives of members of my generation in Europe were badly shaken up by the events of the Second World War, a brief "personal history" may elucidate some aspects of this account.

I was born in Poland, in the city of Plock, and lived in Poland until 1944. Our tranquil family life and my carefree school days were shattered in September 1939 when the Germans invaded our country. My family and I joined the Polish Resistance, barely escaping death in 1943, and I took an active part in the Warsaw Uprising of 1944. Afterwards, I spent half a year in a German prisoner-of-war camp. In April 1945, we were liberated in Bavaria by General Patton's army. Soon afterwards, my friends and I left Germany to join General Anders' Polish Second Corps, then under British command in Italy. Late in 1946 we were evacuated to England and a new life began for me.

I studied telecommunications and mathematics at the University of London, and after graduating worked as a scientist in the ATE telephone company. In 1950, I was married; my wife Barbara used to live almost next door in Warsaw, but we first met many years later in London. In April 1960, I visited the USA at the invitation of the Office of Naval Research (ONR). On 31 December 1960, my wife and I, together with our four children, arrived in Maryland; I am still here—with six children and six grandchildren.

My first return visit to Poland took place in 1980, at the invitation of the Warsaw Technical University, and two subsequent visits took place in 1982 and 1984 at the invitation of the Mathematical Institute of the Polish Academy of Sciences. My only visit to Germany after the war took place in 1970 when I attended a congress in Munich. My only visit to Moscow was in 1984, at the invitation of the Telecommunication Institute of the USSR Academy of Sciences.

2.2. When my father retired from his law practice in 1937 because of failing health, our family settled down to a country life in the vicinity of Pultusk, a small historic town north of Warsaw. I became a student at the local high school, housed in a sixteenth-century building which is still in use today. My mathematics teacher was the headmaster himself and, to my surprise, I got an excellent grade on my first test. That is how my interest in mathematics began. There was another mathematics teacher at the school with whom I had no contact until we met briefly at the school's reunion in 1980. Then, to my even greater surprise, I discovered that he was the father-in-law of the editor of the Dutch publishing company which is preparing my book (on teletraffic) [25] for republication.

Just before the Warsaw Uprising, we lived in Warsaw under assumed names; my father had been killed in a Russian air attack on the city. Life under the German occupation was harsh, and we were in constant danger. At that time I taught myself enough calculus to be able to sign up for a "course in technical drawing" at one of the vocational schools allowed to function by the Germans. The school was intended for artisans but its teachers, in most cases, taught their students much more. All high schools and universities were closed by the Germans, but the underground courses flourished despite constant danger. In our "technical drawing" course, we had regular lectures in real analysis, analytic and descriptive geometry, and indeed, some technical drawing, all taught by the surviving faculty of the Warsaw Technical University. I recall warmly the lectures of Professor W. Pogorzelski, famous for his textbooks on analysis and his work on integral equations [13], [14]. It was with great emotion that 36 years later, on my first return to Warsaw, I lectured in the same rooms in which I had attended classes during those dark days of the German occupation. Besides my studies, I pursued my underground activities and attended military courses, achieving my highest military rank, that of corporal. My wife Barbara, who had similar experiences, reached the rank of sergeant, as she likes to remind me on occasion.

In a book [39] published in England after the war, Captain Zagórski, the commander of my unit in the Warsaw Uprising, described in the form of a diary the day-by-day activities of his soldiers. The book made extensive use of first-person accounts provided by those of us who were in the West. Let me describe one of my stories.

In a look-out in the ruins of a burnt house, I found several books scattered on the floor, among them the Polish edition of Goursat's *Analysis* [8] and some works of Plato. I took them with me, later carrying them on my back to the prisoner-of-war camp and again during the evacuation when the German–Russian front moved westward. Plato and other philosophy books ended their life somewhere in the snow, but Goursat's book sits proudly on my bookshelf here, bearing the University of Warsaw library stamp and the German "geprüft". One of these days (or perhaps in my will) I shall return it to the university.

Speaking of combat, I once took part in the attack on a German stronghold located in a building housing a telephone exchange. In later days, when working with telephone engineers in England, I teased them about lacking my experience in "burning exchanges".

In Italy I served in an artillery unit, which was the most mathematical branch of the armed forces at that time. I was soon transferred to the cadet officer school where my calculus and trigonometry were indeed useful. When the opportunity arose, I dropped out of the school and abandoned my military career; this turned out to be a wise move. I enrolled at the Technical University in Turin to continue my studies while still formally in the army. Lectures and examinations were in Italian; I did well in mathematics, physics and chemistry, but poorly in technical drawing!

After demobilization in England, I became a student at the Polish University College (PUC), a technical college for Polish ex-service men and women having a rather loose association with the University of London. PUC was staffed with Polish faculty and had a program modeled on that of Polish prewar technical universities. I graduated with a degree in telecommunications, having written a thesis on the non-linear equation of a synchronized oscillator; I then spent a year at Imperial College on postgraduate work (gaining a DIC, the College's diploma). I switched to mathematics completely, and after taking a job with the ATE telephone company, I continued my studies at the University of London, gaining first a B. Sc. and then a Ph. D. in mathematics.

2.3. Information theory was gaining popularity in the 1950s. I was fascinated by Wiener's *Cybernetics*, read Shannon with great interest, and attended splendid lectures by two outstanding professors, D. Gabor and C. Cherry. I was struck by the analogies between information theory and the field in which I worked, namely congestion theory ("teletraffic theory" in modern terminology). I pointed out my observations at the Information Theory Symposia [10] p. 528, [2] pp. 25, 109, [26], and expanded my ideas in a paper presented at the First International Teletraffic Congress held in Copenhagen in 1955 [21]. I had great faith in the power of the entropy concept and was deeply impressed by ergodic theorems. My interest diminished when I sensed that ergodic theory was becoming a big business developed by empire builders, and when information theory changed into information systems analysis and processing.

I was, and still am, highly enthusiastic about measures on function spaces and all extensions of the Daniell–Kolmogorov theorem. My Ph. D. thesis dealt with this topic, and some applications to congestion processes. What I had in mind at that time was later accomplished in a much better way by Skorohod. As for overenthusiasm, V. Beneš told me much later that he was flabbergasted upon receiving my letter on applications of the Daniell–Kolmogorov theorem. I had better luck with I. J. Savage, who found my comment useful enough to include in his monograph on subjective probability (perhaps as a warning to others) [18], p. 186. And now, when teaching probability courses at the University of Maryland, I always include the Ionescu Tulcea theorem and related results in my lectures.

Among mathematics lectures in my student days at Imperial College, I still remember those of Barnard (based on Halmos' *Measure Theory*), those of Lloyd on statistics, and those of Anis on random walks. Once we were instructed to attend a seminar talk on the theory of dams by a visitor from Australia, Joe Gani. I also attended meetings of the Royal Statistical Society in London; in particular, I was present when Professor J. Neyman on his first return to England in 1957 presented a paper (jointly with Professor E. Scott) on cosmology [22], and I took part in the discussion in 1958 on a paper by W. Smith on renewal theory and its ramifications [19].

2.4. When, in 1952, I joined the newly formed laboratory at the Automatic Telephone and Electric Co. Ltd. (ATE), I was given the task by our chief, Colonel A. Krzyczkowski, of investigating the theoretical (i.e. probabilistic) aspects of telephone traffic. This laboratory, the brainchild of Colonel K., as he liked to be called, was located in London at Hivac, a subsidiary of ATE. My colleague and friend, R. Sidorowicz, who is now a lecturer in electronics at Imperial College, also joined the laboratory (Colonel K. by the way, was our professor at PUC).

From the start I was involved in the study of stochastic processes, and queueing in particular. The similarity between (birth-and-death) equations in the famous books by Th. Fry and W. Feller came as a revelation to me (see [20]). I read the excellent monograph on A. K. Erlang by Arne Jensen and his colleagues, the thesis of C. Palm, and the papers of Pollaczek, Vaulot, Kendall, Foster, Lindley, Smith and many others. I can say without exaggeration that I knew practically all that was written on queueing theory in those early days. At the same time, I acquainted myself with technical literature on teletraffic. My studies oscillated from the very beginning between abstract formulation and concrete analytic examples. My interest in Markov processes led me to the Hille–Phillips treatise on semigroups and to G. Birkhoff's *Lattice Theory*. The classical books of J. L. Doob, M. Loève, and W. Feller have been with me ever since, to be joined later by the treatises of Dynkin, Meyer, Chung, and Blumenthal and Getoor. Let me mention in passing an amusing incident: a selfrighteous gentleman, in a letter to the editor, warned me graciously that I might fall into a Soviet trap by studying Markov processes (see [24])!

Personal contacts have always played an important role in my life. Over the years, I have had numerous acquaintances and friends, and have always enjoyed a good friendly chat. Let me mention here four close friends from those early days. I met Wim Cohen in 1955, although we had some earlier correspondence, and our friendship has continued uninterrupted to this day. In 1958 at the Hague congress I met Theo Runnenburg, and since then we have seen each other on numerous occasions. At the same congress, my wife and I were introduced to Mr and Mrs Pollaczek—they visited us at Maryland later, in 1963. Lajos Takács arrived in London in 1958 and subsequently emigrated to the USA. We, too, had been in correspondence earlier. Meeting with Tom Saaty, who was with the ONR office in London, was a turning point in our life—he was instrumental in our emigrating to the United States.

At Maryland, I continued with my interest in queueing for some time—writing, lecturing, and attending meetings. I began to lose interest when the field grew too widespread and the papers thinner in new ideas. I turned my attention instead to stochastic processes and potential theory. Later, the situation improved, as I remarked in my comments at the TIMS meeting in 1973 [29], and queueing theory enjoyed a revival due principally to new approaches through point processes and martingales [1]. Nevertheless, I have always remained faithful to my first love, Markovian queues.

3. Later Life

3.1. While working on teletraffic problems at the ATE laboratory, I became convinced that there was a need for a text which would present all known results in terms of recent developments in queueing and stochastic processes. The book on teletraffic used by engineers at the time was that of Berkeley. Valuable as it was, it was rather old and inadequate for modern needs so I intended to write a new book presenting a unified treatment of teletraffic theory, and not merely a collection of known results. Thus, I began my long study of teletraffic literature. The actual writing of the book took about two years, followed by almost a year of production technicalities. This was my full-time job. Presentation, arrangement of material, even notation were all new. The book contained a lengthy chapter on probability and stochastic processes, which was really needed at that time, and a rather extensive mathematical appendix. I received sympathetic help from the ATE traffic engineers, especially from N. H. G. Morris who wrote a chapter on telephone systems and helped generally in shaping the book. J. W. Cohen (then at Delft) read the whole manuscript and suggested many useful improvements.

At the opening session of the First International Teletraffic Congress in Copenhagen in 1955 I had suggested that such a book should be written, but I was not certain that I should be the one to do it. At the second congress in the Hague in 1958 leaflets announcing my forthcoming book [25] were distributed. The book was published in 1960, and the first copy reached me in April just a few days before my departure for a visit to the United States. I proudly carried the book with me, showing it off. In my mind, I always refer to it by the initials CGT.

The book was well received. Initial reviews were favorable (except for one or two from bewildered engineers), and throughout the following years it has gained widespread acceptance, as I have had many occasions to observe. At the Tenth Teletraffic Congress in Montréal in 1983, it was announced that the Council would sponsor republication of the book in their new series on teletraffic. What more could any author desire?

3.2. It may be of interest to say a few words about my association with the International Teletraffic Congress (ITC). Out of ten congresses, I have attended all except one in Australia (1976). I believe that at this moment there are perhaps not more than four of us who have missed at most one meeting. In those early days Wim Cohen and I served on the Bibliography Committee.

(Wim is now a member of the Council.) The fifth congress was held in New York (1967), and I served on the organizing committee under the able chairmanship of Roger Wilkinson.

As the ITC meetings became more and more technical and I drifted further away from teletraffic towards Markov processes, I attended the congresses to meet old and new friends, and to see how things were progressing. ITC meetings have a unique atmosphere, pleasant and friendly, and it was rewarding to chat with friends. I was frequently honored by invitations to chair a session, or to be a discussion leader. As time progressed, it was a pleasure to see the new generation of teletraffic researchers entering the field. After the Torremolinos meeting (1979), a questionnaire was distributed among participants asking for their views on the future of ITC; I expressed hope that there would be more and more papers devoted to theoretical studies of teletraffic. This hope is being fulfilled.

One of the regulars who attended all meetings is the originator of ITC and its chairman, Arne Jensen. The Paris Congress in 1961 was organized by Pierre Le Gall, a friend of mine since those days. The Torremolinos meeting was particularly significant for me, as J. Szczepanski who attended it, invited me to visit Poland. At the Montréal meeting, V. Neiman invited me to an ITC special seminar on teletraffic theory held in the summer of 1984 in Moscow.

3.3. In the summer of 1971, I was invited by Julian Keilson to attend what turned out to be the first Conference on Stochastic Processes and Their Applications, held in Rochester. Julian's idea was to have all applied probabilists meet together to discuss the future of the field. In his opinion, there was a need for more concrete applied work, although he did not shun abstract probability theory. The result was a successful series of conferences held practically every year, the latest (the fourteenth) being held in the summer of 1984 in Göteborg. I have missed only three. The third conference was organized by Joe Gani in Sheffield in 1973, and the fifth by Grace Yang and me at the University of Maryland in 1975. I am perhaps the member with the longest uninterrupted tenure on CCSP—the Committee for Conferences on Stochastic Processes.

At Rochester, we started discussions on the new journal, also called *Stochastic Processes and Their Applications* (or *SPA* for short). The first issue appeared in 1973, Wim Cohen, Julian Keilson, Uma Prabhu and I being its "founding fathers". Through the years, thanks to the editorial efforts of Julian, Uma and now Chris Heyde, the journal has gained firm scientific standing.

At present, both the conferences and the journal are strong and developing. Although both are now out of our hands, they are doing well under the auspices of the Bernoulli Society. From my point of view this has been a success story, and I have described it as such in my article [35]. The work on both the conferences and the journal gave me great satisfaction, and it is a real pleasure to see that it was not in vain.

3.4. Much of my academic life at Maryland has run along the usual path: research, teaching and committee meetings. As these activities, which occupy a substantial part of my time, are not directly related to the main theme of this essay, I shall not discuss them.

I wish, however, to make a few comments. My long association (extending over a decade) with colleagues in our statistics group, Peter Mikulski, Grace Yang and Paul Smith (in the order of their joining Maryland), has always been harmonious, even if we have not always agreed on some issues. I have carried out my research in the areas of Markov processes, martingales and stochastic equations, turning later to potential theory. I maintained personal contacts with Wim Cohen, Lajos Takács, Uma Prabhu, Julian Keilson, Ben Epstein, Micha Yadin, Jef Teugels, Pierre Le Gall, Jan Ostrowski, Theo Runnenburg, and many other workers. We have had numerous visitors in our department and, in 1975–6, we organized our Special Year in Probability and Statistics. I traveled a lot, meeting old friends and making new ones. The places where I have spent weeks or months include Amsterdam, Haifa, Lisbon, Louvain, London, Utrecht and Warsaw. Many interesting new books appeared in this period, and my library grew. I greatly enjoyed the excellent treatise of Cohen on the single-server queue, the combinatorial methods of Takács, Keilson's rarity and exponentiality as well as Chung's boundary theory, Revuz's Markov chains, and Williams' Markov processes and martingales. And, of course, Blumenthal and Getoor followed by Dellacherie and Meyer.

I taught courses on stochastic processes at various levels and, from time to time, I gave a special course on Markov processes, martingales, queueing, and related topics. In the 1960s Tom Saaty and I (subsequently, I alone) taught extension summer courses at UCLA on queueing and stochastic processes; once Norbert Wiener was teaching a course next door to mine.

Among my former Ph. D. students I wish to mention Lou Blake who is now well known for his extension of martingales, Dick Shachtman who now works in biological applications, and Pete Kolmus who spent several years with Choquet in Paris.

After spending about eight years as chairman of a statistics group, struggling with statistical courses, programs, and students, I resigned this position and took my first sabbatical (at Technion, Haifa). Since then, I have lost all interest in administrative work at the university. I now prefer to work in peace and quiet on problems which interest me, and enjoy a good conversation with friends, as well as reading.

I was never an empire builder, and I hated to write proposals—I abhor begging, and especially organized begging. I do not think it is right to measure success by the amount of contract and grant money generated.

3.5. Visiting Poland in 1980 after 36 years of absence came as a shock to me. All places familiar to me either do not exist any more or have changed beyond recognition; all family members and friends have become much older than they used to be!

The most striking effect, perhaps, was to see the new generation which has grown up since I left the country. Often when looking at my niece, I pictured in my mind her mother, whom I remembered when she was her daughter's age. At universities and research institutes I met with bright young scientists, all complaining about their isolation from the West and eager to visit the United States to further their scientific development.

I greatly enjoyed seeing again friends whom I had met earlier at various meetings in the USA and in Europe. I am very grateful to Genek Fidelis from the Mathematical Institute for his hospitality and for the unforgettable experience of mountain climbing in Zakopane.

In 1984, having spent one week in Moscow (a nice place to visit, but ...) I am ready to join the ranks of experts on Russian affairs. I remember the pleasant evening when Professor B. Gnedenko (whom I last saw over 25 years ago at the Edinburgh congress) invited four of us, all authors of books on queueing theory, to the Russian dinner in his apartment at Moscow University.

Vividly embedded in my memory are my three visits to the Technion, Haifa (fall semester 1970, summer 1973, and the academic year 1976–7). I enjoyed my stay there tremendously, due to the great hospitality of my hosts at the Faculty of Industrial and Management Engineering. I was greatly impressed by their work in probabilistic modeling, and had rewarding discussions with Ben Epstein, Micha Yadin (we have published a joint paper), Micha Rubinovich, Igal Adiri and others, Regrettably, during my first visit Paul Naor, whom I met earlier and admired greatly, lost his life in a tragic air accident; I was asked to take over his queueing theory course. Later, in 1976, I was honored by the award of the Paul (Pinhas) Naor Visiting Professorship at the Technion; part of my duties involved the presentation of the Memorial Lecture. Among graduate students who attended my lectures, and who are now established workers in their own right, were Peter Kubat (who now works in industry in Boston) and Micha Hofri (now in the Computer Department, Technion); Eli Rubinstein wrote his master's thesis with me.

In my lectures, of course, I emphasized potential theory and its applications to Markovian queues, I am happy to see that a new generation of excellent young mathematicians now works at the Technion in this area, combining theoretical investigations with real practical applications. And I would like to add that during my stay in Israel, I visited almost every part of the country, even worked in a kibbutz for one month, and became an unofficial guide for Polish pilgrims to the Old City of Jerusalem.

Finally, one more item along memory lane. I participated with great pleasure in the moving ceremony at which Mr Pollaczek was presented the John von Neumann Theory Prize (for 1977), awarded to him by ORSA/TIMS. As Mr Pollaczek was unable to travel to America, the informal ceremony took place (in July 1977) among close friends in the hospitable home of Christiane and Pierre Le Gall, in the scenic countryside of Normandy. In his testimonial speech, Wim Cohen presented Mr Pollaczek with the gold medal on behalf of the societies, and offered congratulations to Mr and Mrs Pollaczek. It was with great sadness that later, in 1981, I paid my last respects to Mr Pollaczek at his grave in Paris.

4. Forever Markovian

4.1. As I have already mentioned, my interest shifted gradually from general queues to Markovian queues, then to Markov processes, and consequently to probabilistic potential theory. I came to the conclusion that potential theory provides a proper framework for the unification and systematic description of Markovian queues. It came to me as a revelation, though it must be obvious to many people, that formally queueing problems provide a splendid concrete illustration of potential theoretic concepts, the typical example being the Dirichlet problem and all its ramifications.

I became interested in Markovian queues almost immediately after completion of my CGT book [25]. At the Chapel Hill Symposium (1964) I stressed the importance of the interpretation of a waiting time as the first-entrance time to a taboo set [27]. This idea, of course, has been used in queueing from the very beginning, but it was the structure of the Markov model and its generality for applications which I found fascinating. I would like to mention here the applicability of these methods to concrete examples in the papers by P. Kühn [12] and myself [30], [31]. I may add that at the same symposium I was privileged to present Pollaczek's paper in his absence. Later at the Lisbon Conference (1965), I was invited to present an expository paper on the Pollaczek method [28]. In both cases, the underlying passage-time ideas were stressed [15]. However, the full power of the Pollaczek method was demonstrated by J. W. Cohen in his famous treatise [5].

Subsequently, I immersed myself deeper into potential theory, struggling with the works of Hunt, Doob, Blumenthal and Getoor, and others. Most of the time I worked with Markov chains with a discrete state space and a continous parameter ("the Chung process"), see [3]. Such chains have a very rich theory and enjoy plenty of applications. At the same time their "intuitive simplicity" serves well as an illustration of abstract mathematical concepts. As before, my work oscillated between abstract topics (boundary theory, potentials, duality) and concrete analytic examples of passage times (first entrance, last exit, etc).

I began to "preach" potential theory, finding it at the center of everything, and not paying much attention to anything else. I lectured about it at Maryland and at other universities where I was invited, as well as at various meetings. On the one hand, such activities were very pleasant and personally gratifying; on the other hand, people began avoiding me like the plague.

And I fell into a most natural trap—I wrote a rather extensive text on passage times in Markov chains in which I attempted to show the applicability of abstract theory (potentials, excessive functions and measures, balayage, killing and reversing, random time change, and so on) to Markovian models, and Markovian queues especially. At present no publisher is willing to look at my manuscript until its size has been reduced drastically, to say nothing of whether the topic itself has "marketable value". I am afraid that it may suffer the fate of my manuscript on advanced undergraduate probability which was found to be "too advanced for undergraduates but too elementary for graduates." In the meantime, I took the opportunity of writing up my ideas on Markovian queues in summary form in the additional chapter of the forthcoming reprint of my CGT book [38] and in an invited paper delivered at the ITC Moscow Seminar [37].

4.2. For what it is worth, let $X = (X_t, 0 \leq t < \infty)$ be a Markov chain with a discrete state space Π and a continous time parameter, describing a state of a queueing system at time t. Assume that X is governed by a standard transition matrix $P(t) = (p_{ij}(t))$, with an intensity matrix $Q = (q_{ij})$, and a resolvent (α-potential) $U^\alpha = (u_{ij}^\alpha)$ for $\alpha > 0$; that both Kolmogorov equations hold, and that sample functions of X are right-continous with left limits. In applications, the coefficients q_{ij} describe the structure and operation of the system. Now select a fixed subset H of Π, and define for any fixed $s \geq 0$, the first-entrance time T_s to H after s, and the last-exit time L_s from H before s, by

$$T_s = \inf(t: t > s, X_t \in H), \qquad L_s = \sup(t: 0 < t \leq s, X_t \in H),$$

with $T_0 = T$ and $L_\infty = L$. Let $X(T_s)$ and $X(L_s)$ be the corresponding positions of the first entry to H and the last exit from H (by right-continuity they take values in H and H^c, respectively). The random interval (L_s, T_s) is the excursion interval straddling s, and the corresponding excursion chain $(X(L_s + t), 0 < t < T_s - L_s)$ takes values in H^c only. It is of great importance that joint distributions of these random variables can be expressed in terms of taboo probabilities defined by

$$H^{p_{ij}(t)} = \mathbb{P}^i(X_u \in H^c, 0 < u < t, X_t = j) = \mathbb{P}^i(t < T, X_t = j),$$

which satisfy Kolmogorov equations (restricted to transitions on H^c) and correspond to the killing of the chain X at the first entrance to the taboo set H.

The α-balayage matrix is $B^\alpha = (d_{ij}^\alpha)$, where d_{ij}^α is the Laplace–Stieltjes transform (with respect to t) of the joint distribution $\mathbb{P}^i(T \leq t, X_T = j)$. The fundamental equations for d_{ij}^α are easily derived from the taboo equations. For $\alpha = 0$, write $B = (D_{ij})$ where $D_{ij} = \mathbb{P}^i(X_T = j, T < \infty)$ is the first-entrance distribution, with the first-entrance probability $D_i = \mathbb{P}^i(T < \infty)$ where by definition $D_i = 1$ for $i \in H$.

The Dirichlet problem consists of finding a (positive bounded) function f on Π which is invariant (harmonic) on the set H^c and equal to the "boundary function" φ on H:

$$Qf(i) = 0, \quad i \in H^c, \qquad f(i) = \varphi(i) \quad \text{for } i \in H.$$

The unique (minimal) solution of the Dirichlet problem is given by the balayage of φ on H:

$$f(i) = B\varphi(i) = \mathbb{E}^i(\varphi(X_T), T < \infty).$$

It follows that the first-entrance probability $D = B1$ is the (minimal) solution of the Dirichlet problem with the boundary function φ identically 1 on H.

Moreover, the function D has the unique Riesz decomposition (implied by the fact that it is excessive, by the Poisson equation):

$$D_i = Ub_i + s_i, \quad i \in \Pi,$$

where Ub is the purely excessive component which is also a potential of a charge b (escape rate), and s is invariant (harmonic). Probabilistically,

$$Ub_i = \mathbb{P}^i(0 < L < \infty), \quad s_i = \mathbb{P}^i(L = \infty),$$

and s is the probability of infinitely many visits to H by the jump chain associated with the chain X.

These results have been used in queueing theory, but the unification provided by the potential theoretic formulation is not only of theoretical interest, but produces a practical method for calculations. With a chain specified by its intensities (q_{ij}) and with proper choice of the taboo set H depending on the interpretation of T, equations for d_i^α can be written down by inspection, and the Dirichlet problem yields the probability that T is finite. Typical cases include situations where T represents the busy period, the waiting time, the time to absorption, overflow time, output, etc.

This is just a beginning, but it should suffice to give a picture of what the subject is about. Is it too abstract? I do not think so. Anybody who has read Karlin and Taylor, or Volume 2 of Feller (a must for an applied probabilist) should not be lost.

4.3. I should like to mention here some other potential theoretic topics which I consider important. The first is the compensation technique introduced by J. Keilson in his approach to potential theory [11]. I employed it in my "perturbation model" which essentially changes a generator of a Markov chain by some external mechanism. (This happened to be related to the "second resolvent equation" in semigroup theory.) This model has quite a wide applicability and, in particular, it includes the "phase-type distributions" of Marcel Neuts. For a summary of these results, see [36]. Another type of transformation which has numerous applications is the random time change. It is determined by the inverse τ of an integral functional A of a chain (X_t), with a new chain (Y_t) formed by $Y_t = X(\tau_t)$. My attempt to obtain a generator for Y was superseded by the beautiful paper of H. Gzyl [9]. As an amusing application, let $X = (X_t)$ be the usual branching process (with absorbing state 0). Then, the integral

$$A_\infty = \int_0^\infty X_t \, dt$$

and the busy period T in the $M/M/1$ queue with group input, have the same distribution (with proper identification of parameters).

Another concept which I think has a great future is that of energy. Following the classical definition (as an integral of a potential with respect to a charge), energy for a Markov chain may be defined by the inner product $p \cdot \mu$, where $p = Ua$ is the potential of a charge a (a function) and μ is a charge (a measure), the dual of a. Another definition uses the integral functional (A_t) determined by charge a, and defines energy as $\mathbb{E}A_\infty^2/2$. I suggested in my communication at the SPA meeting in Montréal in 1974 that the energy concept may serve as a useful characterization of Markovian systems, somewhat as entropy used to be (see [32], [33] and remarks in [37]). Recently, the notion of energy has received considerable attention in the literature, but it has not yet been fully employed in queueing theory; (see [4], [7]).

4.4. In this essay, I have expressed my faith in general models. I believe that those aspects of mathematics which emphasize structure, are essential for applications to model building. Criticism is often expressed that such an approach is too abstract (a polite way of saying useless). Then the question might be asked, "How much abstraction is good for your health?" Obviously, I would not suggest that users of models be familiar with recent advances in the "general theory of processes". Much of this material, after all, is inaccessible to nonspecialists. On the other hand, it is an unfortunate sign of our times that books, papers or courses, based on solid mathematics (irrespective of its level) are frequently regarded by practitioners as "too difficult, too abstract and too fancy".

In my opinion, the abstract material just mentioned is useful and important, but it should be "translated" into practice and properly applied. After all, we all know that yesterday's abstract concepts are today's practical tools; recent examples are martingales and predictable processes.

This brings me to education, both for our students and future researchers, and for current and future practitioners (statisticians, engineers, scientists). Apart from the regular literature on mathematics, stochastic processes, and statistics designed for specialists and students in these areas, there is a need for good expository books and courses designed for users of mathematical models. Basic ideas should be explained in a convincing and logical fashion, pointing out highlights as well as pitfalls, and omitting excessively technical details. However, the cookbook approach and fun stories (so popular nowadays) are definitely harmful and out of place in serious work.

I have offered many courses on stochastic processes for interested nonmathematical users. My approach—through logical, general formulation and particular, illustrative examples—has been appreciated in most instances. My little booklet [34] written in this spirit and for this type of audience, presented "correct and moderately precise" mathematics. At the other extreme, my expository paper on stochastic processes, which intended to explain modern abstract developments and their usefulness, was rejected because it was thought that it might scare readers; it was later accepted after surgery to make it more palatable. However, as I have described in this essay, I have always tried to bridge the gap between abstract theory and practice, "translating" advances in Markov processes into practical tools.

Indeed, on the one hand, tremendous progress in the theory of stochastic processes, especially Markov processes and martingales, has led to applications of probabilistic methods in such classical fields as potential theory, differential equations, harmonic analysis, representation theory, and even the theory of numbers. The recent book by Doob [6] illustrates this trend. It is with some malicious satisfaction that I observe my friends in pure mathematics bewildered by the intricacies of Brownian motion infiltrating their fields.

On the other hand, "thinking with models" is widespread in applied fields, see [17]. In particular, Markovian models are entering new areas like networks of queues, computers, biology (survival analysis), psychology (learning), social sciences and even the stock market and geology. This healthy trend is not without effect on the development of theory: random fields, boundary problems, compensation and perturbation techniques are some recent examples.

Although Markovian models are only approximate, more elaborate models involving semi-Markov or point processes are just important extensions of the Markov structure.

Hence, as has always been the case in history, theory and practice influence each other. And for myself, I feel that Markovian models still have great potential.

Acknowledgment

I wish to thank my daughter-in-law Helena for un-Polishing my English in this essay.

References

[1] Brémaud, P. (1980) *Point Processes and Queues (Martingale Dynamics)*. Springer-Verlag, Berlin.

[2] Cherry, E. C. (ed) (1956) *Information Theory*. Butterworth, London.

[3] Chung, K. L. (1967) *Markov Chains with Stationary Transition Probabilities*, 2nd edn. Springer-Verlag, Berlin.

[4] Chung, K. L. and Rao, M. (1980) Equilibrium and energy. *Prob. Math. Statist.* (Wrocław) 1, 99–108.

[5] Cohen, J. W. (1969) *The Single Server Queue* (2nd edn 1982). North-Holland, Amsterdam.

[6] Doob, J. L. (1984) *Classical Potential Theory and its Probabilistic Counterpart*. Springer-Verlag, New York.

[7] Glover, J. (1983) Topics in energy and potential theory. In *Seminar on Stochastic Processes*, Birkhauser, Basel, 195–202.

[8] Goursat, E. (1914) *Kurs analizy matematycznej*, Vol. 1 (Polish translation). Warsaw.

[9] Gzyl, H. (1980) Infinitesimal generators of time changed processes. *Ann. Prob.* 8, 716–726.

[10] Jackson, W. (ed) (1953) *Communication Theory*. Butterworth, London.

[11] Keilson, J. (1979) *Markov Chain Models—Rarity and Exponentiality*. Springer-Verlag, New York.

[12] Kühn, P. (1973) The impact of queueing theory on the optimization of communications and computer systems. *Proc. XX TIMS, Tel Aviv*, ed. E. Shlifer, Jerusalem Academic Press, Jerusalem, 2, 554–568.

[13] Pogorzelski, W. (1946) *Analiza matematyczna*, Vols 1–4. Warsaw.

[14] Pogorzelski, W. (1953) *Rowania calkowe i ich zastosowania*. PWN, Warsaw.

[15] Pollaczek, F. (1963) Summary of *Théorie analytique*. University of Maryland Technical Report BN-323.

[16] Prabhu, N. U. (ed) (1975) Applied probability: Its nature and scope. *Stoch. Proc. Appl.* 3, 253–255.

[17] Saaty, T. L. and Alexander, J. M (1981) *Thinking with Models*. Pergamon Press, Oxford.

[18] Savage, L. J. (1962) *The Foundations of Statistical Inference*. Methuen, London.

[19] Smith, W. L. (1958) Renewal theory and its ramifications. *J. R. Statist. Soc.* B 20, 293–294.

[20] Syski, R. (1953) The theory of congestion in lost-call systems. *ATE Jnl* 9, 182–215.

[21] Syski, R. (1955) Analogies between the congestion and communication theories. *ATE Jnl* 11, 220–243 (Summary in *Teleteknik*, Copenhagen 1 (1957), 124–125.

[22] Syski, R. (1958) Professor Neyman in London (in Polish). *J. Inst. Polish Engrs in Great Britain* 18, January, p. 23.

[23] Syski, R. (1959) On the role of mathematics in technology (in Polish). *Technika i Nauka* (London) 6, 1–3.

[24] Syski, R. (1960) Reply to a letter (in Polish). *Technika i Nauka* (London) 8, 45–46.

[25] Syski, R. (1960) *Introduction to Congestion Theory in Telephone Systems*. Oliver and Boyd, Edinburgh. (Revised reprint 1985, North-Holland, Amsterdam.)

[26] Syski, R. (1962) Congestion in telephone exchanges. [2], pp. 85–98.

[27] Syski, R. (1965) Markovian queues. In *Proc. Symp. Congestion Theory*, ed. W. L. Smith and W. E. Wilkinson, University of North Carolina Press, Chapel Hill, 170–227.

[28] Syski, R. (1967) Pollaczek method in queueing theory. In *Queueing Theory*, ed. R. Cruon, English Universities Press, London, pp. 33–60.

[29] Syski, R. (1973) Queueing theory symposium: Introduction and summary. *Proc. XX TIMS, Tel Aviv*, ed. E. Shlifer, Jerusalem Academic Press, Jerusalem, 2, 507–508.

[30] Syski, R. (1973) Queues and potentials. *Proc. XX TIMS, Tel Aviv*, ed. E. Shlifer, Jerusalem Academic Press, Jerusalem, 2, 547–554.

[31] Syski, R. (1973) Potential theory for Markov chains. In *Probabilistic Methods in Applied Mathematics* 3, ed. A. T. Bharucha-Reid, Academic Press, New York, 213–276.

[32] Syski, R. (1975) Energy of Markov chains (abstract). *Adv. Appl. Prob.* 7, 254–255.

[33] Syski, R. (1979) Energy of Markov chains. *Adv. Appl. Prob.* 11, 542–575.

[34] Syski, R. (1979) *Random Processes: A First Look*. Dekker, New York.

[35] Syski, R. (1980) Stochastic processes and their applications (in Polish). *Technika i Nauka (London)* 46, 33–42.

[36] Syski, R. (1982) Phase-type distributions and perturbation model. *Applicationes Math.* (Wrocław) 17, 377–399.

[37] Syski, R. (1984) Markovian queues in teletraffic theory. In *Fundamentals of Teletraffic Theory*, USSR Academy of Sciences, Moscow, 430–440.

[38] Syski, R. (1985) Markovian queues. Chapter 11 of the reprint of [25].

[39] Zagórski, W. (1957) *Wicher wolnosci* (English translation: *Seventy Days*). Muller, London.

N. U. Prabhu

Professor Prabhu was born in Calicut, India on 25 April 1924 and received his university education at Loyola College, Madras, where he studied mathematics. After teaching for two years in the University of Bombay, he enrolled in its statistics programme and obtained his MA there in 1950. He began lecturing at Gauhati University, Assam, and soon moved to the Department of Statistics at Karnatak University in 1952 as reader and head of department, a position he held for nine years.

In 1955, he visited the University of Manchester, England, as a British Council scholar, and was awarded an M.Sc. in 1957 for his research on dam theory. In 1961 he was appointed to a readership in mathematical statistics at the University of Western Australia, where he completed the writing of his two books.

He emigrated to the USA in 1964, and after a year at Michigan State University, he moved to Cornell University, where he has been a professor of operations research since 1967. He has been closely involved in the development of stochastic processes, as editor of *Advances in Applied Probability* (1973–82) and as principal editor (1973–9) and editor (1980–4) of *Stochastic Processes and Their Applications*. He has written over 40 research papers and four books.

Professor Prabhu has received recognition for his many contributions to the field of applied probability. He is a fellow of the Institute of Mathematical Statistics and an elected member of the International Statistical Institute. He also served as chairman of the Committee for Conferences on Stochastic Processes between 1975 and 1979.

He married his wife Sumi in 1951 and has two daughters, Vasundhara and Purnima. He is interested in literature (including detective fiction), cinema, music and travel.

Probability Modelling Across the Continents

1. Academic Background

I was born at Calicut, Kerala State, India and attended high school and (two-year) intermediate college there. I wanted to study mathematics, but in those days there was supposed to be no future for arts and science graduates, so I applied for admission to an engineering college. As it turned out, I failed to get this admission, and so I joined the Loyola College of Arts and Science, Madras, where I studied for a bachelor's degree (with honours) in mathematics.

The programme at this college consisted of courses in pure mathematics (analysis, algebra and geometry) and applied mathematics (statics, dynamics and astronomy); in addition I took two optional papers—potential theory and complex analysis. A strong feature of the programme was the interconnection between various branches of mathematics that the instructors stressed constantly. A typical example was the manner in which the proof of the following statement in astronomy:

"The equation of time vanishes four times a year"

was reduced analytically to that of the statement in geometry:

"From a point within the ellipse, four normals can be drawn to it."

The professor of mathematics at the college was Fr C. Racine, a young energetic Jesuit priest who had recently arrived from France with new (for the 1940s) mathematics. He taught analysis out of the book by De la Vallée Poussin. This course and the one on complex variables (based on the book by Goursat) were an important part of my training in mathematics.

Mathematics in India at that time (but probably much less now) was influenced by British mathematics, and consisted mainly of classical pure and applied mathematics with emphasis on problem solving. However, the training in mathematics that I received taught me to appreciate fully its conceptual foundations and to use its techniques skillfully and wisely. In addition I developed a perspective on the discipline of mathematics that has moulded my attitude towards the craft of probability modelling.

After receiving my BA (Hons.) degree in 1946 I taught mathematics for two years at colleges affiliated with the University of Bombay. I found teaching interesting enough, but the position itself did not hold many prospects. Therefore when in August 1948 the University of Bombay opened its postgraduate department of statistics, I gave up my position and enrolled in that programme. The curriculum included courses in topics such as statistical inference, multivariate analysis, experimental designs and sample surveys, but was rather weak in probability theory. Some training in the handling of statistical data was also a part of the programme. I was truly impressed with the vastness of the conceptual framework of mathematical statistics, but probability theory was going to be my chosen field of interest.

I received the MA degree in 1950 and worked as lecturer in mathematics and statistics at Gauhati University, Assam State, for two years and then as reader in statistics and head of the Department of Statistics at Karnatak University, Karnataka State.

2. Visits to England and Australia

During my tenure at Karnatak (1955) I was awarded a British Council scholarship for higher studies and went to work with Maurice Bartlett at the University of Manchester, England. It turned out that my choice of university was perhaps not ideal, as Bartlett's probability proved to be too heuristic for my taste. However, that year Joe Gani came to Manchester as a Nuffield Fellow from Australia and I started to do research on the probability theory of dams under his direction. During the limited duration of my scholarship I was able to fulfill the requirements of the M.Sc. degree at Manchester.

The statistics courses taught at Manchester did not contain any material that was new to me, but I did take a couple of mathematics courses, one of which was on functional analysis and Markov processes taught by Harry Reuter. I enjoyed this course immensely and it influenced my later work on Wiener–Hopf factorization. Also I became acquainted with David Kendall who expressed interest in my work on the theory of dams.

My own academic temperament had much in common with Joe Gani's, and our early student–advisor relationship rapidly developed into successful research collaboration (at the University of Western Australia which I visited during the 1957 academic year) and also led to a close personal friendship.

I returned to my position at Karnatak University in 1958. The Department of Statistics that I had started in 1952 continued to expand, and a modest attempt was made to start a Ph.D. programme. My research continued on an active basis. However, circumstances forced me to leave my position in 1961 and accept the position of reader in mathematical statistics at the University of Western Australia.

In 1961 there was still only moderate activity in the area of probability in Australia; however, since then a group of very fine probabilists and statisticians has emerged. The group in Western Australia, started earlier by Joe Gani, was very active in terms of research publications and graduate students. This activity continued under my tenure as reader. I completed work on two books—on queueing theory [21] and stochastic processes [22]. Also I became acquainted with Pat Moran, who pioneered the probability theory of dams. The academic climate in Australia was indeed to my liking, but unfortunately the political climate of those times was not favourable to my continued stay in that country.

3. Migration to the USA; Some Reflections on my Career

I came to the USA in 1964 as a visiting associate professor of statistics at Michigan State University. This was my first experience of American academe, and it was not an entirely happy one. Since 1965 I have been at Cornell University—as associate professor in 1965–7 and as professor since 1967. In 1972 my wife, our two daughters and I became naturalized U.S. citizens, and it looks as though my wanderings have come to an end.

The applied probability group at Cornell is a part of the School of Operations Research and Industrial Engineering, which is a unit of the College of Engineering. I have tried to develop this subject area according to my philosophical inclinations. My experience is that the environment provided by the engineering college is not entirely conducive to the growth of applied probability to its full stature. I am completing my twentieth year at Cornell, and this is perhaps an opportune time to reflect a little over some aspects of my career.

The scientific community is a truly international one, sharing its concerns over matters of mutual interest and participating in cooperative ventures such as conferences and journals. Membership of this vast international community is one of the privileges of our profession. This is an aspect that I have enjoyed most in my career; it has been my objective to play my part in international activities in the domain of probability.

Teaching and research are important parts of an academic career. I like teaching, but derive less satisfaction from it here in the USA, than I have in Australia and India. I have had excellent rapport with my research students, many of whom have become my close friends and associates. My own research has been in the area of what I have designated as stochastic storage processes [23]. This area was in its developing stage when I started my research career, and my contributions to it have brought me immense enjoyment.

In this section I describe my views on various aspects of the craft of probability modelling in general, and the modelling of queues and storage systems in particular. My views have evolved through various stages, being influenced by my experiences in India, England, Australia, and the USA, and are therefore somewhat personal. In particular, I have found that in most parts of the world academics consider themselves to be members of an élite class, enjoying social if not financial privileges. I am ill-at-ease with this notion, and would not claim for myself any special status. In a society where jealousy is almost institutionalized I consider myself an uncompetitive person.

4. Mathematical Models

Let me begin with some comments on mathematical modelling. The term "mathematical model" is used to describe a quantitative approach to various phenomena. Such models abound in classical physics and applied mathematics (in the British sense of the term). They are all deterministic; the physicists were apparently slow in recognizing the role of chance in model-building. The paper by Kac [9] contains interesting references to some of the controversies that raged between the classical and modern physicists. Neyman [18] in his address over 25 years ago to the American Statistical Association drew statisticians' attention to the role of indeterminism in science and the consequent demands on them. To make the perspective somewhat broader, one might perhaps also emphasize the concept of stochastic control in this connection.

Probability models may be characterized as mathematical models that involve a random element. An older term is *statistical*, used in connection with statistical physics and related areas. At the early stages of development of mathematical statistics considerable attention was paid to fitting curves to observed data. (Kendall's book [14]) contains several examples of this.) Even that may be viewed as probability modelling, but the curve-fitting was carried out rather uncritically, with no attempt to explain the possible *a priori* reasons why a certain curve and not some other might have been fitted to the data. Probability modelling in the current sense of the term emerged during what Neyman [18] calls the era of *dynamic indeterminism*, starting with Mendelism, statistical mechanics, epidemiology and other areas, and now extending to all branches of the natural, physical and social sciences.

5. The Scope of Applied Probability

The subject areas that probabilists seek to model are diverse, and each has its own technical background. Applied probabilists, on the other hand, are usually trained in mathematics, probability and in some cases, statistics. Their intended audience consists of theoretical experts in their various areas, or practitioners (the consumers of applied probability). This is very different from the classical situation when the boundaries across disciplines such as mathematics and physics were not rigidly drawn. Probability modelling consists of the following important steps:

(i) Describe the phenomenon under investigation in fairly non-mathematical terms.

(ii) Set up reasonable hypotheses (assumptions) to translate the above description into mathematical (probabilistic) terms. This constitutes the probability model.

(iii) Ask appropriate questions concerning the phenomenon, and formulate these questions in terms of the stochastic process that arises from the model.

(iv) Test the appropriateness of the assumptions made in (ii). This involves the testing of the model.

(v) Communicate the results of the investigation to the scientist or the practitioner who first proposed the problem, and to the wider audience of all applied probabilists.

The ideal environment for applied probabilists' work is provided by organizations such as the scientific and industrial research organizations of Australia, India and the United Kingdom, and Bell Laboratories and IBM in the USA. The environment provided by a university for collaborative research between probabilists and biologists (for example) has its limitations. In any case, whether or not an applied probabilist is able to perform his chosen task successfully in terms of the steps (i)–(v) described above will depend entirely on his environment.

First and foremost an applied probabilist has to understand the subject matter of his study thoroughly and grasp its technical nature, with a view to explaining the problem to a wide audience. Unfortunately the very important step (i) above is neglected by several authors, who begin their analysis of the model with a rather uncritical mathematical description of the situation and thereby raise questions concerning the genuineness of the model itself. In all branches of applied mathematics the practitioner is usually prepared to allow the mathematician considerable latitude in introducing sophistication in his modelling effort for the sake of mathematical maneuvrability, but it is essential that the real-life features of the situation should be carefully described in step (i) before introducing the necessary sophistication in step (ii).

A probability model gives rise to a stochastic process and the analysis of the model reduces to solution of problems within the theoretical framework of this process. Thus in his pioneering papers on queueing theory, Kendall [12, 13] uses discrete-time Markov chains, while in his recent book on the subject Brémaud [3] uses martingale dynamics. It very frequently happens that in order to carry out step (iii) described above, namely to answer questions concerning the phenomenon under investigation, the applied probabilist expands his theoretical framework considerably, and discovers new properties of known processes or even finds a new class of stochastic processes. Thus the work of Lindley [15] and Smith [31] on queueing theory opened up new vistas on random walks (fluctuation theory, Wiener–Hopf factorization, etc.).

Incorporating a large number of real-life factors into a model usually makes it complex to the extent that the existing theory of stochastic processes becomes inapplicable. In such situations computer simulations of the model might be the only recourse and might lead to broad tentative conclusions.

The questions asked of the model are in the first instance solved by calculating the system characteristics, and at an advanced level involve statistical inference, design and control. The analytical techniques used by applied probabilists are drawn from several branches of mathematics such as real and complex analysis, linear algebra, functional analysis and even mathematical programming and game theory. Thus Rouché's theorem has many uses in applied probability, and properties of linear operators in Banach spaces are also needed. For problems that defy analytical treatment, computers are being used fairly extensively. Exactly which of these techniques should be used depends on the problem in hand, and to indulge in a concerted effort to discredit any one set of techniques amounts to an anti-intellectual activity.

Except perhaps in a few classes of probability models, real-life data are hard to obtain, and testing of models as suggested in step (iv) above is not always possible. The theory of statistical inference for stochastic processes has made great advances in recent years (see Basawa and Prakasa Rao [1]), and should prove useful to applied probabilists.

6. Communication Problems in Applied Probability

Initially, applied probabilists published the results of their research in mathematics and statistics journals; this was very natural, as their academic background was in these subject areas. It is doubtful whether the readership of these journals was the authors' intended audience, and in any case the journals became increasingly reluctant to publish papers on applied probability. The founding of the *Journal of Applied Probability* in 1964 and *Advances in Applied Probability* in 1969 by Joe Gani was a most timely and welcome development, and these two journals have since provided a major venue for the publication of applied probability research.

In the USA the main concerns of applied probabilists were the directions in which the field of applied probability was developing, and the status of applied probabilists in the general scientific community. Efforts to address these problems led to the starting of a series of conferences on stochastic processes and their applications (SPA) in 1971, and to a journal of the same title in 1973. The committee that was set up to plan the conferences was affiliated in 1975 with the International Statistical Institute's Bernoulli Society for Mathematical Statistics and Probability as a subject area committee. The journal *Stochastic Processes and Their Applications*, published by the North-Holland Publishing Company, became an official publication of the Bernoulli Society in 1980. In [25] and [26] I have recorded brief histories of these developments, with which it was my privilege to be associated from the very beginning.

7. The Status of Applied Probabilists

Some applied probabilists seem to feel that they do not always get the recognition they deserve for their work. It is tempting to blame this on the continuing reluctance of classical scientists to recognize the role of chance in physical and natural phenomena. However, there are other, more valid reasons. It is possible that an applied probabilist is viewed as a technician, a problem-solver, rather than as a scientist in his own right. This is clearly a mistaken view. It is true that a considerable part of an applied probabilist's work consists of problem-solving, and I believe this amounts to a significant contribution, of importance to all parties involved. However, applied probabilists are also basically probabilists; they are inclined to ask whether the results obtained in specific models have implications concerning a larger class of stochastic processes, and very often find that they do. Thus they are able to make significant contributions to the theory of stochastic processes, which entitle them to the status of scientists in their own right.

Are applied probabilists trying to "reach the moon without learning Newton's laws of gravitation"? The early applied probabilists emerged from among mathematicians and statisticians with a good background in pure and applied mathematics and mathematical statistics. It was appropriate at that time to designate them as probabilists. In recent times most applied probabilists have been graduates of departments of operations research or of mathematical sciences. Their training includes mathematical programming, game theory and combinatorics (which perhaps constitute modern applied mathematics), besides probability and statistics. This training is broader in the sense that it is appropriate to the needs of current times, but it may perhaps lack depth in mathematics, probability and statistics. Is a person with this background an applied probabilist *per se*, or rather an applied mathematician specializing in probability models?

In the 1950s and early 1960s when applied probability was emerging, it was subject to the criticism of pure mathematicians and mathematical statisticians, who did not fully appreciate the significance of the new discipline. Against this criticism the small community of applied probabilists built a common defence, and the applied probability journals which started their publication at this time provided them with a strong sense of identity. This professional camaraderie was, however, shortlived. As the community grew larger the differences in the academic backgrounds of its members (as explained above) became more evident, and resulted in disparities in their approach to the craft of probability modelling. Trends and fashions emerged, as is so common in other branches of science—the very same factors that applied probabilists had earlier felt they were victims of. Thus events have completed a full cycle.

In the USA financial support from the national research agencies for research projects on applied probability has not been forthcoming to the extent that this area deserves. One cannot blame the agencies for this deficiency, because they do not have an accurate perception of the role of applied probability in the general domain of scientific endeavour, and it would be futile to look to them for any leadership in this matter. In my opinion the factors responsible for this lack of support are the continuing reluctance of mathematicians and statisticians to give applied probability its due place, and the prevailing attitudes of applied probabilists themselves in making some areas of research less fashionable than others. This competition for financial support is of course an essential feature of American scientific effort; one might take some comfort from the fact that the quality of the research accomplished is not always positively correlated with the extent of support received from the research agencies.

The social unrest of the late 1960s and the 1970s in the USA and the resulting challenge to mathematicians (and other scientists) made considerable impact on their subject areas, specifically prompting a heightened awareness of real-life problems. Thus mathematicians have become increasingly interested in problems of biology, operations research, economics and other subject areas. The availability of high-speed computers has made problem-solving a less tedious and perhaps even an interesting exercise. In the broad spectrum of applied science, applied probability has a well-deserved place; perhaps it is not as large as applied probabilists in their pioneering enthusiasm once claimed for it, but it is undoubtedly a significant place.

8. Probability Modelling of Queueing and Storage Systems

In his pioneering survey paper, Kendall [12] stated that the theory of queues has a special appeal for the mathematician interested in stochastic processes. The truth of Kendall's statement has been borne out by the developments of the last 34 years, and queueing theory continues to fascinate mathematicians at least as a source of convenient examples for various concepts of stochastic processes. It provides motivation to applied probabilists to seek new directions for their research, and poses problems of inference and control to operations researchers. Indeed the richness of structure of queueing systems is shared by only a few other areas of applied probability.

The stochastic processes arising from simple queueing models (those with Poisson arrivals and service times having exponential density) turn out to be birth-and-death processes, and the standard properties of these processes are used to answer questions concerning these systems. In somewhat more advanced models (such as those with group arrivals or bulk service), the processes are still Markovian, but not of the birth-and-death type. Their analysis is still standard, attention being concentrated on the limit behaviour of the processes. When non-Markovian processes were encountered, dire warnings were at first issued as to the complications that occur in their analysis, and later, two remedies for the situation were offered. One remedy is A. K. Erlang's method of phases for the system $M/E_k/1$, where one is asked to investigate $Q_1(t)$, the number of service phases present in the system, there being k for each arrival; it turns out that the process $\{Q_1(t)\}$ is Markovian, while the queue-length process $\{Q(t)\}$ is not. However, Erlang's method results in loss of information on $Q(t)$, and it is in fact quite unnecessary to use it. The appropriate procedure for the $M/E_k/1$ system is to consider $\{Q(t), R(t)\}$, where $R(t)$ is 0 if the system is empty, and the residual number of phases of the customer being served otherwise. This two-dimensional process is Markovian, and its analysis is no more difficult than that of $Q_1(t)$. A second remedy for the non-Markovian situation is the technique of imbedding proposed by Kendall [12, 13]. Here, instead of the given continuous-time process one investigates a suitable Markov chain imbedded in it. I used to think that the concept of imbedding had retarded the progress of queueing theory by at least 10 years, because attention was diverted from continuous-time non-Markovian processes (such as point processes and martingales). However, my more recent experience has convinced me that imbedded chains are perhaps the natural processes to observe in control procedures, where controls are usually (but not always) imposed at certain special points of time, such as arrival or departure epochs. The end of this Markovian era is marked by the appearance of the important paper by Takács [35] on the virtual waiting time in the system $M/G/1$.

More general queueing systems in which the independence assumptions on interarrival times and service times are suitably weakened give rise to Markov renewal processes. However, no significant results seem to emerge from this class of models, qualitatively different from those of the standard models.

Early investigations of the general single-server queueing systems were concerned with the waiting time of an arriving customer (as opposed to virtual waiting time). In the pioneering paper of Lindley [15] the basic process is again a (discrete-time) Markov chain, and its limit distribution is of main interest. Here Wiener–Hopf techniques were used by Smith [31] in a non-probabilistic context, and later by Spitzer [32], [33] in a probabilistic context. This led to the surprising discovery of the close connections between queueing problems and random walks. In particular, it turned out that the problems concerning the waiting time and accumulated idle time reduce to those concerning the maximum and minimum functionals of the associated random walk. Combinatorial techniques used in random walks became a standard tool of queueing theory, and were used by Bhat [2] to investigate bulk queueing systems.

The continuous-time analogue of the random walk is of course a process with stationary independent increments (Lévy process). However, the fact that Lévy processes are basic to queueing models was not easily recognized. The compound Poisson process lurking behind the virtual waiting-time process $\{W(t); t \geq 0\}$ of the $M/G/1$ system was used by Reich [28], [29] to formulate his integral equation for $W(t)$, and I obtained the distribution of the busy period of that system by recognizing it as a first-passage time for this compound Poisson process [19]. The theory of dams, which reached a vigorous state of development about this time, gave considerable impetus to the study of continuous-time stochastic processes arising in queues. In particular it led to appropriate formulations of single-server queueing models with static and dynamic priorities (Hooke and Prabhu [8], Goldberg [7]) and also to a new approach to the insurance risk problem (Prabhu [20]). Unification of the theories of queues, dams and insurance risk was also achieved, and it became evident that it would be appropriate to include all of these models under the broad title of stochastic storage models, characterized by inputs constituting Lévy processes. In [24] I presented a comprehensive treatment of these models.

The probability theory of dams had a modest beginning, with Moran's [17] discrete-time, finite-capacity model with additive inputs. Models with correlated inputs were proposed by Lloyd [16]. Continuous-time models with inputs forming a Lévy process were first investigated by Gani and myself [6]. This area of probability models developed very rapidly. In particular a study of continuous-time storage models with Markovian inputs was undertaken by Çinlar [4]. Insurance risk models with claims occurring in a Markov renewal process have recently received considerable attention from European actuaries.

My research on queueing problems within the framework of random walks and on continuous-time dam models motivated my interest in the fluctuation-theoretic aspects of Lévy processes. This led to the theory of ladder phenomena (Rubinovitch [30]) and Wiener–Hopf factorization for Lévy processes (Prabhu [23]), for Markov additive processes (Kaspi [10]) and for Markov processes (Prabhu [27]). This was a most rewarding experience, and it confirmed my belief that applied probability research can indeed lead to significant contributions to the theory of stochastic processes.

In other work on continuous-time storage models the net input process is assumed to be a Brownian motion. Now since the net input equals gross input minus amount demanded, both quantities being nonnegative, it is clear that the net input process is necessarily of bounded variation with probability 1, and so it cannot be Brownian. I therefore question the validity of such models. In the Brownian setting, control problems in storage reduce to familiar problems in stochastic control theory and other problems reduce to those involving entrance or exit times for the Brownian motion (in one or more dimensions) into or from certain regions. These problems are difficult, and presumably they are interesting and even important. However, it would be more honest to formulate them directly as problems on Brownian motion, rather than as problems arising from genuine storage models.

Research on queueing networks started along conventional lines, using the standard theory of Markov processes, most of the work being motivated by computer applications. This subject area has become increasingly important, and more advanced concepts and techniques such as time-reversibility and properties of Poisson flows have been used (see, for example, Kelly [11]). The other newly emerged areas of research use the theory of point processes and martingales (Brémaud [3], Franken et al. [5]) and concepts of stochastic ordering (Stoyan [34]).

A due concern with the practical applications of queueing and storage models has forced applied probabilists to come to grips with the complexities of the stochastic processes arising from the models, and their labours have been amply rewarded. These models have indeed made significant contributions to the general theory of stochastic processes in terms of concepts and techniques; I am happy to have been among the contributors.

References

[1] Basawa, I. V. and Prakasa Rao, B. L. S. (1980) *Statistical Inference for Stochastic Processes*. Academic Press, London.

[2] Bhat, U. N. (1964) Imbedded Markov chain analysis of bulk queues. *J. Austral. Math. Soc.* 4, 244–263.

[3] Brémaud, P. (1980) *Point Processes and Queues: Martingale Dynamics*. Springer-Verlag, New York.

[4] Çinlar, E. (1973) Theory of continuous storage with Markov additive inputs and a general release rule. *J. Math. Anal. Appl.* 43, 207–231.

[5] Franken, P., König, D., Arndt, U. and Schmidt, V. (1981) *Queues and Point Processes*. Akademie-Verlag, Berlin.

[6] Gani, J. and Prabhu, N. U. (1963) A storage model with continuous infinitely divisible inputs. *Proc. Camb. Phil. Soc.* 59, 417–429.

[7] Goldberg, H. M. (1977) Analysis of the earliest due date scheduling rule in queueing systems. *Math. Operat. Res.* 2, 145–154.

[8] Hooke, J. A. and Prabhu, N. U. (1971) Priority queues in heavy traffic. *Opsearch* 8, 1–9.

[9] Kac, M. (1947) Random walk and the theory of Brownian motion. *Amer. Math. Monthly* 54, 369–417.

[10] Kaspi, H. (1980) On the symmetric Wiener-Hopf factorization for Markov additive processes. *Z. Wahrscheinlichkeitsth.* 59, 179–196.

[11] Kelly, F. P. (1979) *Reversibility and Stochastic Networks*. Wiley, New York.

[12] Kendall, D. G. (1951) Some problems in the theory of queues. *J. R. Statist. Soc.* B 13, 151–185.

[13] Kendall, D. G. (1954) Stochastic processes occurring in the theory of queues and their analysis by the method of the imbedded Markov chain. *Ann. Math. Statist.* 24, 338–354.

[14] Kendall, M. G. (1943) *The Advanced Theory of Statistics*, Vol. I. Griffin, London.

[15] Lindley, D. V. (1952) Theory of queues with a single server. *Proc. Camb. Phil. Soc.* 49, 277–289.

[16] Lloyd, E. H. (1963) Reservoirs with serially correlated inputs. *Technometrics* 5, 85–93.

[17] Moran, P. A. P. (1954) A probability theory of dams and storage systems. *Austral. J. Appl. Sci.* 5, 116–124.

[18] Neyman, J. (1960) Indeterminism in science and new demands on statisticians. *J. Amer. Statist. Assoc.* 55, 625–639.

[19] Prabhu, N. U. (1960) Some results for the queue with Poisson arrivals. *J. R. Statist. Soc.* 22, 104–107.

[20] Prabhu, N. U. (1961) On the ruin problem of collective risk theory. *Ann. Math. Statist.* 32, 757–764.

[21] Prabhu, N. U. (1965) *Queues and Inventories: A Study of Their Basic Stochastic Processes*. Wiley, New York.

[22] Prabhu, N. U. (1965) *Stochastic Processes: Basic Theory and Its Applications*. Macmillan, New York.

[23] Prabhu, N. U. (1972) Wiener-Hopf factorization for convolution semigroups. *Z. Wahrscheinlichkeitsth.* 23, 103–113.

[24] Prabhu, N. U. (1980) *Stochastic Storage Processes: Queues, Insurance Risk and Dams*. Springer-Verlag, New York.

[25] Prabhu, N. U. (1982) Conferences on stochastic processes and their applications: a brief history. *Stoch. Proc. Appl.* 12, 115–116.

[26] Prabhu, N. U. (1986) *Stochastic Processes and their Applications. Encyclopedia of Statistical Sciences* 8. (To appear.)

[27] Prabhu, N. U. (1985) Wiener-Hopf factorization of Markov semigroups-I. The countable state space case. *Proc. 7th Conf. Probability Theory* ed. M. Iosifescu, VNU Science Press, Utrecht, 315–324.

[28] Reich, E. (1958) On the integro-differential equation of Takács I. *Ann. Math. Statist.* 29, 563–570.

[29] Reich, E. (1959) On the integro-differential equation of Takács II. *Ann. Math. Statist.* 30, 143–148.

[30] Rubinovitch, M. (1971) Ladder phenomena in stochastic processes with stationary independent increments. *Z. Wahrscheinlichkeitsth.* 20, 58–74.

[31] Smith, W. L. (1953) On the distribution of queueing times. *Proc. Camb. Phil. Soc.* 49, 449–461.

[32] Spitzer. F. (1957) The Wiener-Hopf equation whose kernel is a probability density. *Duke Math. J.* 24, 327–344.

[33] Spitzer, F. (1960) The Wiener-Hopf equation whose kernel is a probability density. *Duke Math. J.* 27, 363–372.

[34] Stoyan, D. (1983) *Comparison Methods for Queues and Other Stochastic Models*. English trans. ed. by D. J. Daley. Wiley, New York.

[35] Takács, L. (1955) Investigation of waiting time problems by reduction to Markov processes. *Acta Math. (Budapest)* 6, 101–129.

Lajos Takács

Lajos Takács was born in Maglód, Hungary, on 21 August 1924. He attended elementary school in Maglód, and secondary school in Budapest. In 1943 he enrolled at the Technical University of Budapest where he studied mathematics and physics. He obtained his doctorate in 1948 with the thesis "On a Probability-Theoretical Investigation of Brownian Motion". In 1957 he received the title of Doctor of Mathematical Sciences from the Hungarian Academy of Sciences for his thesis "Stochastic Processes Arising in the Theory of Particle Counters".

From 1945 to 1955 he worked as a mathematician at the Tungsram Research Laboratory (Telecommunications Research Institute), Budapest. From 1950 to 1958 he was staff member of the Research Institute for Mathematics of the Hungarian Academy of Sciences, Budapest. Between 1953 and 1958 he was associate professor (adjunktus) at the Department of Mathematics of the L. Eötvös University, Budapest. During the academic year 1958–9 he was a visitor at Imperial College and the London School of Economics of the University of London. From 1959 to 1966 he was assistant professor and later associate professor in the Department of Mathematical Statistics, Columbia University. Since 1966 he has been professor of mathematics and statistics in the Department of Mathematics and Statistics at Case Western Reserve University, Cleveland, Ohio.

He won second prize in the Eötvös mathematical competition for high-school graduates in Hungary in 1943, and first prize in the Schweitzer mathematical competition for university graduates in Hungary in 1949. He also won the second Grünwald Prize awarded by the János Bolyai Mathematical Society in 1952, for his paper "Investigation of waiting time problems by reduction to Markov processes".

He married Dalma Horváth in 1959. They have two daughters, Judith and Susan. His recreations include reading, classical music, the opera and the theater.

Chance or Determinism?

1. Early Life

I was born on 21 August 1924 in Maglód, a little town 16 miles from Budapest. From a very early age I was fascinated by numbers. I must have been about four when a neighbor, Mrs Kéry, told my mother that she had sold her pig for 40 pengös. At that age I was more familiar with fillérs than pengös, and I asked my mother how many fillérs made up a pengö. She told me, 100. After thinking this over, I spoke up: "Mrs Kéry has 4000 fillérs." From that time my reputation as a mathematician was established—at least in Maglód. Everyone who knew me wanted to try out my mathematical skills, and I was delighted to oblige.

I went to elementary school in Maglód, but school had very little new to offer me in numbers. I was more interested in working on my own projects. We had 12 apricot trees in our yard and I meticulously worked out an "unbiased" estimate of the weight of the next apricot harvest by counting the buds on one tree in the spring. I learned to play chess. I heard the story about the man who invented the game and his modest request that as a reward he be given one grain of wheat for the first square, two for the second, and for each subsequent square, double the previous number. I calculated the exact amount of his reward in grains, and estimated it in kilos and sacks. Since I did not discover how to calculate the sum of a geometric series, I simply added the numbers for all 64 squares. Probably the young Gauss would have found a simpler way to do it. The cylindrical water barrel beside our well provided me with another mathematical adventure. I estimated its volume by pouring a couple of buckets of water into the barrel and measuring the increase in the water level. I was still in elementary school when I found out the magic property of π in calculating the area of a circle. Using my new-found knowledge, I was able to calculate the exact volume of the cylindrical barrel without using a bucket.

2. Secondary Education

I commuted to Budapest by train to attend secondary school. By now I had notebooks in which I wrote down everything I learned in technology and arithmetic. I had a special interest in electronics and radio technology, and studied many books on these subjects. Every month I bought *Rádió Technika*, a radio technology monthly journal, and I constructed many different kinds of electronic equipment. I was interested not only in the design, but also in the theory behind the design, and especially in the relevant mathematical formulas. The journal of radio technology included a mathematical column which was more advanced than my school mathematics. From it, I learned about logarithms, trigonometry, complex numbers and more. My daily train journey gave me a chance to talk to older students and borrow their textbooks and notes on mathematics.

I was in constant search of knowledge, and pursued anybody who would give me useful information. When I was 15, I approached my new mathematics and physics teacher, Dezső Vörös, with various questions. He suggested that I read Euler's *Algebra* [5]. This seemed a formidable task because the *Algebra* was written in complicated German syntax, and old-fashioned Gothic letters. But Dezső Vörös persisted and I bought the book. I spent the Christmas vacation of 1939 studying Euler. Once I mastered his syntax, I was amazed to find in the book such wonders as the solutions of cubic and quartic equations, Diophantine equations, and the solutions of Fermat's problem for $n = 3$ and $n = 4$. I also bought the booklet *Differential and Integral Calculus* by Manó Beke [2]. He was an outstanding university professor, and what was even better, he wrote lucidly in Hungarian.

When I went back to school after the Christmas break, I told my teacher that in addition to Euler's *Algebra*, I had also learned differential and integral calculus. He wanted to find out what I had really learned and wrote a few integrals on the blackboard for me to evaluate. When I gave him the correct answers, he was so impressed that the next day he lent me his copies of *Differentialrechnung* [20] and *Integralrechnung* [21] from the Sammlung Göschen series. From then on Dezső Vörös was a constant source of information on higher mathematics, of which I made a systematic study in the following years. He lent me his notes on university courses which he had taken at the beginning of the century. I built up a small library of Hungarian and German books. I also used the library of the Technical University of Budapest. It was my good fortune to be acquainted with Dr Ernő Fónyad, a writer, who was a university librarian. Although I was still in high school, he gave me free access to the whole library. He even allowed me to browse in the stacks—a privilege granted to very few people. Throughout my high-school years, I spent my vacations and much of my free time studying mathematics intensively. I learned differential and integral calculus, determinants, analytic geometry, set theory, algebra, number theory, complex variables, differential equations, projective geometry, and special topics such as gamma functions and algebraic functions.

I was particularly interested in number theory. I had already studied Diophantine equations in Euler; I read Vörös's notes on Gusztáv Rados's lectures on number theory as well as several other books on the subject, including works by Wertheim [19], Dirichlet [4] and Dickson [3]. I personally felt that number theory was the most exciting of fields. I made small discoveries. In one of my school textbooks I had read that the number of branches on a tree increases yearly according to the sequence 1, 2, 3, 5, 8, 13, ..., and I wanted to find a general formula for the number of branches in the nth year. At the time I had no background in the calculus of finite differences, but I still found the solution in a roundabout way by observing that the numbers in the sequence are the positive integer solutions in y of the Diophantine equation $x^2 - 5y^2 = \pm 4$, which I could solve. I was happy to find an explicit expression for the nth element of the sequence, and I discovered many interesting properties of this sequence which I studied in detail. At the time I did not know that my discoveries were already known in the literature of Fibonacci numbers.

In Wertheim's book I read Meissel's method for finding $\pi(x)$, the number of prime numbers less than or equal to x, and I generalized his formula for finding the number of primes less than or equal to x in any arithmetic progression. I have encountered a recent paper giving the same solution.

I was a high-school student during the Second World War, and the war effort of the country had an influence not only on our daily lives but on many other areas. One casualty of the war was a mathematical journal for high-school students. This journal was started by the Loránd Eötvös Mathematical and Physical Society in 1894 and it not only contained instructive articles, but also had a problems section. Outstanding solutions and the names of students who submitted correct solutions were published. During the war years the journal was suspended, and I missed the benefits of such a publication.

In my high-school years I was familiar with combinatorics and solved various probability problems, but I considered probability as a branch of combinatorics, not an independent discipline. For me the physical world was deterministic and I believed that sufficient knowledge of mathematics and the natural sciences made it possible to calculate various measurable quantities. It was only much later that I recognized the importance of probability theory and its role in describing our physical world.

Although I spent most of my time on mathematics, I also studied books on classical and modern physics, and atomic physics in particular. I also enjoyed reading literature and philosophy. I had an active school life and participated in sports: I was particularly proud of my accomplishments in high jump, long jump and running. I had many friends, and enjoyed our swimming expeditions in the summer, skating in winter and card games on rainy days. I frequently entertained my friends with mathematical puzzles and tricks. One of my source books was *Mathematische Mussestunden* by Hermann Schubert [11]. Every year I had students whom I tutored in mathematics. One year, when a mathematics teacher in my school was drafted, I was given a mathematics class to teach for several months until the end of the school year.

In addition to my other activities, I had to help my mother in her business. My father had died when I was 14, and my mother, who had a general store in Maglód, was left to run the business by herself. My daily routine included running errands for her in Budapest. During my high-school years I became an expert in ordering, buying and other administrative procedures.

Fortunately, the Hungarian school day ended at 1 p.m. Although we had a sizeable amount of homework, there was still enough of the day left for extracurricular activities for both students and teachers.

3. The Eötvös Competition

Each autumn the Loránd Eötvös Mathematical and Physical Society arranged a national competition in mathematics, and awarded two prizes for students who graduated during that year. These competitions were started as early as 1894. Among the winners were many illustrious mathematicians and physicists such as L. Fejér (1897), T. Kármán (1898), D. König (1902), A. Haar (1903), M. Riesz (1904), G. Szegő (1912), T. Radó (1913), L. Rédei (1918), L. Kalmár (1922), E. Teller (1925) and others. In 1943 I finished high school and participated in the competition. There were three problems to be solved, and we had four hours to answer. I solved all three problems, and won second prize. There was only one other student, István Ádám, who solved all the problems: he was the first prize winner. The solution of the problems did not require a deep knowledge of advanced mathematics; instead, the problems tested skill and ingenuity. I had a good knowledge of higher mathematics, but made practically no use of it. The solutions of both prize winners appeared in the December 1943 issue of the *Matematikai és Fizikai Lapok*; they were published unedited, as submitted for the contest.

4. University Years

In the autumn of 1943 I became a student of the Technical University of Budapest. I attended lectures on mathematics and physics at both the Technical University and the Pázmány University. I became a student assistant to Professor Géza Huszár, and devoted all my time to attending lectures and to continuing my own studies. At the end of the school year the war reached the borders of Hungary. Disorder and devastation followed, and it was not until the autumn of 1945 that regular classes resumed. Among the courses I took were those of Professor Charles Jordan on probability theory and mathematical statistics. These courses turned out to be decisive in my career as a probabilist. I thoroughly enjoyed Professor Jordan's lucid lectures on the classical material. I made small contributions to his lectures. For example, when he discussed the problem of matching, he told us that if $Q(n)$ denotes the number of permutations of $1, 2, \ldots, n$ in which no match occurs, then $Q(n) = (n-1)[Q(n-1) + Q(n-2)]$ for $n \geq 2$ where $Q(0) = 1$ and $Q(1) = 0$. This result was found by P. R. Montmort in 1708, but he did not prove it. I gave a proof based on the observation that $Q(n)$ is also the number of terms which contain no diagonal element in the expansion of a determinant of order n. Professor Jordan had already finished the manuscript of his book on probability theory, but he inserted a short note about my contribution.

In late 1945 I gladly accepted an offer to become student assistant to Professor Zoltán Bay, the professor of atomic physics. He was also the head of the Tungsram Research Laboratory which belonged to one of the most prestigious Hungarian companies, the United Incandescent Lamps and Electric Co. (Tungsram). I was appointed a consultant in Professor Bay's laboratory where all the research was done. This appointment not only provided me with monetary support, but enabled me to participate in various research projects. I was the only student in this position; the majority of the researchers had teaching positions at the university. I participated in Professor

Bay's major project, an exciting and highly publicized experiment, detecting microwave echoes from the moon. He had started it in 1944 but could not finish it because of the war. My job was to calculate the position of the moon at 15-minute intervals and participate in the nightly experiments. On the night of 6 February 1946 the experiments resulted in success. It is interesting to add that simultaneously and independently, American scientists working on the same project anticipated our successful experiments by a few days. These were also a triumph of probability theory. The signal/noise ratio was very small, but by averaging out the noise, the effect of the signals could be detected. These experiments were perhaps the first which pointed out clearly for me the importance of probabilistic thinking. The moon experiments caused a sensation. Newspapermen visited the laboratory, and several newspaper and magazine articles appeared in which my name was always included among the participants.

Professor Bay conducted interesting experiments concerning the nature of the photon. He used the principle of secondary electron multiplication in atomic counting, and by using electron multiplier counters he worked out the first nanosecond coincidence circuit. I became interested in his experiments, and his research provided me with numerous problems. There was an outstanding problem which occupied the minds of several people: in an electron multiplier individual electrons multiply in a series of several plates by secondary electron emission. We can observe the distribution of the heights of the pulses in the output. What is the probability that a given electron gives rise to exactly $k(k = 0, 1, 2, \ldots)$ secondary electrons? In Professor Jordan's lectures I learnt about binomial moments and I tried to apply this notion to the problem. I observed that if we knew the first m binomial moments of the output, then we could calculate the first m binomial moments of the probability distribution in question, the distribution itself being determined by the binomial moments. This method (extended in various ways) gave not only a theoretical solution of the problem, but a practical method for calculations. This was an initial success. Particle counters provided me with problems for years to come.

Dr Tivadar Milner, the head of the chemical division of the laboratory, gave me some problems concerning Brownian motion. From my physics courses I was familiar with the notion and with Einstein's formula, but that was all. In order to be able to work on the problems proposed, I started reading R. Fürth's article "Wärmeleitung und Diffusion" [6] in the second volume of *Differentialgleichungen der Physik* edited by Ph. Frank and R. von Mises. By using this material I could solve Milner's problems, but I felt that a more refined model was needed to study the phenomenon of Brownian motion. So I formulated a realistic three-dimensional model. The displacements of a particle are independent random variables each having an exponential distribution, and the changes in direction are also independent random variables each having a uniform distribution. I was interested in finding the distribution of the total displacement of a particle in an interval of length t and its asymptotic behavior as $t \to \infty$. I also considered the same problems in several dimensions. I liked various mathematical aspects of this topic so much that I chose Brownian motion as the topic of my doctoral dissertation. In the spring of 1948 Professor Bay left Hungary and settled in the United States in Washington, DC, continuing his work at George Washington University. In 1955 he joined the U.S. National Bureau of Standards. He is now 84 and is still active in research.

I received my doctorate in 1948. My thesis, "On a Probability-Theoretical Investigation of Brownian Motion" was refereed by Professor Charles Jordan. During the year that followed I was still a registered student at the university, studying education for a possible career in teaching. In 1949 the Miklós Schweitzer mathematical competition was introduced for recent university graduates. I participated in the first competition and won first prize.

During my student years Professor Jordan was one of my mentors. I paid frequent visits to his home and I admired his wonderful library which contained most of the classical books in probability and mathematics in original editions. For instance, he had a copy of *Summa del Arithmetica* by Lucas da Borgo Pacioli, printed in Venice in 1496.

After Professor Bay's departure there were many changes in the Tungsram Research Laboratory. Several of his close associates left the laboratory; his projects were discontinued and his laboratory later became a part of the newly created Telecommunications Research Institute. I had several offers for academic positions, but I decided to continue my mathematical work in the laboratory, maintaining close contact with problems of the real world.

5. Research in Hungary

Until 1955 I remained at the Tungsram Research Laboratory (later the Telecommunications Research Institute) and continued with my mathematical research. In 1950 the Hungarian Academy of Sciences created the Research Institute for Mathematics, and I became a staff member of this institute too, dividing my time between the two.

I continued my research in probability, which was strongly related to the problems in physics that I had encountered earlier. I worked on two mathematical models of particle counters which were typified by the electron multiplier and the Geiger counter. I considered different input processes such as the Poisson process and the recurrent (renewal) process. I was interested not only in the asymptotic behavior of the process of registrations but in the transient (time-dependent) behavior too. I considered coincidence problems (several counters used simultaneously) and successive transformations with scaling devices used in series. No ready-made mathematical theory was available, and the problems were challenging. I was intrigued by problems that could not be solved by using standard techniques and required the invention of new mathematical methods. I used the theory of stochastic processes, Laplace–Stieltjes transforms and Tauberian theorems in my investigations. A simple new idea which I called the method of average functions turned out to be very successful in tackling various point processes. For many practical purposes it is sufficient to consider counter processes as point processes, but a closer look reveals that we are actually observing signals and not merely random points. Thus a refined mathematical description of counter processes led me to the investigation of secondary stochastic processes generated by a point process. Amplifiers and other physical devices are frequency-dependent, and their effect on various processes directed me to the study of harmonic analysis of stochastic processes. One result of particular importance was the study of the spectrum of noise in vacuum tubes, which led to a refinement of Schottky's formula.

In the Research Institute of Mathematics we had constant contact with people working in science and industry, and these contacts provided us with plenty of opportunities to study their problems. I solved some telephone congestion problems by using the same method as I had used in solving coincidence problems for particle counters. I also gave a generalization of Erlang's formula for telephone traffic congestion. In a textile factory, the problem arose of allocating machines among several repairmen in an efficient way. All these problems required building a good model, formulating the mathematical problems, and working out some methods for solving the equations. It turned out that many important and interesting problems could be reduced to problems in stochastic processes. The theory of stochastic processes was then not as widely known as it is now, and many results and methods which are now standard had to be invented.

I wrote several papers on waiting time and queueing problems. The models discussed had a wide variety of applications in telephone traffic, in storage, in the theory of dams and reservoirs, and in insurance. My research also included recurrent processes, point processes, sojourn-time problems, and a generalization of Markov processes which was essentially the semi-Markov process later introduced by Paul Lévy.

From 1953 I was also on the faculty of Loránd Eötvös University (formerly Péter Pázmány University) where I taught courses in probability theory. The vehicles for exchanging ideas among Hungarian and foreign scholars were the seminars of the Research Institute of Mathematics and the colloquia of the János Bolyai Mathematical Society. I participated in these meetings and gave numerous talks. In 1955 I visited the universities of Warsaw, Wrocław, Poznań and Kraków. This was my first trip abroad, and I enjoyed meeting many famous Polish mathematicians in person. In September 1956 I attended the Mathematics Congress in Vienna; this was my first trip to the West.

6. Years in the United States

The year 1958 was a turning point in my life. I made the decision to leave Hungary permanently and continue my work in the United States. The transition period was made easier for me by invitations from Imperial College and the London School of Economics in England where I gave lectures on queueing theory and stochastic processes.

In 1959 I received an offer of an assistant professorship from Columbia University, New York, which I accepted. For the next seven years I taught probability theory and stochastic processes at Columbia. I had Ph.D. students working with me every year, a consulting job with Bell Laboratories and, later, with IBM. I spent a summer at Stanford University, teaching and writing my book *Introduction to the Theory of Queues* [16]. Another summer was spent in Seattle at the Boeing Research Laboratories. I presented papers at several congresses and meetings in Europe and in the United States, and continued my research, dealing mostly with queueing problems, ballot theory and order statistics. In 1966 I spent part of my sabbatical at Stanford University, working on my book *Combinatorial Methods in the Theory of Stochastic Processes* [17].

Early in 1966 I had received an offer of a professorship from Case Institute of Technology. I found the offer attractive: the Institute had a good library, and the Department of Mathematics, under the chairmanship of Melvin Henriksen, was flourishing. I accepted the position and have been teaching there ever since. Over the years there have been many changes. Case Institute merged with Western Reserve University, and the department has become a Department of Mathematics and Statistics. My courses include probability theory, stochastic processes, queueing theory and combinatorics. My research is also concerned with these areas. My most recent work is in the field of algebraic methods in probability theory and probabilistic number theory.

It might be of some interest to compare my working conditions in Hungary and the United States. In Hungary my primary work consisted of research, and my duties at the university were restricted to lectures without any additional tasks. Thus I could spend more time on research and my output was accordingly higher. On the other hand, in the United States I have enjoyed personal freedom, and material advantages such as photocopying machines, interlibrary loans, and—in better days—research support.

I have found the process of scientific publication in the United States often frustrating. In Hungary the referee's name was always made known to the author, who was given an opportunity to discuss his paper with the referee and answer possible objections. In the United States the referee, under the veil of anonymity, can allow himself greater arbitrariness and sometimes even malice in his decision. He has the power to delay or even prevent the publication of an important paper, which hardly helps the progress of science.

A procedure that I have found most puzzling in the United States is the choice of criteria for the award of research support. In Hungary, scientists were rewarded with a financial bonus for the research which they had already accomplished. In the United States it is the promise of future research plans that is evaluated and rewarded—if one is lucky. I have never been comfortable with the art of making extravagant promises: perhaps that is why my research support has dried up during the past decade, despite my sustained productivity.

7. Conclusion

It would be impossible in this short account to describe my research in detail. I have confined my discussion to my formative years and mentioned only the factors that contributed to my development as a probabilist. I now have about 180 publications, including six books, with many projects still pending.

My own research is always problem-oriented. To begin with, I try to choose a mathematical model which describes a real-world situation as closely as possible and then formulate the problem and search for a solution. I have always been fascinated by problems which cannot be solved by using standard techniques and require the invention of new mathematical methods.

My aim has always been to find the simplest and clearest possible solution to a problem. Sometimes I found a gem such as the following result which I discovered in 1960:

Suppose that a box contains n cards, each marked with a non-negative integer so that the sum of the n numbers is k where $k \leq n$. We draw all n cards one by one from the box. The probability that the sum of the first r numbers drawn is less than r for all $r = 1, 2, \ldots, n$ is simply $(n - k)/n$.

It turned out that this simple theorem was the key to solving many problems in queueing theory, the theory of dams, order statistics, graph theory, mathematical logic and many other areas. My book [17] is devoted entirely to the ramifications of this theorem. Actually, the theorem is a generalization of the classical ballot theorem which was formulated by J. Bertrand in 1887.

The area that I have found most exciting and versatile is queueing theory. Apart from its practical importance, queueing theory is an excellent proving ground for new mathematical ideas. It is hard to find a probabilistic method which has not had its impact on queueing theory. It is not surprising that É. Borel, A. N. Kolmogorov, A. Ya. Khintchine, D. G. Kendall, M. Kac and many other mathematicians have had an interest in the theory of queues and have contributed to the subject.

The importance of probability models cannot be sufficiently emphasized. Every aspect of our daily lives depends on well-designed systems, among them telephone exchanges, transportation networks, air traffic, road traffic signals, hospital administration and factory production lines, not to mention computers and military defense systems. To design any workable system, we must find realistic models and accurate solutions to the problems arising from them.

With the advance of science and technology and the ever increasing complexity of modern life, the number of new systems will increase; they will require new models, new solutions and new theory.

Probability models play a significant role in unexpected areas of mathematics such as number theory. Even using the fastest computers, the calculations needed to factorize a very large integer would take many years of computer time. However, using probabilistic methods, in many cases, numbers as large as 10^{60} can be factorized in a short time. Another interesting example concerns Riemann's conjecture, which is beyond the reach of a conventional mathematical approach, but can be shown to be plausible using probabilistic methods.

In studying and interpreting the physical world, the role of probability theory is likely to increase rather than diminish. It will play a key part in designing the systems essential to maintaining modern life efficiently.

Publications and References

[1] Alexits, Gy., Kalmár, L., Rényi, A. and Turán, P. (1950) The Miklós Schweitzer mathematical competition of 1949. *Mat. Lapok* 1, 294–303.

[2] Beke, M. *Differenciál-és Integrálszámítás*. Budapest.

[3] Dickson, L. E. (1929) *Introduction to the Theory of Numbers*. Chicago.

[4] Dirichlet, P. G. Lejeune [Dedekind, R.] (1871) *Vorlesungen über Zahlentheorie*. Friedrich Vieweg und Sohn, Braunschweig.

[5] Euler, L. (1770) *Vollständige Anleitung zur Algebra*. St. Petersburg.

[6] Fürth, R. (1935) Wärmeleitung und Diffusion. In Ph. Frank and R. von Mises: *Die Differential-und Integralgleichungen der Mechanik und Physik*. II, 2nd edn. Friedrich Vieweg und Sohn, Braunschweig, 526–626.

[7] Hajós, Gy. (1943) Report on the 47th mathematical competition in 1943. *Mat. Fiz. Lapok* 50, 384–388.

[8] Hajós, Gy., Neukomm, Gy., and Surányi, J. (1957) *Matematikai Versenytételek*, II. (1929–55. évi versenyek). Tankönyvkiadó, Budapest.

[9] Jordan K. (1972) *Chapters on the Classical Calculus of Probability*. Akadémiai Kiadó, Budapest. [English translation of the Hungarian original published in 1956.]

[10] Kürschák, J. [Hajós, Gy., Neukomm, Gy., and Surányi, J.] (1955) *Matematikai Versenytételek*. I (1894–1928. évi versenyek). Tankönyvkiadó, Budapest. [English translation: *Hungarian Problem Book* I, II. New Mathematical Library, Random House, New York, 1963.]

[11] Schubert, H. (1898) *Mathematische Mussesstunden*. Leipzig.

[12] Szász, G., Gehér, L., Kovács, I., and Pintér L. (1968) *Contests in Higher Mathematics, Hungary, 1949–1961*. Akadémiai Kiadó, Budapest.

[13] Takács, L. (1955) Investigation of waiting time problems by reduction to Markov processes. *Acta Math. Acad. Sci. Hungar.* 6, 101–129.

[14] Takács, L. (1960) *Stochastic Processes. Problems and Solutions*. Methuen, London.

[15] Takács, L. (1961) Charles Jordan, 1871–1959. *Ann. Math. Statist.* 32, 1–11.

[16] Takács, L. (1962) *Introduction to the Theory of Queues*. Oxford University Press, New York.

[17] Takács, L. (1967) *Combinatorial Methods in the Theory of Stochastic Processes*. Wiley, New York.

[18] Turán, P. (1953) Géza Grünwald memorial prize. *Mat. Lapok* 4, 221–222.

[19] Wertheim, G. (1902) *Anfangsgründe der Zahlenlehre*. Friedrich Vieweg und Sohn, Braunschweig.

[20] Witting, A. (1936) *Differentialrechnung*, 2nd edn. Sammlung Göschen, Bd. 87. Berlin.

[21] Witting, A. (1933) *Integralrechnung*. Sammlung Göschen. Bd. 88. Berlin.

Motoo Kimura

Motoo Kimura was born on 13 November 1924 in Okazaki, a small but historic city in central Japan. He studied botany at Kyoto University and after obtaining his M.Sc. in 1947 worked there as an assistant for two years. In 1949, he was appointed to the National Institute of Genetics at Mishima, where he has since spent most of his working life. In 1953 he went to the USA for graduate study first at Iowa State University and later at the University of Wisconsin, where he worked under Professor James F. Crow to gain his Ph.D. in 1956. He returned to the National Institute of Genetics in 1956 and became head of its Department of Population Genetics in 1964.

While he was still a student, Professor Kimura became interested in genetics and evolution, particularly their mathematical aspects. He is the author of numerous papers and five books which have contributed to significant developments in population genetics. In 1968 he proposed the theory of neutral evolution; his book on the subject, *The Neutral Theory of Molecular Evolution*, was published in 1983.

Professor Kimura has received several honors in recognition of his work. He was awarded the Weldon Memorial Prize by the University of Oxford in 1965 and received an honorary D.Sc. from the University of Chicago in 1978. He is a member of the Japan Academy and a foreign associate of the U.S. National Academy of Sciences. He has been vice-president of the XIth International Congress of Genetics (The Hague, 1963) and president of the Genetics Society of Japan (1980–4). In 1976, he received the prestigious Order of Culture from the Emperor of Japan, a recognition reserved for very few.

Professor Kimura is married and has a son. His hobby, which he has pursued enthusiastically for 25 years, is the improvement of hybrid Paphiopedilum (lady's slipper) orchids.

Diffusion Models of Population Genetics in the Age of Molecular Biology

1. Early Education

When I was a small boy, I was fascinated by the flowers my father loved and raised, and imitating him I grew various plants in the garden at home. The main reason why I was attracted by the flowers was their beauty. In addition, I was fascinated by the mystery of development; I wondered how a beautiful tulip, for instance, could emerge from a mere bulb. My interest in plants continued through elementary school. Then, in the second year of middle school, I met an interesting teacher who was a devoted naturalist. He encouraged my interest in plants, and gradually I became absorbed in botany; as a result, I made up my mind to become a botanist.

I also liked mathematics, although my ability in it was quite limited. I struck up a friendship with a young mathematics teacher. My interest in mathematics was aroused by Euclidean geometry. It so happened that I had to be absent from school for a few weeks owing to severe food poisoning. While recovering at home, I tried to catch up with the geometry course, and I soon found that by following the textbook step by step I could proceed without difficulty. My interest in mathematics was further stimulated by the fact that I could solve all the problems which no one else in my class could. The mathematics teacher eventually suggested that I might become a mathematician, but I did not take his suggestion seriously; I was sure that I was going to be a botanist, and I could not think of any connection between botany and mathematics.

After graduating from middle school (which then consisted of five years' study), I entered the Eighth National High School in Nagoya to pursue my aim of becoming a botanist. The two and a half years of my life in this high school (17–19 years of age) had a decisive influence in forming my subsequent professional career.

Immediately after entering this high school I asked the late Professor M. Kumazawa, a noted plant morphologist, to become my tutor, and he consented. He kindly allowed me to use his laboratory and suggested that I investigate the chromosome morphology of lilies. I found this work most intriguing and I pursued it intensively. I also attended his course in general biology, and was particularly impressed by his introductory account of cytogenetics. For example, the explanation that Mendel's first law could be understood in terms of the random assortment of paired chromosomes was most revealing. It was also in this course that I first learned about Kihara's work on the "genome analysis" of wheat which, according to Kumazawa, was an outstanding contribution that Japan could be proud of. Through these experiences, my interest in cytogenetics was aroused, and, gradually my desire to become a botanist turned into that of becoming a plant cytogeneticist.

It is often remarked that hearing of the achievements of great men and women influences young students to choose the direction of their pursuits; my experience corroborates this. At that time, cytogenetics was the most fashionable field in genetics, and there were many workers in this field in Japan.

In his biology course, Kumazawa included a topic in biometry, and I was fascinated to find that even such things as the length and width of leaves of a tree could be characterized by the mean and standard deviation, and that leaves on the sunny, south side of the tree differed "significantly" from those on the shady, north side. We were also taught introductory theories of probability and statistics in one of the mathematics courses, and I was much interested in them.

The other topic which made a deep impression on me was thermodynamics, taught by a young physics teacher. Although he was not particularly outstanding, he explained in easily understandable terms the basic concepts and experiments supporting thermodynamic concepts. What really impressed me was the fact that one could describe natural phenomena in mathematical terms starting from only a few basic principles. I still remember my fascination when I heard the teacher explain that the law $PV = RT$ could be derived by considering the random motion of "ideal gas molecules" impinging on the walls of a containing box.

An additional incident which enhanced my interest in this type of world view was that I came across the articles of Hans Reichenbach, professor of natural philosophy at the University of Berlin. In one of the German language courses, we were assigned to study, as a supplementary reader, a booklet entitled *Über die Materie* consisting of a few chapters taken from the popular writings of Reichenbach. These writings explained in elementary terms how ordinary physical phenomena could be understood by the behavior and structure of atoms and molecules.

Gradually the idea occurred to me, through my interest in biometry and cytogenetics, that it would be interesting to combine these two fields. Of course, I never imagined that the mathematical theory of population genetics, which had been created by R. A. Fisher and J. B. S. Haldane in England and Sewall Wright in the USA, already existed. At that time, our hero in science was Hideki Yukawa, then a young theoretical physicist who was later awarded the Nobel Prize for predicting the existence of the meson. Although I could not understand his research, nevertheless I felt a yearning toward such work. This was during the Second World War, and because of the pressing circumstances, we graduated after two and a half years of high school rather than the normal three.

2. University Studies

I entered Kyoto University (then called Kyoto Imperial University) as a student of botany in the Faculty of Science. I was attached to the Cytology Laboratory, and also attended all the genetics courses that were available in

Kyoto, including the advanced cytogenetics courses offered by H. Kihara who was professor of genetics in the Faculty of Agriculture.

Before my first year in Kyoto had ended, the atomic bomb was dropped on Hiroshima, and this was followed quickly by the surrender of Japan to the Allied Forces. While listening to the broadcast of the Emperor announcing the surrender, I was relieved that the terrible war had come to an end; my opinion then was that Japanese military power and totalitarianism were the origins of our misery. In fact, I welcomed the end of hostilities and looked forward to a new, more liberal world.

Immediately after the war, however, living conditions deteriorated, and became even worse than they had been during the war. Shortage of food was especially pressing. Many students like myself, who were away from home, suffered a great deal from short rations. Fortunately, one of my cousins-in-law, Matsuhei Tamura, was living in Kyoto, and almost every Sunday I used to visit his home to alleviate my hunger. It so happened that he was a mathematical physicist working on quantum mechanics, serving as associate professor under the famous Yukawa. I used to talk at great length with him, discussing various topics in the natural sciences. Since my ambition was to do something in genetics like what the theoretical physicists were doing in physics, I listened to his stories with intense interest. At that time, most biologists around me in Kyoto could not understand my ambition and could not think of any merit in applying mathematics to biological phenomena.

Although I was a student of botany majoring in cytology, I became more and more absorbed in the mathematical treatment of genetical phenomena. So, almost every day, I visited Kihara's laboratory to discuss various topics relating to genetics with the staff of the laboratory. There, foreign journals such as *Genetics* and the *Journal of Genetics*, and foreign textbooks that were more advanced than contemporary Japanese textbooks on genetics, were available. I found these very interesting and useful. I also profited greatly by attending the course on animal genetics given by the late Taku Komai, then a professor in the Department of Zoology at Kyoto University. The library in this department also had books and journals relating to genetics which I found useful. At that time Kyoto was the center of genetics in Japan, and I was fortunate to become personally acquainted with the major figures in this field.

The first subject that I found interesting to work on was the mapping function which gives the relationship between the map distance of genes on a chromosome and the observed recombination fraction between them. Through this work I became acquainted with the papers of J. B. S. Haldane on various topics in mathematical genetics. Gradually, I was attracted by his work; his papers were not only lucid and easy to understand, but also treated problems which were biologically significant.

Two books which I came across at that time, which helped to widen my views on genetics and evolution were C. H. Waddington's *An Introduction to Modern Genetics* [35] and Th. Dobzhansky's *Genetics and the Origin of Species* [8], both of which were available in pirated editions. By reading these, I was led to the work of Sewall Wright who investigated mathematically the effect of random drift in small populations. I was very interested in Wright's work, but his original papers were too difficult for me to understand at the time.

After graduation from Kyoto University, I left the Cytology Laboratory in the Department of Botany, and moved to the Genetics Laboratory in the Faculty of Agriculture, where researchers were actively at work on the cytogenetics of wheat under the direction of Kihara. In the first year, I was unable to get a definite position in Kihara's laboratory, but he kindly arranged financial support for me from the Kihara Institute of Biology, then situated at Mozume in the suburbs of Kyoto.

Having no specific obligations, I began to study Wright's mathematical papers hoping to understand his work. I particularly felt the need to understand his 1931 paper entitled "Evolution in Mendelian populations", which was more than 60 pages long [37]. At the beginning, the paper was extremely difficult for me to understand; it appeared almost insurmountable. I still remember my resolution to continue studying this paper even if it took me 10 years to understand it. To increase my knowledge of mathematics, I attended some courses given in the Department of Mathematics, and received advice from its members. More importantly, I obtained various books on mathematics and studied them by myself. Gradually, my love of mathematics, which had been dormant for a few years, came back to me.

As to Wright's paper, it proved not to be as formidable as I had first thought, and I was glad that within a year or so, I could understand the main part of it. I continued to read Haldane's papers, but gradually became more absorbed in Wright's work. I also tried to read Fisher's papers, but most of them were too difficult for me to understand.

At this time, Kihara was carrying out "chromosome substitution" experiments with wheat for the purpose of investigating the effect of foreign cytoplasm on gene expression. He asked me to estimate how much chromosome material of the maternal species still remained when this was repeatedly backcrossed by pollen of a different (but related) species for a given number of generations. I formulated a finite difference-integral equation which yielded a nice solution, giving the frequency distribution of the length of the chromosome segment in question in a given generation. The result satisfied Kihara, and I published a paper on this in *Cytologia* [12].

Meanwhile, I was appointed to the position of research assistant in his laboratory. Gradually, news reporting the progress of genetics in the United States and England during the war came to our attention through Kihara, and we were much impressed. New scientific literature from the USA also became available in the main library of the University of Tokyo, and I took the opportunity to go there on several occasions to copy by hand some interesting papers, including those of Wright.

3. The National Institute of Genetics, Mishima

In 1949, the National Institute of Genetics was newly established in Mishima as one of the research institutes under the direct jurisdiction of the Ministry of

Education in Japan. Through Kihara's recommendation, I was able to get a job as a research member of this institute in the fall of that year. Mishima proved to be a small city completely lacking in the cultural atmosphere to which I had become accustomed in Kyoto.

The institute building was a crude wooden structure previously used as the administration quarters of a factory producing airplanes during the war. Almost no scientific literature, let alone new foreign journals, was available, and we had to travel to either Tokyo or Kyoto if we wanted to read them. So, for about a year after my appointment in Mishima, I still used to spend much of my time in Kihara's laboratory in Kyoto. I felt rather lonely in Mishima, and I spent most of my time reading papers on mathematical population genetics and also studying mathematics and probability theory from newly published textbooks written in Japanese.

Through reading Wright's papers, I became more and more absorbed in problems of random genetic drift in finite populations. Two papers which made a deep impression on me were Wright's 1949 paper [40], in which he referred to the "Fokker–Planck equation", and his previous paper [39]. Reading these, I was fascinated by the fact that the process of random drift could be treated by an elegant partial differential equation without resorting to the complicated integral equations given in Wright's 1931 paper. Pondering over the rationale of this equation, I found one day that an elementary explanation could be given by approximating the probability distribution by a histogram (see [6], [16]). Although it is not mathematically rigorous, I still think that this explanation is very helpful for most biologists who wish to grasp the meaning of the Fokker–Planck equation as applied to population genetics. Soon, I learned from advanced textbooks on probability theory, particularly Kunisawa's book [25], then newly published in Japan, that there were two types of partial differential equations, called Kolmogorov forward and Kolmogorov backward equations, which described continuous stochastic processes, and that the Fokker–Planck equation corresponds to the forward equation. I also learned, from mathematical books on analysis and more specifically books on differential equations, of the existence of eigenvalues and the expression of the solution for linear second-order differential equations in terms of eigenfunctions. These were fascinating subjects, and I was deeply impressed by the ingenuity of mathematicians who had discovered such properties.

As to my own research project, its main incentive came from Wright's papers and his debates with R. A. Fisher and his school on the significance of random genetic drift in evolution. Open debate started at this time when Fisher and Ford published a paper [10] claiming that the population sizes of most species in nature were generally too large for the random fluctuation of gene frequencies due to sampling of gametes to have any significance in evolution. Instead, they asserted that the fitnesses of mutant alleles at all loci fluctuated from generation to generation due to random fluctuation of environmental conditions and that this overrode the effect of sampling drift. Wright [41] made a rebuttal, but at that time the argument on Fisher's side appeared to have more force. So, one of my research projects was to clarify the effect of random fluctuation of selection coefficients on the genetic composition of populations.

I was also attracted by Wright's idea that individually deleterious but jointly advantageous mutants can be utilized in evolution when the population is subdivided; the advantageous gene combination will become fixed in some local populations through sampling drift. For this, I formulated the two-dimensional Fokker–Planck equation, but I found it too difficult to handle. In addition, I felt the need to consider a subdivided population structure in which migrants come from nearby colonies. Previously, Wright [37] had used the "island model" in which migrants to each colony were assumed to come in each generation as a random sample from the "gene pool" of the species as a whole. In contrast, I formulated a "stepping-stone model" in which migrants came only from adjacent colonies. I published this as a short note in the Annual Report of our institute [13], although a fuller and more satisfactory mathematical treatment of the model had to await my collaboration with George Weiss [24], [36].

One of the problems which I considered most intensively then was how the gene frequencies change under random fluctuation of the selection coefficient. I formulated a partial differential equation for the stochastic process involved. Fortunately, I was able to solve the equation so that I could reconstruct the entire process starting from an arbitrary gene frequency. I wanted to do the same for the case of pure sampling drift for a selectively neutral mutant in a finite population. Although I found a way of obtaining formulas for the process of change of the moments of the gene frequency distribution step by step starting from the first moment (the mean), I could not find the solution of the relevant equation. For a selected mutant, the problem appeared to be formidable.

At about this time, I came across the paper by William Feller [9] entitled "Diffusion processes in genetics" in the *Proceedings of the Second Berkeley Symposium on Mathematical Statistics and Probability*, possibly in the main library of the University of Tokyo. From this paper I learned that the Fokker–Planck or Kolmogorov forward equations which describe the stochastic processes in population genetics are of singular type and that no arbitrary boundary "conditions need, or can, be imposed". I was puzzled and asked one of the mathematics professors in Kyoto who specialized in classical differential equations about this. He replied that such a question was due to my lack of knowledge of the standard theory of differential equations.

In addition to mathematics, my head was filled with the papers of Wright. I was so obsessed by his work that, one night, Wright appeared to me in a dream; he asked me how many of his papers I had studied. He seemed to be a mild and kindly person, as was later fully confirmed when I got to know him in real life.

My life in Mishima at that time was rather lonely, with no one around me to understand my work. My colleagues were puzzled about what I was doing. So,

I used to tell them how wonderful I felt describing the living world by mathematics in a manner similar to that of the theoretical physicists for the nonliving world. One exception was the late Dr Taku Komai, who knew of Wright's work and who thought that such a line of research was interesting and worth pursuing. Komai came to Mishima as one of the department heads after his retirement from Kyoto University in order to help make the newly established institute into a good research organization. Even if he could not understand the details of my work, he gave me constant encouragement and help. He was an eminent zoologist and was probably the first to introduce modern evolutionary genetics to Japan.

When he was young, Komai had been to the United States to study genetics under T. H. Morgan. He suggested that I should also go to the United States or Britain to extend my work and said that he would be glad to help make my visit possible. At this time, a group of American doctors and biologists (including human geneticists) were stationed at ABCC (the Atomic Bomb Casualty Commission) in Hiroshima to investigate the genetic effects of atomic radiation. One of the biologists, Duncan McDonald, was very kind, and through Komai's recommendation, made arrangements for me to be awarded a scholarship to the United States. Originally, McDonald considered Wright in Chicago as the best person for me to study under, and he wrote a letter to Wright. In reply, Wright wrote that he was not taking any students because he was near retirement. He suggested that Iowa State College in Ames would be a suitable place for me to go, because quantitative and statistical genetics were being actively pursued there. So, McDonald obtained a scholarship for me in the graduate school of Iowa State College. This enabled me to extend my studies under Jay L. Lush of the Department of Animal Husbandry, where the applications of population genetics principles to animal breeding were being intensively investigated. Furthermore, he obtained for me an additional grant to supplement the scholarship so that I could study in Ames without financial worries. As to traveling expenses, I applied for a grant from the Fulbright Fund and this was approved.

4. Visit to the USA

In the summer of 1953, as one of the Fulbright grantees, I left Yokohama for Seattle on board the *Hikawa Maru*, the sole passenger ship left in Japan after the war. The voyage, which took two weeks, was most memorable. The pleasant and enjoyable days on the boat, playing deck golf every day, being served excellent meals, taking naps in the afternoon, soothed with the comfortable vibrations from the engine, seemed to have a most salutary effect on me; from this time on, my chronic stomach trouble, which had annoyed me for years and which might have been caused by nervous tension, disappeared. This was one of the important factors which made my life in the USA and subsequently much happier. During the voyage, I wrote up a manuscript reporting my solution of the stochastic process of gene frequency change under random fluctuation of the selection coefficient around the neutral point. This paper was later published in *Genetics* [14] with the help of James F. Crow of the University of Wisconsin. Our boat arrived in Seattle on 10 August, and I still remember my excitement at setting foot in a far-away land for the first time in my life. After some three weeks of a pleasant orientation course during August on the campus of the University of Washington in Seattle, we departed to our destinations. Compared with the miserable life after the war in Japan, life in the United States seemed to me like paradise; everything I experienced in Seattle was marvellous.

In Ames, however, my life became harder; I was registered as a foreign graduate student in Lush's laboratory, and I was forced to take graduate courses in full, which I found a very trying experience. One difficulty was my poor training in English conversation; at the beginning, I found it almost impossible to understand spoken English except for simple words when spoken slowly. So, I chose to take as many mathematics and statistics courses as were allowed to fill the requirement, so that poor English would not be a great handicap to me. Another disappointment was the study of population genetics. In Ames, activities in this field were represented by Lush and Oscar Kempthorne. Both were working intensively on quantitative genetics. The most fashionable topic there was the expression of correlation between relatives in terms of various components of variance. I could not find much interest in such work. My research project on the stochastic processes of gene frequency changes was temporarily interrupted. I had to use all my spare time to do homework problems, including plotting binomial distributions on graph paper. However, not all these exercises were meaningless; my formal training in more advanced mathematics and statistics must have been of some help in my subsequent work.

After nine months at Ames, my scholarship expired. I wished to continue my studies in the USA at the University of Wisconsin under James Crow; I knew him through correspondence and also met him once in Madison, and I had been greatly attracted by his warm personality and keen mind, in addition to the fact that his academic interests were much closer to mine than most people whom I knew then. For this purpose, I applied for a scholarship from the Japan Society in New York, and this was granted. So, in the early summer of 1954, I came to Madison, Wisconsin to study under Crow. I was happy to find that, compared with Ames, the city was much larger and more entertaining, the campus was more beautiful and situated by a large lake, and the academic atmosphere was much more liberal and congenial. During that summer, before the fall semester started, I was able to work out the complete solution of the process of random drift of a neutral allele in a finite population; this was later communicated by Wright to the *Proceedings of the U.S. National Academy of Sciences* [15]. The summer was pleasant, and I even had time enough to think of the fact that my twenties would soon be ending. Incidentally, a book on theoretical physics [28] which I happened to find in a campus bookstore proved extremely useful for my work; I learned the properties of some orthogonal polynomials, among which the Gegenbauer polynomials just fitted my need to express the solution to my problem.

Although I was enrolled as a graduate student for the Ph.D., I had ample time to pursue my research work. I discussed problems with Crow almost every day, and I was able to do much work on the stochastic process of gene frequency changes. Towards the end of this year, the Third Berkeley Symposium on Mathematical Statistics and Probability was held, to which Crow was invited [5]. He suggested that I might join as a co-author, and I was happy to accept his offer. This was the beginning of a long and fruitful collaboration with him. And additional incident which made my stay in Madison still more memorable was that Dr Wright, who had just retired from the University of Chicago, joined the Department of Genetics in Madison as a professor a few months after my arrival there.

In June 1955, a Cold Spring Harbor Symposium was held under the title "Population Genetics", and I was happy to be one of the invited speakers. To attend the meeting in Long Island, Crow gave me a long ride in his car. I presented a paper [16] which summarized the entire work which I had done so far regarding stochastic processes and the distribution of gene frequencies in finite populations. This was the culmination of my student life in the United States. During the symposium I was able to meet and talk with many eminent scholars in the field from various parts of the world whom I previously knew only by name.

After Cold Spring Harbor, Crow took me to the Brookhaven Symposium on mutation. During the session, we met H. J. Muller, to whom Crow introduced me. Later, after I returned to Japan, Muller regularly mailed me reprints of his papers, and soon I became captivated by his masterly writings on genetics, evolution and more general scientific subjects. Indeed, it was mainly through reading his papers that I gradually came to think that the new knowledge of molecular genetics must be eventually incorporated into the theoretical framework of population genetics.

In the fall semester of 1955, Wright gave his first (and last) course in population genetics in Madison. On the first day of his lectures, the room was filled with an audience that wanted to see this eminent scholar, but the number in the audience soon dwindled. In the ordinary sense, it was not a good course, nor was he an exciting lecturer. Probably, among the students, I was the only one who could follow the course. Crow also attended, and after the lecture, we three used to have lunch together in a nearby restaurant. I was happy to become acquainted with this great scientist, without whose work on random drift I could never have been able to make a career as a population geneticist. Wright was a shy, modest and absent-minded man, and one could only begin to understand his extraordinary ability when one tried to work out the same problems as he.

Toward the end of 1955, I was invited to give a talk at the New Year Meeting of the Institute of Mathematical Statistics. The outcome of this talk was a paper published in the *Annals of Mathematical Statistics* [17]. This paper took a long time from submission to publication. The most important part of it was the section giving the formula for the ultimate fixation probability of a mutant allele having an arbitrary degree of dominance in a finite population. This result was obtained by applying the Kolmogorov backward diffusion equation, and was inserted into the manuscript at the revision stage in order to replace a corresponding, but much less useful, branching-process type treatment which the referee considered to be rather weak. I feel proud that my result on the probability of gene fixation in this paper was later used by Alan Robertson to develop his theory of selection limit [34], one of the greatest achievements in the theory of animal and plant breeding. I was fortunate, at the New Year Meeting, to meet and talk with William Feller, who was much interested in problems of population genetics.

In June 1956, I was awarded my Ph.D., majoring in genetics and minoring in mathematics, for my thesis entitled "Stochastic Processes in Population Genetics", and returned to Japan the next month. Later, in 1964, the contents of this paper, with some revision and extension, were published under the title "Diffusion models in population genetics" [18] in the newly launched *Journal of Applied Probability*. Looking back, I think that my two years' stay in Madison proved to be one of the most productive periods in my scientific career and served as a real basis for my subsequent activities as a scientist. The treatment of gene frequency changes as a stochastic process using the diffusion equation method (now called "diffusion models in population genetics") was not a popular subject at that time, and its biological importance was by no means obvious. I could never have imagined that through the subsequent rise of molecular genetics and the application of its concepts and methods to studies of evolution, a new field would be opened up where the diffusion models of population genetics play a central role in elucidating the mechanism of evolution and intraspecific variation at the molecular level, particularly in the framework of the neutral theory.

5. Return to Mishima

After I returned home to resume my research work at the National Institute of Genetics in Mishima, my collaboration with Crow continued. We visited each other a number of times, and sometimes spent a fairly long time working together; we co-authored many papers, among which the one published in 1964 [22] on the "infinite allele model" or "Kimura–Crow model" as it was later called, turned out to be most influential in the field of population genetics (see also [26]). Although I was mainly responsible for the mathematical treatment, the original idea of considering the occurrence of an infinite sequence of new, not pre-existing alleles in relation to DNA structure of genes came from Crow, and much credit should go to him for it.

In 1968, I proposed the neutral theory of molecular evolution [19]. I believe that this is the most significant contribution which I have made to science. Since I have recently published a book [21] on this subject, explaining in detail how I came to propose the theory, I shall not repeat it here. In essence, the neutral theory states that the great majority of evolutionary changes at the molecular (DNA) level are caused not by Darwinian selection acting on advantageous mutants but by the random fixation (by sampling drift) of selectively neutral or nearly neutral mutants. It also asserts that most of the intraspecific variability at the molecular level is essentially neutral so that most polymorphic alleles are maintained in the species by the balance between mutational input and random extinction.

Since the publication of the neutral theory, I have been involved in prolonged debates with the anti-neutralists (i.e. "selectionists"). The so-called "neutralist–selectionist controversy" has been extensively documented in such writings as Calder [1] and Crow [3], [4], as well as in my book [21], and interested readers are invited to consult these.

Although the theory looked precarious at the beginning, supporting evidence has accumulated as years have gone by. I believe that it has gained much strength now, particularly following the outburst of DNA data in the last few years; the most remarkable piece of evidence is the rapid evolutionary change observed in the pseudo-genes ("dead genes") which have lost their function as genes. They show the maximum evolutionary rate as predicted by the neutral theory (see p. 314 of [21]). Before such DNA data became available, I received invaluable support from the work of two groups. One is the group led by Terumi Mukai of Kyushu University who did careful selection experiments on enzyme polymorphisms using the fruit fly *Drosophila melanogaster*. The other is the group headed by Masatoshi Nei of the University of Texas at Houston who has done extensive statistical analyses of enzyme data.

Since 1968, I have been much concerned with the development of "molecular population genetics" (see [20]) using the diffusion-equation method. This was done in collaboration with my colleagues Takeo Maruyama and Tomoko Ohta. Maruyama has done much work on subdivided population structure [27] while Ohta has developed diffusion models which can treat "linkage disequilibrium" or nonrandom association of linked genes due to random drift in finite populations (see [26]). In addition, Ohta has contributed greatly to making the neutral theory more realistic by incorporating the selective constraint (negative selection) imposed by the tertiary structure of proteins, and also by pointing out the possibility that very slightly deleterious mutations as well as neutral ones play an important role in molecular evolution and polymorphism [29], [30]. We have co-authored a number of papers and one book [23], and I believe that without her collaboration I would never have been able to develop the neutral theory in its present state. One paper which deserves to be mentioned in the context of this article is the one on the stepwise mutation model [32], which was later called by such names as "ladder model", "charge-state model" or "Ohta–Kimura model". Because of its intrinsic interest as a problem in probability theory, this model has been investigated by a number of authors including such an eminent mathematician as J. F. C. Kingman (see [7] for a review).

It is 40 years since I first came to know the world of theoretical population genetics created by Fisher, Haldane and Wright; the main part of my scientific life has been devoted to exploring this world. Although I made some discoveries, progress was rather slow until about 20 years ago. Then, introduction of the concepts of molecular genetics suddenly permitted us to proceed very rapidly into a wider territory, with the result that our horizon has been enormously enlarged; we now have a new field waiting to be fully exploited. Furthermore, during the past few years, molecular biology has begun to make a second round of revolutionary development with new discoveries coming one after the other on the genetic constitution of eukaryotes, i.e., higher organisms having well-organized cell nuclei (and chromosomes). All these new discoveries invite theoretical population geneticists to work on them, but the multigene families which are usually characterized by large scale repetitive structure are of particular interest. They exhibit a new phenomenon called "concerted evolution" (or "coincidental evolution"), in which members of one family evolve in unison. The most plausible explanation for this phenomenon is that it is caused by random spreading of a mutant form on one chromosome through unequal crossing-over or gene conversion. Thus, for the study of evolution and variation of multigene families, we must take into account complicated two-way stochastic processes; random spreading of a new mutant to members of a multigene family on one chromosome, and random spreading of such a chromosome to the entire individuals of the species. Ohta pioneered the development of a mathematical theory that can treat such a complicated system. Readers who are interested in this topic should consult her excellent monograph on the subject [31].

6. Concluding Remarks

In writing this memoir I realize that what I have achieved on a modest scale is based on the numerous achievements of greater minds: I feel that I am standing on the shoulders of those who came before me. In 1865, Gregor Mendel proposed a probability model to explain the laws of inheritance which he discovered by breeding garden peas. During the subsequent 120 years, our probability models, particularly those in population genetics, have developed a great deal. But the course of this development has been a hard and tortuous one.

First of all, Mendel's theoretical approach was so far ahead of his time that it was long neglected until his laws were "rediscovered" at the beginning of this century. Immediately, strong opposition against the Mendelian theory was voiced in England by the biometric school championed by W. F. R. Weldon and Karl Pearson, and this caused acrimonious debates with the

Mendelians led by William Bateson. Soon, however, the victory of the Mendelian theory became evident, being supported by overwhelming facts, and the biometricians were in retreat. It is now evident that the theory of heredity elaborated mathematically by Pearson was wrong, but the statistical methods which he developed, including the use of χ^2, turned out to be of enormous value in later studies of evolution and variation, as pointed out by Haldane [11]. Bateson and many Mendelians of his time doubted that natural selection acting on small continuous variation (as postulated by Darwin) could bring about evolutionary change. Rather, they welcomed the mutation theory of Hugo de Vries, who claimed that new species arise by mutational leaps. In addition, Wilhelm Johannsen's pure line theory seemed to support the mutation theory. The exciting but confusing atmosphere in genetics during the first decade of this century is well described by Provine [33].

Gradually, however, it was understood that Mendelism and Darwinism are mutually compatible. The fusion of these two concepts using biometric methods led to the birth of population genetics; mainly through the efforts of R. A. Fisher, J. B. S. Haldane and Sewall Wright, classical population genetics was essentially complete by the early 1930s. Although this resolved the conflict between the Darwinists and Mendelians, it started a new controversy; this time between Fisher and Wright regarding the role of random genetic drift (due to random sampling of gametes) in evolution. Fisher considered that the natural populations of most species are so enormously large that random sampling drift has only a trifling effect, easily overridden by the most minute natural selection. On the other hand, Wright [38] advocated a theory which he later called "the shifting balance theory", and claimed that, in a subdivided population, local differentiation of gene frequencies is caused by random drift, and through intergroup selection, gene interaction systems can be exploited, thus greatly speeding up evolutionary progress.

I have already mentioned in a previous section the stimulus which I received from this Fisher—Wright controversy. While their divergent opinions were not reconciled, the "synthetic" or "neo-Darwinian" view became progressively dominant in the field of evolutionary genetics. By the early 1960s, it appeared that the panselectionist program armored with classical population genetics was sufficient to explain the mechanism of evolution.

It did not take long, however, before this peaceful state of affairs was disrupted by the emergence of a new field, namely, the study of evolution and variation at the molecular level. In particular, the proposal of the neutral theory of molecular evolution in 1968 started a new round of heated controversy, as I have already mentioned. Although much criticized by the neo-Darwinian establishment, the neutral theory has survived, and has now gained much strength, supported by the wealth of molecular data obtained from recent DNA sequence studies. I am delighted to note that diffusion models of population genetics, which at one time were regarded as being too theoretical to be of actual use, have now been used to make predictions with respect to variation and evolution at the molecular (DNA) level.

The success of the neutral theory motivated a reexamination of the "synthetic theory" of evolution, and promoted the world view that chance plays a much more important role than was previously considered, as is eloquently stated in a recent article by Calder [2]. No doubt, stochastic models will become increasingly important in biology, and I look forward to even greater progress in this area in the future.

References

[1] Calder, N. (1973) *The Life Game*. BBC Publications, London.

[2] Calder, N. (1985) The lottery of life: changing views of evolution and human progress. In *Population Genetics and Molecular Evolution*, ed. T. Ohta and K. Aoki, Japan Scientific Societies Press, Tokyo; Springer-Verlag, Berlin.

[3] Crow, J. F. (1972) The dilemma of nearly neutral mutations: How important are they for evolution and human welfare? *J. Heredity* 63, 306–316.

[4] Crow, J. F. (1981) The neutralist-selectionist controversy: an overview. In *Population and Biological Aspects of Human Mutation*, ed. E. B. Hook, Academic Press, New York, 3–14.

[5] Crow, J. F. and Kimura, M. (1956) Some genetic problems in natural populations. *Proc. 3rd Berkeley Symp. Math. Statist. Prob.* 4, 1–22.

[6] Crow, J. F. and Kimura, M. (1970) *An Introduction to Population Genetics Theory*. Harper and Row, New York.

[7] Dawson, D. A. and Hochberg, K. J. (1983) Qualitative behavior of a selectively neutral allelic model. *Theoret. Popn Biol.* 23, 1–18.

[8] Dobzhansky, Th. (1937) *Genetics and the Origin of Species*. Columbia University Press, New York.

[9] Feller, W. (1951) Diffusion processes in genetics. *Proc. 2nd Berkeley Symp. Math. Statist. Prob.*, 227–246.

[10] Fisher, R. A. and Ford, E. B. (1950) The 'Sewall Wright effect'. *Heredity* 4, 117–119.

[11] Haldane, J. B. S. (1957) Karl Pearson, 1857–1957. *Biometrika* 44, 303–313.

[12] Kimura, M. (1950) The theory of the chromosome substitution between two different species. *Cytologia* 15, 281–294.

[13] Kimura, M. (1953) 'Stepping-stone' model of population. *Ann. Rep. Nat. Inst. Genetics* 3, 62–63.

[14] Kimura, M. (1954) Process leading to Quasi-fixation of genes in natural populations due to random fluctuation of selection intensities. *Genetics* 39, 280–295.

[15] Kimura, M. (1955) Solution of a process of random genetic drift with a continuous model. *Proc. Nat. Acad. Sci. USA* 41, 144–150.

[16] Kimura, M. (1955) Stochastic processes and distribution of gene frequencies under natural selection. *Cold Spring Harbor Symp. Quant. Biol.* 20, 33–53.

[17] Kimura, M. (1957) Some problems of stochastic processes in genetics. *Ann. Math. Statist.* 28, 882–901.

[18] Kimura, M. (1964) Diffusion models in population genetics. *J. Appl. Prob.* 1, 177–232.

[19] Kimura, M. (1968) Evolutionary rate at the molecular level. *Nature* 217, 624–626.

[20] Kimura, M. (1971) Theoretical foundation of population genetics at the molecular level. *Theoret. Popn Biol.* 2, 174–208.

[21] Kimura, M. (1983) *The Neutral Theory of Molecular Evolution*. Cambridge University Press, Cambridge.

[22] Kimura, M. and Crow, J. F. (1964) The number of alleles that can be maintained in a finite population. *Genetics* 49, 725–738.

[23] Kimura, M. and Ohta, T. (1971) *Theoretical Aspects of Population Genetics*. Princeton University Press, Princeton, NJ.

[24] Kimura, M. and Weiss, G. H. (1964) The stepping stone model of population structure and the decrease of genetic correlation with distance. *Genetics* 49, 561–576.

[25] Kunisawa, K. (1951) *Kindai-Kakuritsuron* (Modern Probability Theory). Iwanami Shoten, Tokyo.

[26] Li, Wen-Hsiung (Ed.) (1977) *Stochastic Models in Population Genetics*. Benchmark Papers in Genetics 7, Dowden, Hutchinson and Ross, Stroudsburg, Pa.

[27] Maruyama, T. (1977) *Stochastic Models in Population Genetics*. Lecture Notes in Biomathematics 17, Springer-Verlag, Berlin.

[28] Morse, P. M. and Feshbach, H. (1953) *Methods of Theoretical Physics*, Part I. McGraw-Hill, New York.

[29] Ohta, T. (1973) Slightly deleterious mutant substitutions in evolution. *Nature* 246, 96–98.

[30] Ohta, T. (1974) Mutational pressure as the main cause of molecular evolution and polymorphism. *Nature* 252, 351–354.

[31] Ohta, T. (1980) *Evolution and Variation of Multigene Families*. Lecture Notes in Biomathematics 37, Springer-Verlag, Berlin.

[32] Ohta, T. and Kimura, M. (1973) A model of mutation appropriate to estimate the number of electrophoretically detectable alleles in a finite population. *Genet. Res., Cambridge* 22, 201–204.

[33] Provine, W. B. (1971) *The Origin of Theoretical Population Genetics*. The University of Chicago Press, Chicago, Il.

[34] Robertson, A. (1960) A theory of limits in artificial selection. *Proc. R. Soc. London* B 153, 234–249.

[35] Waddington, C. H. (1939) *An Introduction to Modern Genetics*. Allen and Unwin, London.

[36] Weiss, G. H. and Kimura, M. (1965) A mathematical analysis of the stepping stone model of genetic correlation. *J. Appl. Prob.* 2, 129–149.

[37] Wright, S. (1931) Evolution in Mendelian populations. *Genetics* 16, 97–159.

[38] Wright, S. (1932) The roles of mutation, inbreeding, crossbreeding and selection in evolution. *Proc. VI Internat. Congr. Genet.* 1, 356–366.

[39] Wright, S. (1945) The differential equation of the distribution of gene frequencies. *Proc. Nat. Acad. Sci. USA* 31, 382–389.

[40] Wright, S. (1949) Adaptation and selection. In *Genetics, Paleontology, and Evolution*, ed. G. L. Jepsen, E. Mayr and G. G. Simpson, Princeton University Press, Princeton, NJ, 365–389.

[41] Wright, S. (1951) Fisher and Ford on 'the Sewall Wright effect'. *Amer. Scientist* 39, 452–459.

Julian Keilson

Julian Keilson was born in New York on 19 November 1924. He studied mathematics and physics at Brooklyn College, where, after a three-year interruption during the Second World War, he obtained his B.Sc. in 1947. His graduate studies were in theoretical physics at Harvard University, and a Ph.D. was awarded in 1950. Two years were spent subsequently at Harvard as a postdoctoral fellow in electronics, studying Brownian motion and electrical noise.

The next four years were spent at MIT's Lincoln Laboratory pursuing diffusion in semiconductors, the physics of transistors, and signal detection in radar systems. He then moved to the Applied Research Laboratory of Sylvania Electronic Systems where his interest in stochastic processes, applied probability and operations research developed under ideal research conditions. In 1963 and 1965 he was given leave to visit the University of Birmingham in England as a fellow. Extended corporate support enabled him to write his monograph *Green's Function Methods in Probability Theory*, which appeared in 1965.

In 1966 he was offered a position at the University of Rochester as professor at the Graduate School of Management. He was the first chairman of the Department of Statistics when it started in 1968, and has divided his time since between management and statistics at Rochester.

In 1970 he helped initiate the journal *Stochastic Processes and their Applications* and served as co-editor with N. U. Prabhu until 1979. A series of annual international conferences was launched simultaneously, with the First Conference on Stochastic Processes and Their Applications held at Rochester in 1971.

Professor Keilson has contributed to the areas of physics, engineering, mathematical biology, probability theory, and operations research. He has addressed himself to additive processes defined on chains, asymptotic exponentiality associated with rarity, birth–death processes, the structure of distributions, state-space methods, the influence of boundaries on spatially homogeneous processes, and algorithmic methods. Related tools have been focused on queueing theory, reliability theory and logistics. A second monograph describing some of these ideas, *Markov Chain Models—Rarity and Exponentiality*, was published in 1979.

Extensive effort has been devoted over the years to journal and professional society activity. He has been a member of the International Statistical Institute and the Operations Research Society, and a fellow of the Institute of Mathematical Statistics. He has been an editor of *Advances in Applied Probability* since 1977. He has been a visiting professor at Stanford University and at the Massachusetts Institute of Technology.

He is married and has a son and a daughter. Chess has long been an enthusiasm: in 1953 and 1954 he was Massachusetts State Chess Champion.

Return of the Wanderer: A Physicist Becomes a Probabilist

1. Youth and Early Education

I grew up in New York City in the Depression. Both my parents were immigrants. My father, Jonas I. Keilson, had arrived in his late teens from Lithuania and had built up a small business as middleman between families of modest means and large merchants in lower Manhattan selling suits, overcoats, and furniture. My father provided contacts, expertise, and credit, and his business flourished. My mother, Sarah Eimer, had come as a baby from Austria. I was the third of three boys. The first, Philip, died in infancy in an accident. The second, Sidney, became an accountant and businessman. I was born on 19 November 1924. A girl, Marcia, came some nine years later and is a clinical psychologist.

When the Great Depression hit in the early 1930s, my father's business collapsed. His customers were unable to continue their payments to him, and his credit collapsed in turn. The family then fell on hard times and subsisted on my brother's meager earnings as a hotel clerk and precarious income from a makeshift dress business my mother ran in the front parlor of our apartment.

Somewhat oblivious to all this, I entered Townsend Harris High School in September 1937. Admission to this school was on the basis of a highly competitive citywide entrance examination. It was a boys' school providing a superb academic education in three years and guaranteeing entrance to City College, now CUNY, an excellent free institution aspired to by all the city's poor families. Townsend Harris was abolished a few years after I graduated in 1940, a victim of anti-elitist sentiment and budget cutting.

The school served me well. I commuted from the family home in Flatbush, Brooklyn, an hour (and five cents) away by subway. Despite the burdens of an evening job delivering medicine for a small uptown pharmacy, and a growing preoccupation with chess, I did reasonably well at school. I did especially well in mathematics, which had an almost religious fascination for me, and I won the gold medal in that subject at graduation. I also was a member of the school chess team. Chess and mathematics have an hypnotic appeal to young worshippers, and together soaked up whatever free energy remained in my day.

Despite the family's economic struggles, going to college was never in question. Higher education had the highest family priority. I entered Brooklyn College, which was easier to commute to, in September 1940. I continued to do very well in mathematics, did well in chemistry and French, but neglected other subjects such as history and philosophy which I was too young to appreciate and hence barely passed. I was absent from many classes and spent much of my time in the cafeteria playing chess. The country went to war in December 1941. In anticipation of being called to service, I took a job at the end of my sophomore year and was inducted into the Army in March 1943.

I took Air Force basic training at Miami Beach, Florida, where the many resort hotels had been converted into training barracks and military installations. After ten weeks of testing, calisthenics, rifle instruction and firing ranges, I was shipped to Silver Spring, Maryland, then a small town on the outskirts of Washington, DC, for a three-month course at the Capitol Radio Engineering Institute in radio maintenance. Conscientious study of the books and manuals gave me an excellent theoretical understanding of radio technology, but I was lacking in confidence and manual dexterity, and had difficulty soldering wires to terminals. Nevertheless, I completed the course successfully. After further training as a radio operator at Sault Ste Marie, Michigan, I was sent to Nakina, Ontario, in the dead of winter to serve in a radar platoon helping guard the northern approaches to the United States and lower Canada. Ten weeks there taught me a deep respect for the potential harshness of winter. The temperature dropped on occasion to 40 degrees below zero, Fahrenheit, and the blasts of wind gave rise to incredible wind-chill factors. The platoon was much relieved to be shipped to Tampa, Florida, and then to the Hawaiian Islands for radar duty there. My next two years were spent as a radio maintenance man and radio operator on the islands of Molokai and Maui. Discipline was lax, and life was placid. The war ended and I was sent home and discharged at Fort Dix in early March 1946.

Army life had taught me that I would have no worthwhile future without further education. I rushed back to Brooklyn College, enrolled for the spring term one month late, and resumed my mathematical studies. The war had transformed me from a lackluster student who cut classes to a singleminded scholar who studied with sustained drive. The elective offerings in mathematics were quickly exhausted. I loved mathematical analysis and complex analysis but was turned off by a bad course using Bliss's *Calculus of Variations*, a rather dry book, and a second course on foundations which gave me an overdose of Peano's axioms. Kellogg's *Potential Theory* and Jeans' *Electricity and Magnetism*, however, excited me, and I switched my major in my senior year to physics. In the fall semester, I took four nominally sequential electives simultaneously, with complete success. On the strength of this performance, the physics department backed my application to Harvard's Ph.D. program in physics and I was admitted.

I can attribute much of my subsequent success in probability and operations research to the excellent training in mathematics received at Brooklyn College. Mathematics in America owes a great deal to the quality of that program, which turned out many notable mathematicians. Most notable was Richard Bellman, who contributed enormously to mathematics in his lifetime. I was very proud to win the silver medal in mathematics in my senior year, 1946–7, and to receive a copy of Widder's *The Laplace Transform* [96]. I could not know then how useful that book would someday be.

2. Harvard University and the Lincoln Laboratory

I lost no time, and started my graduate studies that summer. Harvard stimulated me. The students were sharp and the teaching was excellent. I particularly enjoyed the courses in electromagnetic theory, dynamics and quantum mechanics. The latter was taken under Julian Schwinger, a brilliant theoretical physicist who subsequently won the Nobel prize for his work in quantum electrodynamics. His lectures were remarkable for their force, speed and outward clarity. The delivery, without notes, was smooth and seemingly effortless. One took notes furiously to record the rush of ideas, and left with a wonderful sense of euphoria. Only afterward, when one transcribed the notes into a permanent ledger, did one realize that the subject was subtle and difficult and that the feeling of understanding with which one came away from the lecture room was an illusion.

My second year of study was occupied largely with advanced courses and topics courses, and preparation for the qualifying oral exam. Harvard was so demanding in its admissions standards in physics that it did not believe in the need for written qualifying exams, and its professors were too hard-pressed by their national commitments and their preoccupation with their own research to administer them. Very few students failed the oral qualifying examination. The experience, nevertheless, was terrifying, at least for me. To display one's competence in response to searching and slightly tricky questions, one had to be relaxed. I was not, and left the exam convinced that I had done poorly. But I too passed and was grateful for the mercy shown me. At the end of the year, I was accepted by Professor Schwinger as a research student.

My thesis topic was the splitting of l-shells in heavy nuclei by the second-order tensor interaction to account for the Maria G. Mayer "magic numbers". This was a topic, basically, in quantum mechanics. In 1949, the nature of the forces governing the motion of nucleons inside the nucleus was not well understood. The thesis was an attempt to shed a small amount of light on these forces. I was one of a half-dozen students working with Professor Schwinger. He was terribly busy, and meetings with the students were somewhat infrequent. On the afternoon of a collective appointment, the students would form a queue in order of arrival and patiently wait until Professor Schwinger would appear and it would be their turn. The interaction with each student would be brief but intensive. One would leave with many ideas and considerable confusion. Ultimately a critical mass of results would develop and the thesis would be finished. The process was imperfect, but many excellent theses were written over the years. I received my degree in June 1950. The year my thesis was written was a painful one: my mother was critically ill. She struggled to attend my commencement, and died soon afterward.

My exposure to the discipline of theoretical physics was to have a lasting influence on me. It developed in me a strong sense of practicality and the power of approximation. It strengthened my command of the tools of mathematics and my appreciation of operational methods. In particular, the ideas of orthogonal polynomials, spectral methods and Green's function methods learned from Schwinger would have great influence on my subsequent research.

I was not very happy with the thesis. Indeed I was not happy with quantum mechanics, quantum electrodynamics or field theory. These disciplines were based on hypotheses such as the Pauli exclusion principle which defied human intuition. The theories were highly successful in explaining physical observations, that is in establishing a correlation between theoretical predictions and laboratory measurements, and no other framework was available for the explanations needed. But the process of converting mysterious assumptions into experimental predictions via extended calculations did not appeal to me. Much of it seemed mechanical and arbitrary. It seemed to be neither good mathematics nor good physics, and left me feeling empty. I did not know then, and do not know today, whether my lack of enthusiasm stemmed from personal inadequacy or basic esthetic and intellectual failures of the theory. I did know that I had to do more classical kinds of physics with a heavier mathematical content.

Fortunately, I was able to stay on at Harvard as a research fellow in electronics. The pay was modest, but I did not need much, and it was a chance to pursue new directions. The theory of electrical noise had developed strongly in the 1940s under pressure of growing need for better signal detection methods for radar and sonar. David Middleton was active in this area at Harvard, and it was his interest that initiated mine. I was exposed to a very useful compendium of papers on random walks, diffusion of particles and electrical noise edited by Nelson Wax [95]. The mathematical and analytical problems were very interesting, and I began to work in this area.

As a physicist, I felt a need to reconcile the diffusion equation, which portrays continuous fuzzy particle motion, with the Boltzmann integral equation which is the natural vehicle for describing the essentially discontinuous character of the velocity of particles undergoing random motion. A vital tool of diffusion theory and of the theory of electrical noise is the Fokker–Planck equation [94], a partial differential equation capturing asymptotics of the integral equation through a simpler vehicle. There was need to justify and clarify the diffusion approximations of Einstein, Smoluchowsky and Bachelier. The paper [65] that was written with James E. Storer, later to be a colleague at Sylvania, was my first significant publication of a mathematical character. The paper turned out to be quite influential and has since been widely cited by physical scientists.

With the pressure of graduate study behind me I was able to return to my avocation, chess. As a fellow, I was entitled to represent Harvard on the university chess team which participated in Boston's chess league competition. Through this activity, I met Marshall Freimer, then an undergraduate. Acquaintance with Freimer later led to my association with the University of Rochester. I also competed in area chess tournaments and was Massachusetts State Chess Champion in 1953 and 1954.

In the spring of 1952 I became a staff member of Lincoln Laboratory, a research institute run by the Massachusetts Institute of Technology (MIT). I was assigned to the Solid State Research Group whose leader was Benjamin Lax. As a natural outgrowth of my interest in diffusion, I began to study diffusion in transistors, then still largely laboratory toys but widely recognized to be the key to a new era in electronics. Lincoln Laboratory was busily engaged in the design of a large scale semi-automatic air control system for the whole country. The system required heavy use of computers, and the reliability required of such a system virtually ruled out old-fashioned vacuum tubes. I wrote several papers on the sensitivity of the transistor to its geometry, and also attempted to contribute to the physics of the solid state by addressing the modification of the energy level structure in certain compounds by magnetic fields. But my distaste for quantum mechanics reasserted itself, and I found myself quickly drawn to areas of greater appeal to me.

The Solid State Group was good to me in an unexpected way. I enjoyed talking to a library assistant, Paula Lyman, who worked in the group. A few initial dates led to a steady relationship. On 8 March 1954, we were quietly married in Warwick, Rhode Island, and took an apartment in Arlington, a suburb of Boston. A year later, we bought a large old colonial house in Bedford, somewhat farther out, which would be our home for over ten years. There our first child, Julia, arrived in 1955. Our son, David, was born in 1958.

Lincoln Laboratory provided a very flexible environment. My interests in electrical noise were encouraged, and I was able to participate in radar system design. The need for a distant early warning line of radar sites to signal the intrusion of aircraft created special problems. The normal air traffic in northern Canada was negligible, and human radar operators would not be able to maintain alertness. Automatic electronic detection was needed, and radar systems had to be modified appropriately. There were also many radar design problems associated with the imminence of sophisticated radar jamming techniques. I was privileged to participate in a very exciting brainstorming summer study group which gathered experts from key national and British laboratories and universities to think about impending technological change. It was this study group and Lincoln Laboratory that initiated radical radar design changes to cope with electronic warfare.

3. The Years at Sylvania's Applied Research Laboratory

I became somewhat restless after four years at Lincoln. My salary did not meet the demands of a large suburban home and a growing family. When Leonard Sheingold, a scientist I had known at Harvard, was placed in charge of a new Applied Research Laboratory at Sylvania Electronic Systems at Waltham, Massachusetts, I was interested. He had gathered under him a number of bright physicists and engineers whom he had known from Harvard and MIT, and had a franchise to engage in research of quality in the area of electronic systems, with few restrictions. The essential requirement was that the work be excellent and innovative and lead to publications that contributed to the company's visibility and reputation. I was further enticed by a generous salary offer, impressively higher than my Lincoln Laboratory salary. I accepted.

My initial interests at Sylvania were in the area of electrical noise and related radar and communications systems problems. I published in physics journals, the *IEEE Transactions on Information Theory*, and applied mathematics journals. In 1958, Anthony Kooharian, who had recently joined the Applied Research Laboratory, started talking to me about some queueing problems that had arisen in the context of reconnaissance systems. We both found the mathematical problems involved to be fascinating, and a sustained collaboration began. Our earliest work was on the transient behavior of the $M/G/1$ queue. To meet our needs, we developed state-space methods, which I had worked with in the theory of electrical noise, appropriate to the $M/G/1$ queue. The analytical problems there were quite novel to us, and we were very pleased to have our first paper, "On time-dependent queueing processes", accepted by the *Annals of Mathematical Statistics* [51]. The paper implemented ideas of Cox on supplementary variables [3] and overlapped with similar work by Wishart [97]. Kooharian and I tried to extend our results to the transient behavior of $G/G/1$ queueing systems. A huge joint effort gave rise to a second paper [52] which was a tribute to human persistence in the face of adversity but did not really solve the problem, which is still open. The paper taught us that profundity is not always a virtue and cannot compete with simplicity and success.

As with so many other researchers exposed to the challenge of queueing theory, I remained a grateful captive of the subject. I found its difficulty demanding and frustrating, its variety endless, its importance clear and its beauty rewarding. I have never had any patience with colleagues who condemn the subject because of the excess of papers published of low quality, or who claim that the problems have all been solved. It is becoming clear to me that the work needed has hardly begun. The subject grows constantly in practical importance, and few of the needs have been met. In this respect it differs little from most of pure and applied mathematics.

I had the good fortune to encounter Professor H. E. Daniels, head of the Department of Mathematical Statistics at the University of Birmingham, and his colleague, Dr David M. G. Wishart, a recent student of William Feller, at the 1960 Berkeley Symposium. Wishart and I were able to meet again at a session of the International Statistical Institute in Paris in August 1961. After the meeting David drove me to Bristol to participate in a meeting of the Royal Statistical Society, and to present a paper. The drive through France and southern England was exciting, and the Bristol meeting gave me a chance to meet many of the British workers in probability theory and statistics. The possibility of a visit to Birmingham was broached, and I was interested.

Professor Daniels was very supportive of the visit. The University of Birmingham could not provide enough money to cover the air travel and living expenses of a family for a year. But a University Fellowship was arranged providing £1000 as a contribution, and residence standards were relaxed to permit the fellow and his family to arrive at the beginning of March 1963 and to stay about six months. Sylvania was more than supportive. They gave me a leave of absence and a consulting agreement providing the equivalent of my normal income, generating thereby a *de facto* sabbatical from the company. The family arrived on schedule to find that England was in the midst of a terrible winter. We were ushered to the home provided by the university, a furnished flat filling the second floor of a huge Victorian mansion. The flat unfortunately had not been occupied that year, and mold had begun to form on the walls. But the gas fires were set roaring, we were showered with English hospitality, and a totally enjoyable and productive visit began. The 1963 visit and a second family visit in the spring and summer of 1965 were memorable times for us, and we still look back on them with pleasure.

In September 1963, at the end of my first visit to Birmingham, I was invited by Professor J. Th. Runnenburg of the mathematics department of the University of Amsterdam to give a series of lectures on Green's function methods at the Mathematics Centre. I had met Professor Runnenburg at the ISI meeting in Paris and found his company stimulating. The visit to Amsterdam was pure pleasure. The mathematics department was then housed in a charming old building on the Nieuwe Achtergracht. Professor Runnenburg had been appointed to the chair in probability theory vacated by the early death of Professor David van Dantzig. Many fine students were being nurtured by Runnenburg to populate some of the key teaching posts in probability. One student who became a good friend and collaborator subsequently [63], [64] was F. W. Steutel. Runnenburg and I were able to come up with an interesting demonstration of the infinite divisibility of the $G/G/1$ waiting-time distribution that appeared [91] soon after. I had many other enjoyable and rewarding visits to Amsterdam in later years, but that first visit was unhurried and wonderful.

David Wishart and I worked very hard, and a stream of papers on additive processes defined on finite Markov chains flowed forth. Three substantial papers [81], [82], [83] emerged from the collaboration, adding to the pleasure of the visits. Contact with Henry Daniels was also fruitful and many of his ideas on conjugate transformation and saddlepoint methods [4] played a role in subsequent work. During the second visit, I also met Hamish Ross, an energetic young statistician, and began work with him on the first-passage-time distribution for the Ornstein–Uhlenbeck process [59].

Some of the work with Wishart was on spatially homogeneous processes modified by boundaries. I had a deep interest in such problems as an outgrowth of my background in mathematical physics, where Green's function methods are crucial to boundary-value problems. In classical potential theory and electrostatics, the field inside a boundary on which the potential is constant is shown to be equivalent to that of a dipole layer distributed on the boundary induced by sources inside the boundary. While studying the ergodic distribution of bounded random walks at Sylvania, I was delighted to find that Green's function methods were totally relevant and provided structural insight into ergodic distributions unavailable otherwise. The dipole layers had their analogue in source distributions of mixed sign with total mass zero whose potential was the ergodic distribution. The source distributions could be thought of as a generalized signed measure whose influence was equivalent to that of the boundaries. The analogy with physical boundaries was complete. Indeed for homogeneous diffusion in three dimensions modified by a reflecting boundary, the distribution induced was the familiar dipole layer of electrostatics!

My first paper displaying these ideas and their importance as tools was published in 1962 [8]. The induced measures provide an intuitive interpretation of Wiener–Hopf factorization and provide alternatives to such factorization for the analysis of the Lindley process and related random walks. A demonstration of this appeared in the first issue of the *Journal of Applied Probability* in 1964 [17]. The use of Green's functions and the related ideas of compensation had broader scope. They provided a systematic method of treating spatially homogeneous processes modified by boundaries and associated replacement processes, which are often otherwise resistant. The method has been exploited in many papers since [10], [14], [20], [23], [46], [54], [68], [74], [84].

David Wishart was interested in these ideas. He was able to visit the Applied Research Laboratory in the summer of 1964. During that visit he was enormously helpful in the preparation of a paper organizing the ideas in a more general, unified way. The paper was presented at the very excellent Symposium on Congestion Theory at North Carolina organized by Walter L. Smith and William E. Wilkinson, and appeared in the *Proceedings* [20]. The encouragement I received from Henry Daniels, David Wishart, J. Th. Runnenburg and many colleagues motivated me to gather my ideas into a monograph. Again Sylvania was totally supportive, permitting me to write the monograph as part of my research activity and providing me with a full-time secretary to type the many manuscript drafts. The monograph [22] was published by Charles Griffin, London, in their Statistical Monographs Series and appeared in 1965.

4. Academic Life

Sylvania's idealism and generosity to me were made possible by a government policy which specified that a small percentage of each federal contract could be employed to sponsor basic research. This policy changed in the mid 1960s, and the motivation and character of many industrial research laboratories began to change. By then, I had become totally involved in my basic research and felt a need for a university environment, teaching and graduate students. Fortunately, I had acquired some research visibility and there was academic demand nationally for research workers in mathematics, engineering and statistics.

One invitation was from the University of Rochester, where Dean William Meckling had launched an ambitious Graduate School of Management. The school sought to establish management as a hard, scientific discipline built on a solid orientation to economics, finance and operations research. The university under President W. Allen Wallis, a key figure in American statistics, backed the school strongly. Professor Marshall Freimer, whom I had known at Harvard, was aware of my work in probability theory and operations research and introduced me to the school. The vigor of the new program was attractive and the presence of Professor J. H. B. Kemperman in the Department of Mathematics and his interest in me made Rochester inviting. When I was offered a full professorship with tenure, I accepted. The family moved to Rochester on 3 July 1966. My activity with Sylvania continued on a consulting basis for several years. It ended when the Applied Research Laboratory was discontinued.

I was very grateful to be able to enter academic life at an excellent university without having to run the gauntlet of promotion and tenure. There are only a handful of laboratories today at which scientific growth at a basic level is fostered by freedom and a policy of open publication. Many laboratories are hungry for workers of broad scientific competence, but few of these have the foresight to nurture the bright young people they have to the level of strength needed.

Academic life was exciting. Most of my teaching was at a graduate level and I found lecture preparation satisfying. The Graduate School of Management encouraged me to start an interdisciplinary Center for System Science to coordinate university interests in operations research and systems engineering. The center had a good level of research productivity, but had no real teaching function. It required someone less preoccupied with research to generate the external resources needed for its growth.

I was also asked by President Wallis to play a role in the establishment of a Department of Statistics in the College of Arts and Science. The school wanted a department with a strong orientation to statistical inference and biostatistics. My statistical interests were in probability theory and I had little real data orientation. Nevertheless I was asked to get the department started as its first chairman, with the understanding that I would seek a statistician with an orientation to statistical inference for the long pull. With the help of my colleagues, such a person was soon found. Professor W. J. Hall at North Carolina took over, to my great relief, and proved to be a fine colleague over the years.

My first year at Rochester coincided with Kemperman's sabbatical year at Stanford. He had a rather large house which we were happy to sublet. He also asked me to look after the tracer-kinetic needs of Professor Christine Waterhouse in the Medical School, with whom he had been working for some years. I undertook this substitute role gladly, not knowing that I would become deeply involved for almost 15 years. Waterhouse was a professor of medicine, specializing in renal disorders, director of the Clinical Research Unit at the university's Strong Memorial Hospital, and a dedicated researcher. She was preoccupied with the biochemical processes underlying starvation induced by cancer, and conducted many tracer-kinetic studies of metabolic changes on volunteer patients. I helped analyze the data. The analysis was not mechanical. The basic tools of compartmental analysis seemed inappropriate to analyze the data since the models required the volume of distribution of pools which seemed artificial and spurious at times. The models also had difficulty dealing with intermediate metabolites. Rather than continue to work with an untrustworthy, often senseless, format, we finally scrapped compartmental analysis and devised an alternate metabolic framework modelling the body's logistic system as a network of parallel organs across the heart with random organ selection by a metabolite arising from the turbulent mixing in the heart. We called this model a circulatory model for human metabolism, and demonstrated its feasibility and simplicity in a paper published in the *Journal of Theoretical Biology* [86]. Simultaneously a monograph was issued discussing the probabilistic issues in great detail [49]. It was a privilege to be associated with these medical studies. Unfortunately my colleague Professor Waterhouse left the university in 1980 and these activities ended.

The Graduate School of Management under Dean William Meckling encouraged and supported research fully. The school had gathered a very excellent faculty in all its areas of commitment and provided the strongest support for travel, publication, seminars and academic visitors. During the early years of my appointment at Rochester, many colleagues were able to spend one or more years teaching and working with me. I recall with pleasure such visits by Hans Gerber, S. Subba Rao, H. Callaert, F. W. Steutel and H. Ross. Gerber and I worked together on strong unimodality of lattice distributions [44]. With S. Subba Rao, processes with chain-dependent growth rate were studied [67], [68]. The spectral structure and exponential ergodicity of birth–death processes were explored with H. Callaert [2]. Tables for the distribution of the first-passage times for the Ornstein–Uhlenbeck process were completed with H. Ross [59]. The analysis of and tables for a related maximum process of meteorological interest were completed on a subsequent occasion [60], [61]. A long visit by F. W. Steutel devoted to infinite divisibility and mixtures of distributions [63], [64], and shorter visits by E. Arjas, R. Syski [77] and G. P. H. Styan [66] were sponsored by the Department of Statistics as well as the Graduate School of Management.

Academics working in applied probability had a professional identity problem in the late 1960s. They were neither inference-oriented statisticians nor pure mathematicians, and many of them did not feel an affinity for operations research. The Institute of Mathematical Statistics was not interested in organizing many sessions in applied probability, and its conferences seemed correspondingly unappealing. A need was perceived for new professional outlets. Professor J. W. Cohen, then at Delft, Professor N. U. Prabhu of Cornell, Professor Ryszard Syski of College Park, Maryland, and I had many active discussions of this need. With their encouragement, I made contact with Dr Wimmers, the president of the North-Holland Publishing Company and organized a new journal, *Stochastic Processes and their Applications*. A contract had to be negotiated with one editor, and I assumed that responsibility. I was unwilling, however, to have the sole editorial burden, and persuaded Professor Prabhu to edit the journal jointly with me. The four founders of the journal constituted an editorial board.

It was thought desirable simultaneously to organize conferences promoting the same professional objectives. Rather than struggle to find outside financial support for such a conference, I enlisted the aid of the Graduate School of Management, and organized the first Conference on Stochastic Processes and their Applications (SPA) at Rochester. A great deal of secretarial support, printing and postal costs were needed, and some university money for local hospitality as well. An auditorium in the brand-new Medical Education Building was donated, and little else was needed. The conference was held in 1971 and had about 50 attenders and about 20 speakers. It met with an enthusiastic response, and paved the way for the many such conferences which have been held since at Leuven, Sheffield, Toronto, College Park, Tel Aviv, Enschede, Canberra, Evanston, Montreal, Grenoble, Ithaca and Benares. A brief account of the history of these conferences has been given by Professor Prabhu [93].

The journal prospered, and Professor Prabhu and I functioned as joint editors through 1979. By then both the journal and the conference series had matured. The conferences had already passed into the custody of the Bernoulli Society. It became clear that the journal should also be controlled by the same group, and that an editor should be appointed for a fixed term of office. I supported the candidacy of Professor Prabhu for the first four-year term and stepped down.

I was not always in agreement with all my fellow editors. The journal and the conference series had come into being because applied probability was not adequately represented in the professional societies then active. I wanted the term "applied" to be more than a benediction and a hope, and the journal to promote stochastic processes as a discipline contributing to scientific understanding, not as an end unto itself. A mathematics paper may be difficult and even beautiful, but still sterile in the procreative sense if it never relates to ideas and needs outside mathematics. Some of my colleagues believed, perhaps rightly, that mathematical excellence should be the sole criterion for a paper's acceptance and that no one could predict future relevance to applications. The argument was somewhat empty, since the papers submitted were largely academic in their orientation.

My association with Professor Runnenburg of Amsterdam paid off in an unexpected way. Two very strong students were encouraged by Runnenburg to come to Rochester for further graduate study. The first to come was Teunis Ott, an energetic and independent person. He was encouraged by me to address his research to a topic in infinite divisibility, which he handled with dispatch [92]. After two academic stints at Case Western and at the University of Texas at Dallas, he went to Bell Laboratories at Holmdel. His many contributions to queueing theory and to applied probability since have been a source of satisfaction. The second Dutch student was Adri Kester, who worked very closely with me for many years. His thesis topic was in the area of stochastic monotonicity. Two separate papers [48], [60] were written with him at Stanford University when he accompanied me on my sabbatical in 1973–4. The close working relationship I have had with my students has been one of the richer rewards of teaching.

The sabbatical year was a very productive one. For some time I had been preoccupied with time-reversibility and distribution structure arising from it. I was also caught up in birth–death processes and related topics of rarity and exponentiality. As part of my visit to the Department of Statistics at Stanford, I presented an informal series of lectures on some of these topics. In the middle of my sabbatical year, I visited the Department of Mathematics at Helsinki in Finland. The invitation was arranged by Elja Arjas. I had met him at Leuven at the second SPA conference and had since been in interaction with him. With his encouragement I presented a series of 10 lectures on time-reversibility, spectral structure and rarity, and these were recorded and edited by M. Heikkila, a graduate student. The one-month visit to Helsinki with my wife Paula was delightful. The sabbatical year was capped by a summer visit to Berkeley's Industrial Engineering and Operations Research Department at the invitation of Professor Richard Barlow. He made available to me the energetic assistance of D. Robert Smith, then a graduate student, and with his help all the ideas on time reversibility, spectral structure, and rarity coalesced into a monograph, *Markov Chain Models—Rarity and Exponentiality*. The monograph appeared first as a working paper sponsored by the Operations Research Center at Berkeley. A slightly revised version was published by Springer-Verlag in 1975 as Volume 28 of their Applied Mathematical Sciences Series [40].

A topic of great interest to me was expanded into a paper for the Conference on Reliability and Fault Tree Analysis [38] organized by Professor Barlow at Berkeley that summer. It was shown that the concept of system failure time could be made precise and unambiguous for redundant systems with independent repairable components, provided that the system was highly reliable. Basically, the state of the system when working is unimportant since the system relaxation time is negligible in comparison to the time to failure, and the distinction between competing candidates for system failure time fades. The existence of a system failure time permits one to employ the concept as a focus of trade-off studies in system design. Two papers [45], [47] developing these ideas and discussing system design optimization algorithmically were written subsequently with Professor S. E. Graves of the Sloan School at MIT.

For many years I had been struggling with a shortcoming of queueing theory which seemed to be insuperable. Many of the descriptive distributions of queueing theory had a succinct representation only as a Laplace transform or, sometimes, had a characterization as a solution of a functional equation in the Laplace-transform domain. Often, real-domain representations existed in the form of an infinite series of multiple convolutions or as the limit of a sequence of iterative operations in the real domain involving differentiation, integration, convolution, and polynomial multiplication. Such real-domain "answers" were useless, however, because continuum operations were foreign to the discrete character of computers.

The possibility of an escape from such difficulties arose during a visit to the Center for Naval Analysis, when Dr Walter R. Nunn, a friend and colleague of long standing, showed me a curious identity in Abramowitz and Stegun [1] stating that the convolution of two Laguerre polynomials had a simple operational form. He thought there might be a chance of exploiting this identity. With a little effort we were able to establish an operational calculus that was highly successful in pulling back the dreaded "Laplacian curtain" of queueing theory. Functions mapped into sequences, continuum convolution mapped into lattice convolution, and differentiation and integration into lattice operations. The method also had a simple and natural extension to the full continuum that permitted, for example, the evaluation of the ergodic distribution of the Lindley process [69] via simple iteration. We were very pleased with the method. A first paper [56] was finished quickly, and a second [87] with U. Sumita came soon afterward. Sumita, then a research student at Rochester, became deeply involved with the Laguerre transform method, and produced a superb thesis refining and strengthening the method and establishing its full scope.

Ushio Sumita had come from Japan in 1974 as a systems analyst to Xerox Corporation in Rochester. His wife was a skilled violinist then studying in Switzerland. While visiting Rochester, she auditioned successfully for the Rochester Philharmonic. Sumita enrolled in the Ph.D. program at Rochester in September 1976, and began working with me. Much of his dissertation research was conducted at MIT during my sabbatical year, 1980–1, in residence at the Laboratory for Information and Decision Systems, where I had my office. When he completed his dissertation, he spent a year at Syracuse University before being appointed to the faculty at Rochester in the Graduate

School of Management's Computer and Information Systems Area. He and I have collaborated on many research topics, and co-authored many papers. He has also been instrumental in establishing strong ties between the School of Management and Japan.

The sabbatical year at MIT was also very full. My trip was sponsored by Mathematics, Electrical Engineering, the Sloan School of Management, and funding from a grant. My former colleague, Dr Anthony Kooharian, had recently joined GTE Laboratories, the research arm of Syvania's parent corporation, and I had strong activity with his research group, housed in the very building where I had first worked for Sylvania back in 1956. Once again that year I had a sabbatical within a sabbatical. My colleague Professor F. W. Steutel at the Technical University at Eindhoven invited me to be in residence at Eindhoven for a month and arranged for a few lectures in Eindhoven, Enschede and Amsterdam. The Steutels were good hosts, and Eindhoven was a beautiful city.

I had the good fortune of visiting Japan in the summer of 1982, as a guest of Japan's equivalent of Bell Laboratories, the Electrical Communication Laboratories (ECL) of the Nippon Telegraph and Telephone (NTT) Public Corporation at Musashino-shi in the suburbs of Tokyo. In particular, I was a visitor to the Traffic Group at ECL and gave an intensive series of lectures on queueing theory over a period of five weeks, emphasizing topics from my own background. The Japanese people are notoriously hospitable to guests, and NTT upheld these traditions wonderfully. An enormous office was vacated for me, equipped with two desks, a conference table, a blackboard, a china cabinet, a sofa, two easy chairs and a coffee table, all quite uncrowded. I was also provided with a luxurious room at the Keio Plaza Hotel in Shinjuku. Numerous lectures at Japan's leading universities in Tokyo, Kyoto and Osaka were sponsored as well, and for all of these visits I was accompanied by one or more colleagues from the Traffic Group, lest I get lost. The hospitality shown me, private and public, was extraordinary, and the extended visit to a corporate laboratory in Japan was a rare privilege. My Japanese friends and colleagues were hard-working and fun-loving, and I shall not soon forget my visit that summer.

I also had opportunity to meet the family of Ushio Sumita and that of his wife, Mariko, both living in Kamikitazawa on the Keio Line. They too were unstinting in their generosity. I was delighted to meet Ushio's father, Umetaro Sumita, who had a wonderful command of English and was still professionally vigorous in his early seventies.

The ties to Japan continue. In 1983–4, a colleague from NTT, F. Machihara, visited the Graduate School of Management with his family to interact with me [56], [90]. In the fall of 1985 I spent two months at the Tokyo Institute of Technology as a fellow of the Japan Society for the promotion of science.

Other graduate students have come. In recent years, I have been fortunate to work with Ravi Ramaswamy on quasistationarity [57], [58], related spectral topics, and interesting topics in queueing theory. The renewed ties with GTE Laboratories have also led to interesting work with L. Servi on the scheduling of computers [62]. The flow of problems shows no signs of let-up, I am happy to say.

I have often looked back on the crossroads of my career and thought of the wisdom of the paths chosen. In retrospect, graduate study in physics might seem to have been a diversion. But it provided a unique and crucial background for much of the work I have done. The years at Lincoln Laboratory and Sylvania also created very unusual opportunities. I have always been grateful that I wound up in academic life at a research-oriented university. The rewards of a life of teaching and study, of close students and colleagues around the world and of broad opportunities for travel of the most fulfilling kind have been priceless.

Publications and References

[1] Abramowitz, M. and Stegun. I. A. (1965) *Handbook of Mathematical Functions*. Dover, New York.

[2] Callaert, H. and Keilson, J. (1972) On exponential ergodicity and spectral structure for birth-death processes. *Stoch. Proc. Appl.* 1, 187–236.

[3] Cox, D. R. (1955) The analysis of non-Markovian stochastic processes by the inclusion of supplementary variables. *Proc. Camb. Phil. Soc.* 51, 433–441.

[4] Daniels, H. E. (1954) Saddlepoint approximations in statistics. *Ann. Math. Statist.* 25, 631–650.

[5] Keilson, J. (1954) A suggested modification of noise theory. *Quart. Appl. Math.* 12, 71–76.

[6] Keilson, J. (1955) On diffusion in an external field and the adjoint source problem. *Quart. Appl. Math.* 12, 435–438.

[7] Keilson, J. (1961) The homogeneous random walk on the half-line and the Hilbert problem. *Bull. Inst. Internat. Statist., Proc. 33rd Session ISI, Paris* 39(2), 279–291.

[8] Keilson, J. (1962) The use of Green's functions in the study of bounded random walks with application to queueing theory. *J. Math. Phys.* 41, 42–52.

[9] Keilson, J. (1962) Non-stationary Markov walks on the lattice. *J. Math. Phys.* 41, 205–211.

[10] Keilson, J. (1962) The general bulk queue as a Hilbert problem. *J. R. Statist. Soc.* B 24, 344.

[11] Keilson, J. (1962) A simple random walk and an associated asymptotic behaviour of the Bessel functions. *Proc. Camb. Phil. Soc.* 58, 708–709.

[12] Keilson, J. (1962) Queues subject to service interruption. *Ann. Math. Statist.* 33, 1314–1322.

[13] Keilson, J. (1963) A gambler's ruin type problem in queueing theory. *Operat. Res.* 11, 570–576.

[14] Keilson, J. (1963) The first passage time density for homogeneous skip-free walks on the continuum. *Ann. Math. Statist.* 34, 375–380.

[15] Keilson, J. (1963) On the asymptotic behaviour of queues. *J. R. Statist. Soc.* B 25, 464–476.

[16] Keilson, J. (1964) Some comments on single server queueing methods and some new results. *Proc. Camb. Phil. Soc.* 60, 237–251.

[17] Keilson, J. (1964) An alternative to Wiener-Hopf methods for the study of bounded processes. *J. Appl. Prob.* 1, 85–120.

[18] Keilson, J. (1964) On the ruin problem for the generalized random walk. *Operat. Res.* 12, 504–506.

[19] Keilson, J. (1964) A review of transient behavior in regular diffusion and birth-death processes. *J. Appl. Prob.* 1, 247–266.

[20] Keilson, J. (1965) The role of Green's functions in congestion theory. In *Symp. on Congestion Theory*, University of North Carolina Press, Chapel Hill.

[21] Keilson, J. (1965) A review of transient behavior in regular diffusion and birth-death processes—Part II. *J. Appl. Prob.* 2, 405–428.

[22] Keilson, J. (1965) *Green's Function Methods in Probability Theory*. Griffin, London.

[23] Keilson, J. (1966) The ergodic queue length distribution for queueing systems with finite capacity. *J. R. Statist. Soc.* B 28, 190–201.

[24] Keilson, J. (1966) A limit theorem for passage times in ergodic regenerative processes. *Ann. Math. Statist.* 37, 866–870.

[25] Keilson, J. (1966) A technique for discussing the passage time distribution for stable systems. *J. R. Statist. Soc.* B 28, 477–486.

[26] Keilson, J. (1966) A theorem on optimum allocation for a class of symmetric multilinear return functions. *J. Math. Anal. Appl.* 15, 269–272.

[27] Keilson, J. (1967) On global extrema for a class of symmetric functions. *J. Math. Anal. Appl.* 18, 218–228.

[28] Keilson, J. (1968) A note on the waiting-time distribution for the $M/G/1$ queue with last-come-first-served discipline. *Operat. Res.* 16, 1230–1232.

[29] Keilson, J. (1969) A queue model for interrupted communication. *Opsearch* 6, 59–67.

[30] Keilson, J. (1969) An intermittent channel with finite storage. *Opsearch* 6, 109–117.

[31] Keilson, J. (1969) On the matrix renewal function for Markov renewal processes. *Ann. Math. Statist.* 40, 1901–1907.

[32] Keilson, J. (1971) On log-concavity and log-convexity in passage-time densities of diffusion and birth-death processes. *J. Appl. Prob.* 8, 391–398.

[33] Keilson, J. (1971) A note on the summability of the entropy series. *Inf. and Control* 18, 257–260.

[34] Keilson, J. (1972) A threshold for log-concavity. *Ann. Math. Statist.* 43, 1702–1708.

[35] Keilson, J. (1973) Simple expressions for the mean and variance of a generalized central limit theorem and applications. *Proc. XXth Annual Meeting, Institute of Management Science, Tel Aviv, Israel, 24–29 June 1973*, 538–541.

[36] Keilson, J. (1974) Sojourn times, exit times, and jitter in multivariate Markov processes. *Adv. Appl. Prob.* 6, 747–756.

[37] Keilson, J. (1974) Convexity and complete monotonicity in queueing distributions and associated limit behavior. In *Mathematical Methods in Queueing Theory*, Proceedings of a Conference at Western Michigan University, 10–12 May 1974, Springer-Verlag, New York.

[38] Keilson, J. (1975) Systems of independent Markov components and their transient behavior. In *Reliability and Fault Tree Analysis*, SIAM, Philadelphia, 351–364.

[39] Keilson, J. (1978) Exponential spectra as a tool for the study of server-systems with several classes of customers. *J. Appl. Prob.* 15, 162–170.

[40] Keilson, J. (1979) *Markov Chain Models—Rarity and Exponentiality*. Applied Mathematical Sciences Series 28, Springer-Verlag, New York.

[41] Keilson, J. (1981) On the unimodality of passage time densities in birth-death processes. *Statist. Neerlandica* 35, 49–55.

[42] Keilson, J. (1982) On the distribution and covariance structure of the present value of a random income stream. *J. Appl. Prob.* 19, 240–244.

[43] Keilson, J. (1983) Stochastic systems (Lecture notes). Graduate School of Management, University of Rochester, Working Paper Series No. QM 8306.

[44] Keilson, J. and Gerber, H. (1971) Some results for discrete unimodality. *J. Amer. Statist. Assoc.* 66, 386–390.

[45] Keilson, J. and Graves, S. C. (1979) A methodology for studying the dynamics of extended logistic systems. *Naval Research Logist. Quart.* 26, 169–197.

[46] Keilson, J. and Graves, S. C. (1981) The compensation method applied to a one-product production/inventory problem. *Math. Operat. Res.* 6, 246–262.

[47] Keilson, J. and Graves, S. C. (1983) System balance for extended logistic systems. *Operat. Res.* 31, 234–252.

[48] Keilson, J. and Kester, A. (1977) Monotone matrices and monotone Markov processes. *Stoch. Proc. Appl.* 5, 231–241.

[49] Keilson, J. and Kester, A. (1977) A circulatory model for human metabolism (Unpublished monograph). Graduate School of Management, University of Rochester, Working Paper Series No. 7724.

[50] Keilson, J. and Kester, A. (1978) Unimodality preservation in Markov chains. *Stoch. Proc. Appl.* 7, 179–190.

[51] Keilson, J. and Kooharian, A. (1960) On time-dependent queueing processes. *Ann. Math. Statist.* 31, 104–112.

[52] Keilson, J. and Kooharian, A. (1962) On the general time-dependent queue with a single server. *Ann. Math. Statist.* 33, 767–791.

[53] Keilson, J. and Kubat, P. (1984) Parts and service demand distribution generated by primary production. *European J. Operat. Res.* 17, 257–265.

[54] Keilson, J. and Machihara, F. (1985) Hyperexponential waiting time structure in hyperexponential $GI/G/1$ systems. Graduate School of Management, University of Rochester.

[55] Keilson, J. and Mermin, N. (1959) The second order distribution of integrated shot noise. *IRE Trans. Inf. Theory* IT-5, 75–77.

[56] Keilson, J. and Nunn, W. (1979) Laguerre transformation as a tool for the numerical solution of integral equations of convolution type. *Appl. Math. Comput.* 5, 313–359.

[57] Keilson, J. and Ramaswamy, R. (1984) Convergence of quasi-stationary distributions in birth–death processes. *Stoch. Proc. Appl.* 18, 301–312.

[58] Keilson, J. and Ramaswamy, R. (1984) The bivariate maximum process and quasi-stationary structure in birth-death processes. Graduate School of Management, University of Rochester, Working Paper Series No. QM 8405.

[59] Keilson. J. and Ross, H. F. (1975) Passage-time distributions for Gaussian Markov (Ornstein–Uhlenbeck) statistical processes. *Selected Tables in Mathematical Statistics* 3, 233–327.

[60] Keilson, J. and Ross, H. F. (1978) The maximum of the stationary Gaussian Markov process over an interval—Theory, table and graphs. Unpublished.

[61] Keilson, J. and Ross, H. F. (1979) The maximum over an interval of meteorological variates modelled by the stationary Gaussian Markov process. Preprint volume, 6th Conference on Probability and Statistics in Atmospheric Sciences, 9–12 October 1979, 213–216.

[62] Keilson, J. and Servi, L. (1984) Oscillating random walk models for $GI/G/1$ vacation systems with Bernoulli schedules. Graduate School of Management, University of Rochester, Working Paper Series No. QM 8403.

[63] Keilson, J. and Steutel, F. W. (1972) Families of infinitely divisible distributions closed under mixing and convolution. *Ann. Math. Statist.* 43, 242–250.

[64] Keilson, J. and Steutel, F. W. (1974) Mixtures of distributions, moment inequalities and measures of exponentiality and normality. *Ann. Prob.* 2, 112–130.

[65] Keilson, J. and Storer, J. E. (1952) On Brownian motion, Boltzmann's equation and the Fokker–Planck equation. *Quart. Appl. Math.* 10, 243–253.

[66] Keilson, J. and Styan, G. P. H. (1973) Markov chains and M-matrices: Inequalities and equalities. *J. Math. Anal. Appl.* 2, 439–459.

[67] Keilson, J. and Subba Rao, S. (1970) A process with chain dependent growth rate. *J. Appl. Prob.* 7, 699–711.

[68] Keilson, J. and Subba Rao, S. (1971) A process with chain dependent growth rate—Part II, The ruin and ergodic problem. *Adv. Appl. Prob.* 3, 315–338.

[69] Keilson, J. and Sumita, U. (1982) Waiting time distribution response to traffic surges via the Laguerre transform. In *Applied Probability—Computer Science, The Interface* 2, Birkhauser, Boston.

[70] Keilson, J. and Sumita, U. (1982) Uniform stochastic ordering and related inequalities. *Canad. J. Statist.* 10, 181–198.

[71] Keilson, J. and Sumita, U. (1983) Extrapolation of the mean lifetime of a large population from its preliminary survival history. *Naval Res. Logist. Quart.* 30, 509–535.

[72] Keilson, J. and Sumita, U. (1983) A decomposition of the beta-distribution, related order and aymptotic behavior. *Ann. Inst. Statist. Math.* A 35, 243–253.

[73] Keilson, J. and Sumita, U. (1983) The depletion time for $M/G/1$ systems and a related limit theorem. *Adv. Appl. Prob.* 15, 420–443.

[74] Keilson, J. and Sumita, U. (1983) Evaluation of the total time in system in a preempt/resume priority queue via a modified Lindley process. *Adv. Appl. Prob.* 15, 840–856.

[75] Keilson, J. and Sumita, U. (1986) A general Laguerre transform and a related distance between probability measures. *J. Math. Anal. Appl.* 113, 288–308.

[76] Keilson, J. and Sumita, U. (1983) A time-dependent model of a two echelon system with quick repair. Graduate School of Management, University of Rochester, Working Paper Series No. QM 8311.

[77] Keilson, J. and Syski, R. (1974) Compensation measures in the theory of Markov chains. *Stoch. Proc. Appl.* 2, 59–72.

[78] Keilson, J. and Vasicek, O. (1975) A family of monotone measures of ergodicity for Markov chains. Center for System Science 75–03, Graduate School of Management, University of Rochester.

[79] Keilson, J. and Waterhouse, C. (1972) Transfer times across the human body. *Bull. Math. Phys.* 34, 33–44.

[80] Keilson, J. and Wellner, J. (1978) Oscillating Brownian motion. *J. Appl. Prob.* 15, 300–310.

[81] Keilson, J. and Wishart, D. M. G. (1964) A central limit theorem for additive processes defined on a finite Markov chain. *Proc. Camb. Phil. Soc.* 60, 657–667.

[82] Keilson, J. and Wishart, D. M. G. (1965) Boundary problems for additive processes defined on a finite Markov chain. *Proc. Camb. Phil. Soc.* 61, 173–190.

[83] Keilson, J. and Wishart, D. M. G. (1967) Addenda to processes on a finite Markov chain. *Proc. Camb. Phil. Soc.* 63, 187–193.

[84] Keilson, J. and Zachmann, M. (1985) Homogeneous row-continuous bivariate Markov chains with boundaries. *Math. Operat. Res.* To appear.

[85] Keilson, J., Cozzolino, J. and Young, H. (1968) A service system with unfilled requests repeated. *Operat. Res.* 16, 1126–1137.

[86] Keilson, J., Kester, A. and Waterhouse, C. (1978) A circulatory model for human metabolism. *J. Theoret. Biol.* 74, 535–547.

[87] Keilson, J., Nunn, W. and Sumita, U. (1981) The bilateral Laguerre transform. *Appl. Math. Comput.* 8, 137–174.

[88] Keilson, J., Sumita, U. and Zachmann, M. (1981) Row-continuous finite Markov chains, structure and algorithms. Graduate School of Management, University of Rochester, Working Paper Series No. 8115.

[89] Keilson, J., Petrondas, D., Sumita, U. and Wellner, J. (1983) Significance points for some tests of uniformity on the sphere. *J. Statist. Comput. Simul.* 17, 195–218.

[90] Keilson, J., Machihara, F. and Sumita, U. (1985) Spectral structure of $M/G/1$

systems–Asymptotic behavior and relaxation times. Graduate School of Management, University of Rochester, Working Paper Series No. 8414.

[91] Kingman, J. F. C. (1965) The heavy traffic approximation in the theory of queues. Discussion by J. Keilson and J. Th. Runnenburg, in *Proc. Symp. Congestion Theory*, University of North Carolina Press, Chapel Hill.

[92] Ott, T. (1975) Infinite Divisibility, Imbeddability and Stability of Finite Semi-Markov Matrices. Doctoral Dissertation, Graduate School of Management, University of Rochester.

[93] Prabhu, N. U. (1982) Conferences on Stochastic Processes and their Applications: a brief history. *Stoch. Proc. Appl.* 12, 115–116.

[94] Wang, M. C. and Uhlenbeck, G. E. (1945) On the theory of Brownian motion II. *Rev. Mod. Phys.* 17, 323–342.

[95] Wax, N. (1954) *Selected Papers on Noise and Stochastic Processes*. Dover, New York.

[96] Widder, D. V. (1946) *The Laplace Transform*. Princeton University Press, Princeton, NJ.

[97] Wishart, D. M. G. (1960) Queueing systems in which the discipline is 'last-come first-served'. *Operat. Res.* 8, 591–599.

Peter Whittle

Peter Whittle was born on 27 February 1927 in Wellington, New Zealand. After graduating from the University of New Zealand with a B.Sc. in mathematics and physics in 1947 and an M.Sc. in mathematics in 1948, he joined the University of Uppsala, Sweden, for his postgraduate studies. He obtained his doctorate of philosophy there under Professor H. Wold in 1951, and held the post of docent at the university's Institute of Statistics for two years. In 1953 he returned to the Department of Industrial and Scientific Research in Wellington, New Zealand, and progressed to become Senior Principal Scientific Officer by 1958.

In 1959 he was appointed university lecturer in the Statistical Laboratory at the University of Cambridge, England, and in 1961 took up the chair of mathematical statistics at the University of Manchester. He returned to the Churchill Chair of the Mathematics of Operational Research at Cambridge in 1967 and has been director of the university's Statistical Laboratory since 1973. Among his many honours are membership of the International Statistical Institute (1956), the Guy Silver Medal awarded by the Royal Statistical Society (1966), fellowship of the Royal Society (1978) and honorary membership of the Royal Society of New Zealand (1981).

Professor Whittle married Kathe Blomquist in 1951 and they have six children. His hobbies include guitar, jogging and the growing of an ash coppice.

In the Late Afternoon

Modesty, together with an awareness of the quizzical reader, inclines me to write an article less autobiographical than has generously been invited. However, it is true, I realize, that the autobiographical form serves well as a thread upon which to string one's thoughts, observations and prejudices.

1. Early Training

My own interest in probability was awoken during vacation work in the New Zealand Department of Scientific and Industrial Research (DSIR) in the late 1940s. We had statistical problems aplenty, of a particularly enlightened character because they were provided by other scientists, and also because Ian Dick, the director of the Biometrics Section (as it then was), was determined that it should become an elite group. Those who know, think that he succeeded. There was a particular interest at that time and place in the work of the Scandinavians in stochastic processes and time series analysis: Arley, Jensen, Lundberg, Cramér, Wold, Karhunen. It was these works and Maurice Bartlett's papers which brought home to me the particular fascination of the combination of dynamic and probabilistic concepts. I had by now, for a mixture of reasons, decided to do my Ph.D. work in Sweden, and I learned Swedish and spatial processes simultaneously by translating Bertil Matern's work on sampling surveys in forestry [4].

My four years in Sweden (1949–53), with Herman Wold, were fascinating and enriching. Herman's interest had now turned away from time series analysis to econometrics, but he spared no effort to help me. I owe him gratitude for many particular acts of kindness, but also for introducing me to a much more scholarly world than any I had known, a world in which intense ambition was moderated by courtesy, fairness and the scrupulous recognition of credit.

Those were still the days when the Swedish Ph.D. oral was a formal public defence, in white tie and tails. My "first opponent" was Ulf Grenander, formidable in his knowledge of the literature. The "second opponent" was none other than Bertil Matern, by now a friend who added to his amiable, gentlemanly nature a wry humour which is much commoner among his countrymen than is supposed outside Sweden. The three hours of the "defence" were the shortest of my life; the quarter-hours chiming in the corridor seemed to mark rounds of no more than the usual length.

I continued to work in time series analysis over the period 1949–54, [5]–[11], and think I may claim to have constructed the asymptotic likelihood theory for the stationary Gaussian process, essentially if unrigorously. This work was done too early (and, perhaps in unconscious emulation of the admired Bartlett, too gnomically) to reap its due recognition. Strangely, the analogous theory for spatial processes which I published in 1954 [12] has been better recognized. The two MacGonagals of time series analysis, whose motto was "if we understand it, it's ours", took up the time series ideas with much vigour, partial comprehension and no acknowledgement some 15 years later.

One of the very pleasant events during my years in Uppsala was Maurice Bartlett's attendance there at a meeting on factor analysis which Herman Wold had arranged. This was actually the first time I had met Maurice, and I have written of my first impression elsewhere ([27], p. 217): "I recall plainly my first glimpse of him on a drizzly railway platform: an unexpectedly bulky, ruddy and cheery figure, almost farmer-like, with a hat whose floppy brim was etched by the several chromatograms of successive seasons." Posterity will evaluate Maurice as he has always been evaluated in the United Kingdom: as someone ahead of his time in his unique blend of stochastic/statistical/ physical thinking. The many (then) young people who were profoundly influenced by him are in no doubt of that.

2. Theory and Practice in DSIR

In 1953 I returned to New Zealand and to DSIR. The subsequent six years were to prove deeply formative. I continued with time series work for a while, and local applications proved interesting. An analysis of some oceanographic data (from rock channels near my own boyhood home) demonstrated the existence of two interpretable seiches, one much more dissipative than the other, and analysis of evidence of some nonlinear effects produced the first treatment of a threshold model [13]. Shame to say, this was also the completest analysis I ever made. Time series analysts have a particularly low calculation/theorizing ratio—perhaps more excusable in those days, when computation of a single correlogram took the punched-card machines several days.

The other quite thorough analysis was of fluctuations in the numbers of New Zealand rabbits—pests of the first order [14]. This came about through the long and fruitful contact between our group and the Animal Ecology Section of DSIR under Dr Wodzicki—formerly Count Casimir and Polish Consul to New Zealand. The model was necessarily nonlinear (host–parasite interactions being a feature) which made analysis less clean but the continuing challenge of such models all the stronger. I have always found nonlinear effects fascinating, but have never been able to do much about them. However, when Norman Bailey [1] wrote of his stochastic epidemic models in 1953 that he could not discern the stochastic equivalent of a threshold theorem, I was delighted to observe just that in the split of his calculated distributions of total infecteds into J-shaped and U-shaped curves. I formalized the statement of a threshold theorem in a paper [15] held up for a year while the referee wrote his contribution.

The time-series work then led into the study and statistical analysis of spatial models, material published in 1954 [12]. I was puzzled, as was everybody at that stage, by the absence of a definite "direction of conditioning", although I still believe this best resolved by study of the full spatio-temporal process. The data I used was taken from Fairfield Smith's collection of uniformity trial data, which I used again [17] to show that, for these data, the spatial autocorrelation functions behaved as inverse powers, $s^{-\lambda}$, in their tails, and that s^{-1} behaviour occurred definitely and dominantly. A spatio-temporal model explained the s^{-1} law. There it was left: fractals and self-similar processes were as yet unthought of.

However, quite the most significant set of events for me was to encounter, in succession, a number of what one might term "statistical/physical" problems. These were fascinating in their nature, and in their difficulty first challenging and then heartbreakingly frustrating. I remember feeling that I had acquired a burden of problems that I could neither solve nor relinquish, and that would neutralize me for life. Indeed, they did stay with me, but the long-term effect was more beneficial than I thought at the time.

Somehow, and I don't now recall how, I got on to the problem of analysing what was essentially a Markov field on a spatial lattice. I was trying to calculate the partition function, and saw this as the calculation of the maximal "eigenvalue" for a multidimensional matrix with naturally defined multi-dimensional multiplication rules. I ran into all the usual problems, and was, perhaps fortunately, too isolated and immature to realize the tremendous effort that physicists had already put into such problems. The only thing that came of it for me was my derivation of the exact joint distribution of the transition totals in a sample from a Markov chain [16]. This was my first experience of hitting J (I won't say B) in trying to hit A—there was to be another.

Frank Evison of the Geophysics Division then came to me with a problem which he thought would be a back-of-an-envelope matter for a statistician: the deduction of the effective mechanical properties of rock with randomly varying structure. The rock in question lay beneath one of New Zealand's many hydro-electric dams, was volcanic, and flecked throughout with inclusions of weaker material. That started me thinking, as early as 1955, about random partial differential equations and products of random matrices. A similar problem in the same vein that came along was the observation that trees whose bark was attacked by boring insects died when the density of holes exceeded a fairly definite critical value. The general notion was that sap rose through the cortex, and that the flow was sufficient for life as long as damage to the cortex did not exceed a critical value, but fell right away above that value. This was a problem that could be seen either as a matter of conduction in a randomly varying medium, or as what was later to be known as a percolation process. I tried manfully to solve the problem, but to no great effect.

A further problem of this type presented itself, again in 1955. This was the question of critical effects in polymerization, and came out of Hamish Thompson's interest in the agglutination reaction for viruses and antibodies. This seemed as frustrating as any of the others, but in fact was to stay with me and prove tractable some years later [20], [21].

The year 1955 seems to have been singular; in that year I also first became interested in reversibility—just why I cannot recall. This interest led to a paper whose sad fate I have already recounted elsewhere ([27], p. 218). In it I demonstrated that birth and death processes were reversible, showed that the third-order Kolmogorov conditions were sufficient for reversibility under certain conditions and propounded the idea of dynamic reversibility. In any case, reversibility became another bug acquired for life.

3. New Horizons

In 1957 I spent a very happy year at Canberra in a group which made up in quality what it lacked in quantity. Pat Moran, Ted Hannan and Geof Watson then started for me what has always been the very happy and stimulating Australian connection. I continued to think of many things, the abortive statistical/physical problems, curve smoothing (Bayesian), Tchebichev inequalities for vectors and processes, and, still, time series. The time series work was taking a new turn, towards prediction and control. Thus I acquired yet *another* bug—optimization.

Optimality notions have always been much stronger in statistics than in probability, for obvious reasons. They are explicit in prediction theory and also in stochastic control. I saw that the Wiener optimal filtration techniques for prediction could indeed be transferred to control. Others (Newton/Kaiser/Gould and Holt/Modigliani/Simon/Muth) had in fact already done work in this line, but I had my own approach in what were then very early days. I published my book on the subject (*Prediction and Regulation*, [18]) in 1963. The treatment was totally by use of generating functions in the stationary context. As Fate would have it, Kalman's work appeared at just the same time, using recursive techniques in the Markov context. Both approaches have their place, but Kalman's work was quickly appreciated, whereas interest in my own was kept alive largely by a small group of economists who had reason to prefer the first approach.

In 1959 I came to Britain to join the Statistical Laboratory at Cambridge, and since then it has been all go, as they say. I found Britain a kindly and lively country, which has treated me well. The years since 1959 seem to have gone so fast that I also feel inclined to treat them more rapidly than the earlier years in New Zealand and Sweden, during which so much was laid down. Or, perhaps it is more correct to say that I view the post-1959 years as later chapters to which I could not do justice in the same article.

My stay in Cambridge was ended in 1961 by a distinction of which I still feel unworthy: that of becoming Maurice Bartlett's successor in the Statistical Laboratory at Manchester University. Maurice's powerful mind and kindly presence seemed to linger on in that department. For department, university and city *as they were* I have an abiding affection, shared, I believe, by the motley band of Antipodeans, Greeks and Moslems in the department, who often outnumbered the uncomplaining natives.

There were two pieces of work which pleased me most in this period, appearing in 1963–4. One concerned problems in sequential analysis and design [19], [22]: I showed that the minimal cost functions could be regarded as homogenous functions of an unnormalized distribution (the posterior distribution of the unknown state variable, in this case "the true hypothesis"). This enormously simplified the dynamic programming equation, and generated a train of ideas and partial results which remain unexploited to this day. The other piece of work was a partial cracking of the polymerization problem, reversibility arguments playing a very central role [20], [21].

My interest in optimization was developing. About 1963 I signed a contract with the publisher John Wiley for a work on dynamic programming; a contract I was not to honour until 1981, as it turned out [32], [33]. I also kept up an interest in spatial processes. People were still arguing whether these should be defined by specification of joint or of conditional probabilities. A research student, David Brook, produced completely on his own a simple result [2] which, it has only recently been generally recognized, cut right through much of the complicated analysis of this issue that was going on at the time.

The year 1964 must have been, like 1955, a seminal one. In ignorance of Jackson's work I essentially obtained Jackson's results on networks of queues. I published a couple of years later [23], [24], and discovered the preemption yet later again. However, there was one novelty in my analysis: the observation of partial balance, which was to load me with yet another bug for life.

It was not long after this that Joe Gani came to the Sheffield chair, bringing the generous energy and vision that he had already and has since demonstrated in so many parts of the world. He developed a fine department, which was indeed able to buoy up the Manchester department when this later passed through a bad phase. However, the act of truly general significance was his founding of the applied probability journals, still Sheffield-based, which have proved both so needed and so catalytic.

4. The Cambridge years

My increasing interest in optimization made it natural for me to accept the honour of appointment to the Churchill Chair in the Mathematics of Operational Research at Cambridge, when it was offered to me in 1966. The succeeding period represents yet another new and major phase which, essentially, has to be another story. I have always been attracted by the appeal to simple, unifying principles, and in 1970 I published my text on probability, axiomatizing the notion of expectation [25]. For reasons explained there I am convinced that this is the natural approach, so convinced that I prefer to state the case and then not argue it. That the text has been translated into Russian pleases me enormously. In 1971 I wrote a text on optimization [26], more to systematize my own thinking than anything else. Again I appealed to a simple principle, the geometric view of Lagrangian theory. The view was by no means novel, but was novel to me, and I wrote for myself . . . .

Again, I believe that I summarize the succeeding years best by describing the pieces of work that pleased me most. These are three in number and were concentrated in 1979–80, a period which provided another of those surges.

One was the work on the multi-armed bandit problem. John Gittins had been working alone on this problem in his own special way since the 1960s, and spoke to me about his ideas in 1969–70. My attitude to these shifted in a curious way, through several phases. Initially, I could not believe that the problem, as he described it (the rationalization of research programming in the chemical industry) was sensible, and I became indeed quite impatient and discouraging. Then I realized that he really had a problem. Finally, I realized that he also had a solution, a solution to this classic problem of marvellous subtlety and intuition. However, if he had a proof, it was not one that was either easy or, to me, comprehensible. For a long time I tried to find an alternative view and proof. The idea, when it came, was in fact triggered again by John himself. Specifically, in his 1979 Royal Statistical Society paper [3], John described the "forward induction" characterization of his index. When later that year I tried to see how this would be calculated in simple cases, I came quite suddenly on the new approach described in my 1980 paper in the *Journal of the Royal Statistical Society* [28]. I think it can be said that this paper played a critical role: it was small enough, sharp enough and different enough to breach the dam of indifference which had for more than a decade blocked recognition of John's magnificent insight.

The second advance which pleased me was a leap forward in my own work on polymers [29], [30] which brought me much nearer to being able to treat the kind of models I had always regarded as "ultimate": models of internal adaptation and self-organization in systems.

The third piece of work [31] concerned what I have referred to as "risk-sensitive LQG control". This is a question of replacing the quadratic cost criterion of familiar linear/quadratic/Gaussian treatments by the exponential of a quadratic. It is an interesting modification in that one can thereby incorporate a degree of optimism or pessimism on the part of the optimizer into the formulation. Contributions had been made to the subject, notably by Jacobson, but the general case with imperfect state observation resisted analysis. I arrived at an unobvious but in fact "right" generalization of the certainty equivalence principle which solved the problem. However, I mention this mainly because it was again a case of aiming at A and hitting something in between G and M, perhaps. I had in fact been trying to make progress on the problem of adaptive control—of control when one has simultaneously to estimate parameters, and control actions can yield information as well as directly affect the system. I thought I had solved a version of the problem but I had not: I had solved the risk-sensitive LQG problem.

5. Some Thoughts on Probability

If I am now asked what general thoughts occur to me I should perhaps be brief, in the trust that these will not be last thoughts.

In probability, as in anything else, there are established "industries", proper and honoured in their time, but surviving in heavy form for no very good reason. I could mention the study of sums of independent identically distributed random variables, and the measure-theoretic embalming industry. In my view one of the most welcome developments of the last 10 or 20 years has been the fading of these in favour of an interest in the stochastic processes of physics at a reasonably fundamental level: e.g. random fields, systems of interacting particles, and percolation processes. Before that, these matters were left to a few gifted and independent souls: Bartlett, Hammersley and Kac spring to mind. For the rest, one found a thin interest in certain easily idealized examples (e.g. Geiger counters, cascade processes) and a number of self-appointed frontiersmen, but the genuine mandarins of probability theory were certainly happy to leave these regions to the barbarians.

Was this attitude prevalent because so often, in the educational system, a decision to follow probability implies a decision to drop physics and the like? Certainly nobody but a probabilist of a certain type would refer to a spatial process as a process in "multidimensional time"!

There is also now a welcome fluidity in the professional boundaries. There are so many subjects of great practical and intellectual importance for which both the stochastic model and the statistical analysis are difficult to formulate. Not surprisingly, these are cultivated by people who are neither probabilists nor statisticians—partly because these are the people who see the end however lost they may be for the means, and partly because the average probabilist or statistician is exceedingly conventional in his views. Areas that I have in mind are, for instance, stereology, tomography, image reconstruction, pattern recognition, expert systems and the latent variable models of economics. Vigilance and flexibility are the watchwords if probabilists and statisticians are not to end up as the camp-followers in these areas.

Where one sees that probability is going or should go is a matter coloured very much by one's personal tastes and determined very much by one's own perception. I myself see the interesting and challenging area as that of large systems: their functioning, their reliability, their economy, their internal adaptation, their evolution, and their self-organization. Russian interest in these matters has always been strong, for reasons of temperament, scale and structure that one can understand. I cannot see an area more challenging or more fundamental. It has been said, by those who have attempted to study the functioning of neural nets by experiment and dissection, that understanding will in fact never come that way, but only by a mathematical understanding of what phenomena are *possible*. To follow this route requires an appreciation of stochastic effects, critical effects, and of optimality (because of the reconciliation of required performance and of physical constraints), but optimality in some relaxed, asymptotic sense, perhaps even in some strange open-ended sense. Then there will be techniques required that we cannot yet conceive: techniques to determine how a large system of unreliable elements can perform reliably, how it can advance to a new level of self-organization, whether there is another level beyond that ....

Lastly, I would name two personal qualities, which I recognize in myself and which I think have been useful, although some will rightly say that they have their negative obverses. One is laziness. If a piece of work is heavy and complicated then it is wrong. In calculations one will of course persist for quite a way with mounting complication and obscurity, but there has to be a resolution, a simplification at some point if one is on the right path. Of course, the simplicity may be "simplicity" in a subtler sense than one would before have understood. This raises the second aspect, which is almost stochastic in character. One has to keep a particular openness of mind. Solving a problem is like going to a strange place, not to subdue it, but simply to spend time there, to preserve one's openness, to wait for the signals, to wait for the strangeness to dissolve into sense.

References

[1] Bailey, N. T. J. (1953) The total size of a general stochastic epidemic. *Biometrika* 40, 177–185.

[2] Brook, D. (1964) On the distinction between the conditional and the joint probability approaches in the specification of nearest-neighbour systems. *Biometrika* 51, 481–483.

[3] Gittins, J. C. (1979) Bandit processes and dynamic allocation indices. *J. R. Statist. Soc.* B 41, 148–164.

[4] Matern, B. (1947) Metoder att upskatta noggrannheten vid linje- och provytetaxering. *Meddelanden från Statens Skogsforskningsinstitut* 36, 1–138.

[5] Whittle, P. (1951) *Hypothesis Testing in Time Series Analysis*. Thesis, Uppsala University. Almqvist and Wiksell, Uppsala.

[6] Whittle, P. (1952) Some results in time series analysis. *Skand. Aktuarietidskr.* 35, 48–60.

[7] Whittle, P. (1952) The simultaneous estimation of a time series harmonic components and covariance structure. *Trab. Estadist.* 3, 43–57.

[8] Whittle, P. (1952) Tests of fit in time series. *Biometrika* 39, 309–318.

[9] Whittle, P. (1953) The analysis of multiple stationary time series. *J. R. Statist. Soc.* B 15, 125–139.

[10] Whittle, P. (1953) Estimation and information in stationary time series. *Ark. Mat.* 2, 423–434.

[11] Whittle, P. (1954) Some recent contributions to the theory of stationary processes. Appendix (pp. 196–228) to [34].

[12] Whittle, P. (1954) On stationary processes in the plane. *Biometrika* 41, 434–449.

[13] Whittle, P. (1954) The statistical analysis of a seiche record. *J. Marine Res.* 13, 76–100.

[14] Whittle, P. (1955) An investigation of periodic fluctuations in the N. Z. rabbit population. *N. Z. J. Sci. Technol.* B 37, 179–200.

[15] Whittle, P. (1955) The outcome of a stochastic epidemic—a note on Bailey's paper. *Biometrika* 42, 116–122.

[16] Whittle, P. (1955) Some distribution and moment formulae for the Markov chain. *J. R. Statist. Soc.* B 17, 235–242.

[17] Whittle, P. (1962) Topographic correlation, power-law covariance functions, and diffusion. *Biometrika* 49, 305–314.

[18] Whittle, P. (1963) *Prediction and Regulation by Linear Least-Squares Methods*. English Universities Press, London.

[19] Whittle, P. (1964) Some general results in sequential analysis. *Biometrika* 51, 123–141.

[20] Whittle, P. (1965) Statistical processes of aggregation and polymerization. *Proc. Camb. Phil. Soc.* 61, 475–495.

[21] Whittle, P. (1965) The equilibrium statistics of a clustering process in the uncondensed phase. *Proc. R. Soc. London* A 285, 501–519.

[22] Whittle, P. (1965) Some general results in sequential design. *J. R. Statist. Soc.* B 27, 371–394.

[23] Whittle, P. (1968) Equilibrium distributions for an open migration process. *J. Appl. Prob.* 5, 567–571.

[24] Whittle, P. (1968) Nonlinear migration processes. *Proc. 36th Session ISI*, 642–646.

[25] Whittle, P. (1970) *Probability*. Penguin, London.

[26] Whittle, P. (1971) *Optimization under Constraints*. Wiley, London.

[27] Whittle, P. (1975) Reversibility and acyclicity. In *Perspectives in Probability and Statistics*, ed. J. Gani, Applied Probability Trust, Sheffield, 217–224.

[28] Whittle, P. (1980) Multi-armed bandits and the Gittins index. *J. R. Statist. Soc.* B 42, 143–149.

[29] Whittle, P. (1980) Polymerization processes with intrapolymer bonding, Parts I–III. *Adv. Appl. Prob.* 12, 94–153.

[30] Whittle, P. (1981) A direct derivation of the equilibrium distribution for a polymerisation process. *Teor. Verojatnost. i Primenen.* 26, 350–361.

[31] Whittle, P. (1981) Risk-sensitive linear/quadratic/Gaussian control. *Adv. Appl. Prob.* 13, 764–777.

[32] Whittle, P. (1982) *Optimization over Time*, Vol. 1. Wiley, London.

[33] Whittle, P. (1983) *Optimization over Time*, Vol. 2. Wiley, London.

[34] Wold, H. (1954) *A Study in the Analysis of Stationary Time Series*, 2nd edn. Almqvist and Wiksell, Uppsala.

Ralph L. Disney

Ralph L. Disney, born on 27 February 1928 in Baltimore, Maryland, is a scholar of international standing in queueing theory, queueing networks, and random processes in queueing. He is currently Charles O. Gordon Professor in Industrial Engineering and Operations Research at Virginia Polytechnic Institute and State University, a post to which he was appointed in 1977.

He has been president of the Applied Probability Technical Section/College of the Operations Research Society of America (ORSA)/The Institute of Management Sciences (TIMS). He has served on the council of ORSA and been division editor for probability and random processes for the TIMS publication *Management Sciences*.

He was senior editor for the American Institute of Industrial Engineers (AIIE) for nearly 10 years and is a fellow of that society. He is the author or co-author of 60–70 research and review papers and one book; another book appeared in 1985 and a third is due in 1986. He was the co-editor of the 1982 *Proceedings of the Symposium on Applied Probability and Computer Science: The Interface*.

Disney has been OAS Visiting Professor at the Instituto Tecnológico de Aeronáutica in São José dos Campos, Brazil, and Graduate School Distinguished Visiting Professor at Ohio State University. Most recently (1984) he was Visiting Professor at the Instituto de Matemática e Estatística at the University of São Paulo, Brazil. He has given numerous short courses and invited seminars on random processes in queueing theory, and in 1982 was the leading speaker at the symposium sponsored by ONR and the Department of Mathematical Sciences of The Johns Hopkins University.

He has received many honors for his teaching and research, including the David Baker Award of the AIIE. He is a member of most of the professional societies relevant to his work in applied probability, and has served on many of their national and international committees.

The Making of a Queueing Theorist

1. What is a Probabilistic Modeller?

What is an "applied probabilist" or a "probabilistic modeller"? How does one become one? There used to be a cartoon that hung on the walls of many engineering schools. On it was the face of Alfred E. Newman. Above the face were the words. "For years ago I couldn't even spel ingineer", and under it, "Now, I are one". Despite the caricature there is a not-so-obvious message: one is hard pressed to explain whatever it is one is, and how one has become it. Often whatever one is, is not quite what one sees oneself as, and this in turn is not what others see.

I remember a meeting of a committee of mathematicians of which I had been invited to be a member. The other members of the committee had all been formally educated as pure mathematicians, though only one was working in that field. The others worked as applied mathematicians. At one of our meetings where we had to stay overnight, we engaged the hotel clerk in a conversation. He asked if we were all mathematicians. The response of my colleagues was interesting. "No", one said. Then, pointing to me, "He's only an engineer." So maybe I are one. But I've been labelled a computer scientist, a mathematician, an operations researcher, a probabilist, an applied probabilist, an engineer, and some things I cannot repeat in print. In the following account, the reader will see, more or less, where I came from and what I have done. Then the semanticists can decide what I am.

2. A Fateful Meeting

In November 1948, I dropped out of college for many reasons. Over the next two years I occupied myself with a variety of jobs, all supposedly full-time. None was. During that time I met Mrs Bratten, whose first name I do not remember. She was the mother of Tom, one of my high-school friends. She spent many long hours pushing, shoving, cajoling, pleading with me to return to school; she finally wore down my resistance. I don't think I ever thanked her.

In September 1950, I had an appointment with Robert H. Roy, the Dean of Engineering at The Johns Hopkins University in my home town of Baltimore. We were to talk about my re-entering the university after a two-year absence. Roy was late for our appointment, and after waiting 15 minutes I thought, "The hell with it. He won't let me back in anyhow." After all in my first two years I had not shone as a student. My 1.9 grade point average (out of 4.0) proved that.

For years I rationalized my low grade point average as being due to my view of myself as a "hot shot" athlete. I had come out of high school having played on ten varsity teams in football, basketball, and baseball. I found the 154-hour engineering curriculum, which was common then, to be a taxing load to go along with a varsity sports schedule. But, while there are many explanations for that grade point average, it was not due to the added burden of sports. Too many of my colleagues and too many other students have managed to combine sports and engineering education and have done well in both.

Finding me headed for the bus home, after leaving Roy's office, a high-school friend took me back, and the Dean and I talked for a few moments, perhaps the most important few moments in my life. Not only was I allowed to return to school without penalty, but Roy also offered to help pay the seemingly great tuition cost of $1000 a year through a scholarship. To this day I do not know why he did that. But for his generosity, I have no idea what I would be doing today. Most likely I would not be writing a paper on the making of a probabilistic modeller.

3. Education

Being a student again after a two-year lay-off was hard work. Not only had I forgotten a great deal, but I realized that I had not known much to start with. The major problem was to retool myself in mathematics. The pure mathematics of the Hopkins mathematics department was not within my ability or interest at that time so I re-enrolled in the Engineering School hoping to be educated in applied mathematics. Out of respect for Rob Roy, I enrolled in industrial engineering.

Acheson Duncan was a member of the small, three-man faculty of industrial engineering when I returned. I started my statistics work with him. For more than four years he was my statistics mentor. He engaged me in several of his statistics projects, including the writing of the solution manual to the first edition of his widely acclaimed text.

It was Duncan who strongly encouraged me to continue my education in statistics at the graduate level. He suggested several universities that I should attend, but going away from home was not financially feasible. Support for graduate education was not then what it later became, so I stayed on at Hopkins for graduate work in industrial engineering.

One must understand two aspects of The Johns Hopkins University as it existed then, in order to understand some of what follows. First, the university is a private school founded originally on the classical European approach to education. There was a plaque on the wall dedicated to one of the professors which stated "Teacher, scholar, leader of men". The word "researcher" was omitted; but at Hopkins everyone was a researcher, so that mention of research was no special accolade. To be a scholar was more.

Second, as part of the Hopkins philosophy, especially as espoused by Rob Roy, graduate programs had little formal structure and few required courses.

One took those courses which appeared necessary to becoming a scholar in one's chosen field.

During my graduate studies I knew I wanted to do applied mathematics. What I did not know was what I wanted to apply mathematics to. Through formal course work I tried macro-economic theory, mathematical psychology, sociology, operations research, and statistics. I even tried to go outside the usual mathematics-based fields into labor history (where I became a member of the honorary Tudor and Stuart Club for history and English majors). Eventually, I settled into the field of operations research.

As was true of many of my colleagues, I had to work my way through college. I took a job in a private tutoring school. The purposes of the school were to prepare prospective college-bound high-school students for college entrance examinations, and to tutor students having difficulty with their school work. My specific areas were mathematics, physics, and chemistry. Among my students were returning service people, high-school drop-outs, a night-club entertainer and a student who was learning Euclid and Ptolemy from Latin texts. These were the people who taught me how to teach and enjoy it.

Near the end of my second or third year of graduate school Ach Duncan encouraged me to attend a course on linear statistical models given by Bill Cochran at The Johns Hopkins Hospital. The high point of the course, for me, occurred in the second term when Paul Meyer taught a course in probability from the first volume of Feller's book. (The second volume hadn't then been written.) For the first time I began to understand the "why" of the statistics course I had taken. As a result, I edged toward the field of probability though it took several more steps before I landed there.

A thought occurs to me as I look back on my formal education in statistics. As a very young child I was often left in the care of one of my grandmothers because both of my parents were working people. My grandmother scoured the neighborhood to fill a drawer with clocks for me to take apart. I spent hours taking apart old clocks, not to become a watchmaker but simply to learn how they worked. Whether that experience was formative or indicative of my formation, I do not know. It is and has been indicative of my thought process in research and learning.

4. Teaching and Applied Research

In 1955, I left Hopkins as an "all but dissertation" Ph.D. to work for the Operations Research Office (ORO) in Washington, DC. My new wife, Lois, whom I met when we were both graduate students at Hopkins, had been accepted as the first female student in industrial engineering at Hopkins, in 1954. She had been sent by Dean Roy to ORO to seek a job as research analyst to support herself while seeking her master's degree. The relation between Hopkins and ORO was close, and she was taking part of her course work at ORO. This educational aspect of ORO appealed to me also. Since it was forbidden for two people in the same family to work for that company, I took the job away from her.

ORO was a good place to start a career as an applied probabilist. Many excellent people were working there on some hard "real" problems. My group had the job of finding ways of reducing the variability of the army's stock of sized clothing. The problem was immense, as there were inventories of thousands of different items of many different sizes, spread geographically throughout the world. It would have been hopeless to get at the problem without modelling, and the models had to be probability models.

I learned for the first time that such modelling is not a one-person activity. In that respect it is quite different from the research I had been accustomed to in college. The scope alone of our inventory problem guaranteed that one had to know many things, not just probability. I quickly learned how essential team research was, and how necessary it was to learn new things rapidly, to interact with other researchers, to have a free, unencumbered flow of information, and to carry on a continual dialogue with others working on the same problem. That experience and ensuing ones have conditioned much of my approach to research and learning.

I think I also learned that there is life after college, and that the minimum one can hope for from a college education is a basis for learning. This thought has conditioned my approach to teaching for nearly 30 years.

By June 1956 my desire to teach had become overwhelming, and I accepted a position at what is now Lamar University in Beaumont, Texas. We moved there with our first daughter. Here I further strengthened my background. I taught in the Department of Industrial Engineering as well as the Departments of Mechanical Engineering, Electrical Engineering, and Mathematics. I was most excited by the mechanical engineering course. It was called "Similitude", and consisted of building mathematical models of mechanical systems, finding electrical analogs and then programming them on an analog computer. To this day, I enjoy finding analogies between seemingly diverse fields. I have sought out these analogies by spending a considerable amount of my time talking with researchers and reading in diverse fields.

In 1959 I moved to the University of Buffalo, now the State University of New York at Buffalo, with my wife and now two daughters, and took on a consulting job with a conveyor manufacturer whose problems occurred at the interface of design and operations. Conveyors designed for specific purposes did not appear to operate as expected. Designs were based on an assumption of deterministic flow of material on the conveyor, but operations encountered the usual work stoppage, waiting caused by breakdowns, shortages, or production delays. The operating systems were stochastic, and the designs did not take this into account.

We provided the designers with a collection of Monte Carlo simulations and taught them to use these simulations to aid their design practices with "what if" scenarios. This was my introduction to queueing and queueing network theory. With this work I became firmly and permanently an applied probabilist.

In the initial phase of the conveyor study I tried to build analytic queueing models, with no great success. There was scant literature on queueing networks in 1959, the 1957 paper by J. R. Jackson [5] being about the only one. My curiosity was piqued to try to find out how queueing networks worked. Naive optimism led me to lay out a three-year program of basic research into queueing network theory. That research has been on going for 25 years and though I am much further along in my understanding, I have not yet finished my program.

In 1962 I was invited to the University of Michigan as a visiting lecturer. We moved there with two daughters and now a son. I stayed on for 15 years, eventually becoming full professor. In 1964 I finally completed my dissertation work and graduated from Johns Hopkins with a Doctor of Engineering degree. All told, I was enrolled, though not in attendance, at Johns Hopkins for 18 years, surely among the longest-tenured students in that school's history.

There is a story I enjoy telling about our frequent trips from Buffalo to Orchard Lake, Michigan, where Lois's parents lived. Our usual route was across Ontario province in Canada. At one of our crossings, the Canadian customs official asked us the usual question, "Where were you and your family born?" My answer was: I was born in Baltimore, Lois was born in Pontiac, Michigan, Lynn our oldest daughter was born in Bethesda, Maryland, Carrie our younger daughter was born in Beaumont, Texas and Edward our son was born in Buffalo, New York. Upon hearing this the customs official asked, with a startled expression on his face, "You all in the Air Force?". I am not sure that he would have believed the true story of our treks.

My degree was completed with A. B. Clarke as my advisor. Bruce was a mathematics professor at the University of Michigan at the time. He had previously done some fundamental research in queueing theory (see [1]). My topic was concerned with traffic processes in queueing networks and was based on experiences from the Buffalo conveyor work. Through Bruce's influence and my own inclination I became a queueing theorist. I have no regrets about that choice: the field has been extraordinarily good to me.

During the time I spent at the University of Michigan I had several opportunities to work on research with fellow faculty and thereby to develop my skills as a probabilistic modeller. Charlie Lipson of the Department of Mechanical Engineering engaged me to work on some mechanical reliability problems which were my first introduction to probabilistic modelling outside of production problems. Lipson was a consultant to the major automobile companies.

Don Cleveland of the Department of Civil Engineering attended several of my queueing courses. His area of research was road traffic theory. We started working together on some of his problems, and eventually produced two or three Ph.D. students in his department.

In addition, I became engaged in statistical analysis and probabilistic modelling with many other research programs. The statistical analysis of new procedures for treating burn patients, of synthetic time standards, and of short-term memory data furthered my statistical development. Probabilistic modelling of a large-scale public ambulance system, of a collection of psychological and neurophysiological problems, and several computer problems furthered my development as a probabilistic modeller.

In 1971 I returned from a sabbatical leave in Brazil to find that Seth Bonder, one of my departmental colleagues, was going to leave the university to form his own company. He asked me to work part time with this company at what he called a "guru" level. Essentially that meant being the company gadfly, poking my nose into any research project the company had, talking with the researchers, advising on the research and on the direction of the research, and lending whatever assistance I could to people working on research projects as diverse as weapons systems evaluation and mass screening for cancer.

I found these research activities exciting and extraordinarily different. They occurred during the heyday of research funding, so that one could engage in a large number of diverse probability-based problems. It gave me experience that would be difficult to reproduce today. If I am a probabilistic modeller, it is because of that experience.

5. What is "Probabilistic Modelling"

I have misgivings about the words "probabilistic modeller". When one models anything concerned with real problems, it is dangerous to presuppose an approach. One models what there is to be modelled. One does not model only those problems that are "probabilistic" or only "deterministic". The danger lies in trying to force a specific structure on a given problem, and the words "probabilistic modeller" seem to me to do that.

In spite of these misgivings about the title, it is true that most problems that have been brought to me have had probabilistic elements. Einstein's supposed conclusion that God doesn't play dice with the universe, I think, misses the point. Our only scientific means of knowing about the universe is through models. Whether the piece of the universe with which I must deal is really stochastic or not is something I cannot fathom as a modeller. I must take my problems as they come, and as they have come they appear to exhibit enough randomness that one would be remiss in ignoring it.

Thus, it seems to me, probabilistic modelling is no different from any other modelling. Modelling is a process. The process produces an entity called a model. This model may or may not lean on the mathematics of probability theory. In fact, the model may or may not even lean on mathematics. The common thread tying together all models, however, is the process of modelling, the process of abstracting from reality to produce a structure (a model) that can be manipulated with more facility than the reality itself. Thus, one might build a mathematical model or an iconic model. Which is built depends on many conditions, but the intellectual process of arriving at a model is much the same. One has some notion of that which is to be modelled and, importantly, some idea of the purpose of the model.

Learning about the phenomenon which is to be modelled is often a long, tedious process. Here is where interactions with others more knowledgeable in the subject to be modelled is crucial. Seldom have I known enough about an area to be modelled that I could go it alone. At this point one is learning about the phenomenon in every way possible. Talking (and more importantly listening), reading, poring over data sets, sifting through anecdotal data to find common threads takes time and patience, but a model cannot be created without this work.

Based on this background one begins the modelling process. Assumptions are made and tested against data or against others' experience where this can be done. Inputs and outputs are defined. Controllable variables and control variables are identified. Purpose may be established here with greater precision than originally. All of these things are usually done iteratively: seldom is a first assumption the final word. More often it is in need of clarification, amplification, clearer enunciation. Ultimately, one distills from all of this a set of axioms, postulates, definitions, and variables. One has started to abstract from the real world.

From this beginning one then attempts to determine relations among the various relevant elements. Here one must make the final decision on the type of model to build. A clear statement of objectives helps in this decision. Shall the elements be related through a collection of probabilistic statements or shall they be related deterministically? If the purpose of the study is to gain an understanding of the expected behavior one may not need a probability model at all. On the other hand, if one needs to control the tail probability of some distribution, a probability model is demanded at this point. As before, this building of relations among the relevant elements is an iterative process. Seldom is one satisfied with the first approach, and seldom does a first try match the reality of the situation as exemplified by previously collected data. In fact, at this point one might iterate the entire thought process, collect more data, elicit more input, or reconsider any previous step. Ultimately, of course, these reconsiderations must stop. Due to time or budget constraints the modeller must be content with what he or she has.

When a model has been constructed, it may not be much different from one to be found in textbooks as an example or homework exercise. The problem now becomes one of exercising the model, playing with it, learning from it, usually in a formal sense. But one may not be out of the creative woods yet. To exercise the model may require developing new techniques. One may need to develop new numerical methods to solve equations in the model, for example. But, ultimately, what comes from the model is a statement that says, essentially, "If the real world works as I have modelled it, then it should have the following behavior."

The acid test of the modelling process is then the re-examination of the real-world phenomenon being modelled to see if it does in fact behave as expected on the basis of the model's prediction. If it does, at least to a usable level of correspondence, one has a "good" model and can use it for the original purpose. If the model does not give results close enough to the real world, it is a "bad" model. Then the hard work begins. What is going on in the real world that is not being handled correctly in the model? Is there a dependency between random variables that the modelling process assumed to be negligible? Is there a nonlinearity that is modelled as a linearity? Why does the model not give results, at least approximately, like those of the real world? At this point, the whole conceptualization is reconsidered. New models are created. The original model is modified or enriched. This reconceptualization may require several iterations until one has a "good" model.

As I reread this section, several thoughts occur. First, modelling is a process. It is an art. One learns to model, but the teaching of modelling, in any usual sense of the word "teaching", may not be possible. One learns to model by modelling. One doesn't become a mature modeller without experience. One becomes accomplished in the art through modelling many different types of problems; ten different projects teach more than one project done ten times over. Second, modelling is not the straightforward process one reads about in some books purportedly discussing the topic. It is a creative process. There is no how-to-do-it manual. Except in the most trivial cases, one creates, tests the creation both against itself for internal consistency and against the real world for external reliability. Then, often, one repeats this process of creation and testing until it leads to a model that is internally consistent and externally reliable. Modelling seldom proceeds in a straight line. Third, the requirement that a good model be a usable tool for studying the real world places the modelling process outside the world of mathematics. While it is true that nearly every model that I have developed in my professional life has used knowledge of mathematics in general and probability theory in particular, the process of creating a real-world model and that of creating new mathematics are different.

6. Current Research

While I was doing the consulting work described in Section 4, I kept my work in queueing network theory going. Consulting was an important activity in learning how to model. It was great fun to be able to poke into so many areas about which I had no initial knowledge. It was fascinating to see the analogies that popped up in seemingly different areas. It was awe-inspiring to see the power of mathematical modelling. But in my role as consultant, I seldom plumbed the depths of a problem. Since leaving the University of Michigan in 1977, I have essentially dissociated myself from consulting. I find my work in queueing network theory more satisfying.

In 1977, I accepted the Charles O. Gordon Professorship in the Department of Industrial Engineering and Operations Research at Virginia Polytechnic Institute and State University. It was a good move for me to make at the time. The new position has given me the opportunity to concentrate on more basic problems in queueing and queueing networks. I have been able to build a small, but strong, program in applied probability. With the help of university funding we have been able to invite visitors from both the United States and overseas for both short and long-term visits. This gives our group an opportunity to see first-hand how the field of applied probability is developing in many places. The prospects are exciting.

To put the later discussion into context, I will briefly discuss queues, queueing theory, and queueing network theory. On the face of it, queueing theory and its extension into networks is the study of a common phenomenon, waiting lines. Waiting lines or queues occur in nearly every aspect of modern society. They occur in traffic jams when people come and go to work. They occur in the lunch room when people eat. They occur in the bank and the post office. They occur in computer systems and in airplane systems, both at the ticket counters and in the delays experienced in transit. They appear in production facilities as work in process, incoming goods or final goods inventories. It has been suggested that they occur in human psychological systems where they are called memory, and in human physiological systems as neural network phenomena. Though, at first glance, the study of waiting phenomena would appear to be humdrum, its importance is due to the ubiquity of the problems.

To illustrate what queueing problems are, consider the problems of a bank teller system. In a teller system there is a collection of people who provide a service; abstractly they are called servers. There is a demand for service by customers. Now, the important recognition in these banking systems, and, indeed, in every queueing system is that while it takes time to provide the service demanded by the customer, these service times are usually not predictable. Rather, if one observes these systems, it appears that the service times to a sequence of customers are random. Further, if one observes the times of arrival of customers, these also appear to be random. Because of these two sources of randomness, either the formation of a waiting line is inevitable or customers choosing not to wait are lost to the system, though they may try again at a later time. Thus, waiting lines form almost inevitably wherever a random service process interacts with a random demand process. That is why waiting lines in some form or another are ubiquitous.

Suppose we now enlarge our banking facilities to include, say, a mortgage department and a department for opening an account. One can imagine (and can see) a collection of servers whose functions are different and a collection of customers who may use some or all of the service facilities available. If one calls the different service facilities *nodes* and imagines that these nodes are connected by *arcs* then one has the image of a *network*. The service times at these nodes are random. The arrival processes (there may be one for each node) to the network are still random but there is a new element: there are arrival processes to a node from other nodes. Thus, congestion phenomena can occur at each node not only as a result of the randomness of service times and arrivals to that node from outside, but also as a result of congestion phenomena at other nodes in this network. These interconnected queueing systems form a queueing network.

Abstractly, then, queueing theory and queueing network theory are concerned with properties of interacting random processes. Because of the randomness involved throughout, these theories are in the area of probability theory and specifically in the area of random processes. My research has been devoted to this area since I first encountered queueing phenomena in conveyor systems about 25 years ago.

What are the basic problems of concern? In our bank teller example one can think of many problems whose resolution casts light on the quality of service provided. For example, one can look at the problem of the length of the customer queue. This problem arises, as do all of them, from the interaction of the random service and arrival times. The problem is to determine the length of the queue as a consequence of these two random phenomena. Since queue length is also a random phenomenon one must use probability statements such as, "the probability that the queue will be seven customers at 11:00 a.m. is 0.4" to describe it.

In a similar vein, one can consider the problem of waiting times. How long must a customer wait before being served? Again this waiting time is the consequence of randomness, and its solution must be couched in probability terms.

A distinction between a probabilistic model and a deterministic model can be drawn from our banking example. Some years ago I was confronted by a perplexed bank vice-president. He had recently had an expensive redesign made of his teller system but congestion problems, including long waiting lines and long waiting times of customers, were worse than ever. After explaining his new system to me he proceeded to place a box full of data on my desk and asked if I could figure out a solution to his problem, quickly. Now, all of my previous engineering education had taught me that a well-behaved system was one that was "balanced", i.e., the input rate to the system and the production rate of the system were approximately equal. In a system with no random variation this dictum makes sense. In a stochastic system such as this teller one, the system becomes unstable when it is balanced according to this dictum. In looking over the data supplied by the vice-president it became clear that his redesigners had consciously tried to balance the system. The ratio of input rates to service rates at certain times of the day varied from 0.95 to 1.05. It was this "balance" that was causing the problem. My suggestion to him was to reschedule the times his tellers were away from their windows so as to drive this ratio below approximately 0.8, a value that a "quick and dirty model" indicated would alleviate his problems. The man never returned to my office after that. Whether my suggestion solved his problem or put him out of business I still do not know.

When one turns to queueing networks, the problems to be solved become considerably more complex. The problem of the queue length is now transformed into a multidimensional one in which one may want to know the length of the queue at each node. This one problem has occupied a considerable portion of the research literature for many years. The waiting-time problem seems to be quite complex. Here one must determine the waiting time of a customer at each node that the customer traverses, and then sum these nodal times. Complications are caused by the fact that these nodal waiting times depend on each other in curious ways. This problem of waiting times has only recently come under serious study. There is still much we do not know about it.

My own research has concentrated on a rather different aspect of these queueing network problems. I have been more concerned with identifying the properties of the flow of customers on the arcs of the network. In this sense, I am concerned more with a theory of traffic than with a theory of queues, though the two concepts are not independent. One can argue (and I do) that in the study of nearly every queueing network problem, one assumes that the properties of the service system and of the demand process are known. In that sense these are the given premises of the study. What are the consequences of these premises on the traffic processes on the arcs? That seems to me to be the crux of the matter. In the references, I note three recent major works that review the field of queueing network theory and my contributions to it [2], [3], [4].

7. Some Comments on the Field

Through the years I have been privileged to be invited to many places in the United States and abroad. These trips and our program at Virginia Tech have given me an opportunity to experience some of the key research in the world. Internationally recognized journals are available to all queueing researchers and provide the possibility of homogenizing research directions, approaches, and problems to be solved, but such homogeneity has not developed to the degree one might expect.

A large eastern European school of queueing theory and a smaller group of researchers in Japan approach the topic principally through the theory of point processes and marked point processes. A French school, with correlative work in the United States, contributes to the subject through martingale theory. A collection of researchers in West Germany, the Netherlands, and the United Kingdom, with a larger contingent in the United States, consider the topic through Markov processes and their reverses. Small groups in the United States approach the subject through operator-theoretic concepts applied to Markov processes, through computer simulation (though this approach seems to be largely used for "real-life" problem solving rather than basic research), and through sample path analysis and perturbations. Some researchers are concerned with the design and control of networks using decision-theoretic concepts. Basic research on statistical analysis of networks is under way. My own approach to the problems is through Markov renewal processes and various transformations of them.

As one might expect, a large collection of researchers throughout the world work on what can be called basic research problems. I place my own work as well as that of most of the researchers mentioned above in this category. In addition, many are developing queueing models so as to solve specific, real problems. One might call such workers applied researchers, or in some cases engineers.

The picture I have drawn, possibly with some bias, is of many prolific researchers working in the area of queueing theory. The number of papers published annually throughout the world is staggering, and seems to be growing. I recently received a paper to review from a journal that previously published almost no queueing papers. The editor's covering letter bemoaned the fact that he was currently swamped with queueing papers. There is no single journal of queueing theory, and if there were to be one it could not accommodate the number of papers submitted for publication. Thus, researchers publish where they can and that may be in any one of as many as 50 different journals, some in English, some not. They must make a choice of what journals to read and what journals to publish in. Some of these journals are obscure, at least for the general queueing audience, and consequently the great wealth of results is known to only a few. Research directions and results are insulated one from another, with a resultant balkanization of the field.

In a conference held in 1973 participants debated the topic "Queueing theory is dead". It seems to me that today the issue is settled. Not only is queueing theory not dead, it is more alive than ever. An enormous number of new avenues have been opened to researchers in the last 20 years, both in basic research and in real applications. Whereas much of queueing theory prior to, say, 1965 was devoted to the use of Markov-process theory, today the tools of the trade have broadened considerably. Furthermore, while the areas using queueing theory prior to 1965 were primarily concerned with telecommunications, current applications range much further afield. Telecommunications continue to be heavy users (and producers) of the theory, but computer analysis has also become a major component of it, and queueing theory has invaded many other areas. Production theory, one of the early areas of queueing theory, has continued to be important. Some of the biological areas, which developed Markov models of many phenomena and, at least at a mathematical level, overlap queueing theory significantly, continue to be important. New uses in military command, control and weapons systems evaluation have appeared. In fact, it seems to me, that queueing theory as a mathematical discipline has become so involved in the general theory of random processes that it is quite difficult to disentangle any one field and its uses from the others. Queueing theory is definitely not dead. It may be going through a metamorphosis, but as long as there are traffic and congestion problems of any kind, research into them will continue.

8. Summing Up

This is the hardest paper I have ever written. Over a nine-month period, I have written, rewritten and re-rewritten dozens of versions of it, and still find this draft unsatisfactory. How does one write about the making of any professional? Writing about the making of an object is difficult enough: witness the almost uninterpretable instructions one finds on how to assemble a bicycle or how to run a microcomputer program. But the writing of the making of a life's work seems to me to be the most difficult of all.

How does one convey the excitement of working in any field? How does one comprehend the turns of fortune that got one there? How does one explain the luck that put me back in college, or the luck that landed me in a tutoring school, or led to my being asked to consult on an enormous variety of topics? I have tried, but I am sure that the sense of fulfillment occasioned by these last 30 years of work and research is beyond my capacity to convey to anyone.

In trying to make this article achieve some semblance of sense, I have imposed on many people. My wife Lois, who knows more about me than I do, has read drafts of this paper so many times, she knows it by heart. Next time she will write it. My daughter Lynn, who lived through most of these 30 years, also read several drafts of this paper. John Sivinski and Barbara Merkel made the mistake of visiting while it was in draft form: of course, they were forced to read it and comment on it. Peter Kiessler visited to do another job and got caught in this net of readers and commenters. Georgia Ann Klutke and John Barnes, two of my graduate students, have read a draft of the paper: I hope it does not cause them to seek another advisor. Yash Mittal, my colleague, read a draft and found it too impersonal. I had thought it too personal for its purpose. In fact, my only close friends who did not have a hand in this were my daughter Carrie who was busy raising my first granddaughter, Tiffany, and my colleague Bob Foley who was busy tearing up his knee again in a softball game. And there was Paula Davis, my secretary, who not only had to read and comment on it, but also had to type several drafts. To all of these people I owe an acknowledgement: thanks.

References

[1] Clarke, A. B. (1956) A waiting line process of Markov type. *Ann Math. Statist.* 27, 452–459.
This is probably the first paper that determines the time-dependent solution to an $M/M/1$ queue.

[2] Clarke, A. B. and Disney, R. L. (1985) *Probability and Random Processes: An*

Introduction with Applications. Wiley, New York.

This is a heavily revised version of the same authors' 1970 book, *Probability and Random Processes for Engineers and Scientists.* The new edition gives an introduction to queueing (Chapter 13) and the requisite background. It is intended for a third-year undergraduate audience.

[3] Disney, R. L. and König, D. (1985) Queueing networks: A survey of their random processes. *SIAM Rev.* 335–403.

This is an extensive survey of queueing network theory.

[4] Disney, R. L. and Kiessler, P. C. (1986) *Traffic Processes in Queueing Networks.* Johns Hopkins University Press, Baltimore, Md.

This is a research-level monograph discussing some of our research into traffic processes (flows) in queueing networks. It should appear by late 1986.

[5] Jackson, J. R. (1957) Networks of waiting lines. *Operat. Res.* 5, 518–521.

This is the first paper to study queueing networks as vector-valued Markov processes. Much of present-day applied queueing network theory owes a major debt to this paper.

3
THE CRAFT IN DEVELOPMENT

Marcel F. Neuts

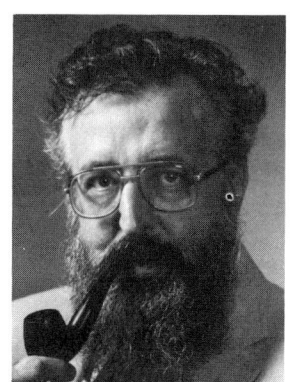

Marcel F. Neuts was born on 21 February 1935 in Ostend, a resort town on the Belgian coast. In high school he was a student in the modern languages and scientific section, rather than the more common classical humanities. This gave him an early exposure to mathematics, for which he developed a passionate love at first sight.

While most mathematically gifted students undertook engineering studies, the appeal of mathematics for him was such that he preferred to enter the Catholic University of Louvain to study the subject. Most courses in the mathematics curriculum then served the needs of physics and engineering students, so that the emphasis on analytic methods was strong and abstract treatments were postponed until the later years. Professor Henri Florin kindled his interest in probability, game theory and statistics and provided a healthy counterweight to the Bourbaki orthodoxy of the time. After graduating in 1956, Marcel Neuts was appointed to a one-year instructorship at the new university in Leopoldville (now Kinshasa, Zaire), where he spent a most interesting and educational year.

A fellowship of the Belgian–American Educational Foundation enabled him to pursue graduate study at Stanford University in 1958; he received his Ph.D. in 1961 with a thesis in game theory under Professor Samuel Karlin, from whom he learned much mathematics and the virtue of unrelenting hard work. Graduate study and thesis research were done under pressure of time, as the obligatory military service in Belgium could no longer be postponed.

In 1962, Professor Neuts became a member of the statistics and mathematics faculty at Purdue University. Following a year as visiting professor at Cornell University, he decided in 1970 to investigate the algorithmic problems associated with stochastic models. This led to the development of a variety of matrix-analytic procedures, whose ramifications and applications continue to be examined. Professor Neuts has also contributed to the theory of extreme values and to biomedical modelling.

In 1976 he was appointed to the Unidel Chair in Statistics and Computer Science at the University of Delaware, where he continued and expanded his research program in algorithmic probability. Since September 1985, he has been professor in the Department of Systems and Industrial Engineering at the University of Arizona. He has written 100 papers in probability and is the author of a textbook and a research monograph. He has supervised 16 Ph.D. theses, several by individuals of unusual research abilities, and regards this as his most significant and lasting contribution to the subject.

Among his honors are a Lester Ford Award, a fellowship in the Center for Advanced Study at the University of Delaware and a Senior U.S. Scientist Award from the Alexander von Humboldt Foundation. He has held lecturing and visiting appointments at The Johns Hopkins University, Cornell University, the Technion, Haifa and the University of Stuttgart.

Professor Neuts married Olga Topff in 1959; they have a son and three daughters. Between mathematical problems he pursues intellectual interests in Jewish learning, psychology and cultural anthropology, and finds relaxation in camping, hiking and (cautious) mushroom hunting.

An Algorithmic Probabilist's Apology

1. A Look Backwards

It is a common observation that the intrinsic time of human experience is a highly nonlinear function of clock time. Anyone whose earliest efforts at scientific computation date back some 30 years or more must now feel separated from them by a very long span of time. The computational environment of the 1950s already belongs to a past that will be unimaginable to the generation for which the personal computer is a childhood toy. It may therefore be appropriate to start this essay with a few personal recollections typical of that era.

Between the later stages of my undergraduate studies at Louvain and the completion of my doctoral work at Stanford (1955–60), I encountered substantial numerical tasks only in courses in actuarial mathematics and astronomy, and occasionally in statistics. In 1955, I had enrolled in an advanced elective course in astronomy. During the first session, the professor explained that we were to participate in one of his research projects by fitting a regression surface of the form $U = k X^a Y^b Z^c$, to a few hundred astronomical observations (X, Y, Z, U). The class was divided into two groups, each equipped with a huge volume of tables and much writing material. Each group was to work out the regression equations independently and by hand. Intermediate results were to be cross-checked for accuracy. As it was clear that work of this nature would fill the entire semester, I discreetly absconded halfway through the first session and switched allegiance to a different elective course.

The laboratory in financial algebra, taken in the actuarial program that same year, was not optional. One afternoon each week was spent using annuity tables and a mechanical calculator to solve the simple equations for the real interest rates of various government bonds. Even square roots, when they arose, had to be computed by hand. An electric calculator with a square-root key was admired from afar in a locked cabinet. It was reserved for use by the faculty members, as its substantial cost of some $1000 (at 1955 values) made it too precious to be entrusted to undergraduates.

During my graduate studies in the United States, numerical assignments were given only in the courses on linear models. Although only small sets of data were involved, each homework would take me two or three times longer than a corresponding assignment in real analysis or stochastic processes. My numerical skills were so unreliable that every operation had to be checked and rechecked, and this sorely taxed my patience. In comparison, I also found the underlying mathematics far less exciting and challenging and I felt relieved to have these courses behind me.

By the time I was deemed ready to enter a career of research, my enthusiasm for numerical work had sunk to an all-time low. The following impressions, gained from my early experiences, were to mold my attitudes for a long time. The mathematical content of the problems for which I had been asked to do numerical work was minimal. In contrast, the drudgery of the work and the time it required were considerable. My minimal computational experience had come from encounters with applied aspects of astronomy, actuarial science and statistics; mathematical disciplines where precise numerical answers have always remained essential. Even in those fields, a definite hierarchy separated those who developed "ideas" from those who merely evolved and implemented numerical methods.

Until 1971, I did not do numerical computations. Although access to the computer had become easy in the mid-1960s, I felt neither need nor incentive to devote time to its use. Besides, there was enough to learn and use in the varied analytic methodology of stochastic models. The incisiveness and specificity of analytic results had always attracted me more strongly than the measure-theoretic aspects of probability. Analytically, there was much to emulate in the courses of Samuel Karlin and Joe Gani at Stanford, in a seminar by Douglas Lampard and James McFadden at Purdue and in the extensive publications of Wim Cohen, Julian Keilson and Lajos Takács, among many others.

During that long period of algorithmic latency, I returned a few times to assess, as it were, the situation of numerical potentialities. Each time, however, my earlier impressions were confirmed and my aversion to algorithmic methods strengthened. A short course, taken in Belgium during 1958, had emphasized machine-language programming and left me with the impression that one kind of drudgery had supplanted another. A FORTRAN course in 1963 was similarly discouraging; we spent hours punching cards and waiting to find that yet another run had aborted because of a missing comma or a mistyped symbol.

2. The Transition

The period since 1960 has been extraordinarily fruitful, not only in the technological development of the computer, but also in the growth in variety and depth of probabilistic models. The analytic results on stochastic models were becoming quite complex, and frequently led to types of equations, such as the matrix–Wiener–Hopf equations of queueing theory and the matrix–Volterra integral equations of Markov renewal processes, which had only sporadically arisen in other branches of mathematics. It was clear that practical implementation of probability models would raise important questions on approximations, efficient algorithmic analysis and feasible, accurate computation. During 1965, I offered a seminar course for operations research students at Purdue and discussed a number of algorithmic approaches to semi-Markovian models. Viewed in retrospect, these were quite sound, but because I preached what I did not practice, some difficulties of numerical computation were clearly overlooked, while others, mostly pertaining to truncations and approximations, were greatly exaggerated.

Toward the end of the 1960s, remote computer terminals became commonly available. Although still noisy and frustratingly slow, they made interactive computation a reality. To scientists with numerical or algorithmic needs and interests, this was, I believe, the true beginning of the computer revolution. The nuisance of punched cards was mostly eliminated and the time involved in debugging and testing a coded algorithm was greatly reduced.

A number of other, more personal, factors led me to revise my attitudes towards algorithmic mathematics. Foremost among these were the statements of respected colleagues in industry that the computer implementation of many of the deeper analytic results, notably in queueing theory, was very difficult and sometimes numerically unreliable. Except for simple models, formal solutions were seen only as analytic existence proofs; practice relied instead on simulation or *ad hoc* numerical approaches in which the existing theory played only a minor part. Within the academic community, I was positively influenced by the writings of Richard Bellman and Ulf Grenander, who had stressed the mathematical challenge of modelling with an algorithmic objective and had set standards and examples by discussing such problems in probability or closely related areas. There can be no doubt, on the other hand, that the *Zeitgeist* of 1970 and the astonishingly negative reactions of many colleagues to algorithmic concerns interacted with a personal streak of cussedness and led to my decision to make computational probability a primary research objective.

This is not the place for a detailed review of the research results which followed that decision, nor for an assessment of their merits for the understanding and applicability of stochastic models. A review may be found in [1], and details in the books and articles cited there. Evaluations are best left to others and to time, though it is gratifying to see that computational probability has now become a vigorous, active field with diverse input streams of which my own efforts form but one.

The remainder of this essay is perhaps better devoted to some reflections on the shift in paradigm from traditional to algorithmic mathematics, on some of its consequences for education in probability and on the role that algorithmic analysis may play in the future of probability.

3. On Becoming an Algorithmic Mathematician

When a mathematician develops a genuine algorithmic interest, revisions both of principle and of practice impose themselves. A revision of principle is needed because of the legacy of history, which until the computer era has totally favored the dialectic over the algorithmic in mathematics. A number of essays, notably by Peter Henrici and by Philip Davis, have discussed that historical fact and its consequences. I will therefore single out only a few points for elaboration.

Many good mathematical problems, particularly those related to applications, have interrelated dialectic and algorithmic aspects which ideally should not be completely separated. The algorithmic aspects have features which, from a traditional viewpoint, may be disconcerting. Foremost among these is that no algorithmic problem can ever be said to be "completely solved". For example, after courses in real analysis and linear algebra, a student may believe that the solution of systems of linear equations or the integration of functions of a single variable are but simple cases of well-understood general constructions. Dialectically that student may be well-educated, but he or she is likely to be astonished by the extent and depth of the ongoing research on how these "elementary" operations can be performed for very large or ill-conditioned systems, or for rapidly oscillating functions. Once certain results of existence, uniqueness and characterization are in place, the only remaining interest of many problems may well lie in addressing their algorithmic issues. In stochastic models, for instance, a theorem of my own states that certain positive-recurrent Markov chains have a matrix-geometric vector of steady-state probabilities and gives a complete characterization of that vector. For some models, the computation of that solution may involve matrices of very large dimension. When dealing with a model of that type, one should not merely note the form of the solution, but examine how the necessary matrix operations can be performed. Failing that, the proposed solution is stillborn.

Equally disconcerting is the fact that equivalent procedures, all equally correct, may perform completely differently as algorithms. Moreover that difference is often inherent, and is unrelated to the state of computer technology; it follows from the sensitivity of algorithms to features that play no role in the proofs of theorems. A nonsingular but ill-conditioned matrix, or a convergent recurrence which suffers from severe loss of significance, can raise utter havoc with a correctly coded algorithm. For algorithms of any generality, it is well-nigh impossible to avoid unfortunate input data for which such difficulties occur.

This requires the revision of practice. Besides mathematical insight and rigor, the choice of a final algorithm requires the highest programming care and extensive testing; these also are the responsibility of the numerical mathematician. The coding of an algorithm should be only reluctantly entrusted to others, but its testing never. Sound algorithmic judgements require a hard-won and broader expertise than the verification of the underlying theorems.

This is not yet widely recognized, and numerate mathematicians too often have to sit through presentations of insufficiently tested algorithms with shortcomings that would be unacceptable if the proof of a theorem were at stake.

These observations are crucial in the supervision of theses and dissertations in this area. Only after three or four years of intense personal involvement with numerical methods did I feel sufficiently competent to assume this responsibility. With few role-models to follow, I had to look to graduate education in the experimental sciences in order to develop an approach suitable for algorithmic mathematics. Next to a sound graduate education in probability and the core areas of mathematics, I now encourage the student to become involved in computational problems of moderate complexity at the earliest possible stage. The discovery of the difficulty of numerical mathematics at the thesis stage can be disconcerting and discouraging, and it is still quite easy to get an advanced degree in applied mathematics, statistics or operations research without ever having to prove that a proposed solution is indeed computable. As the student rarely avoids all the pitfalls of numerical computation, his or her supervision requires much closer attention than I have found to be the case for purely theoretical theses.

Since 1975, algorithmic probability has attracted some of the best students to come my way. Among my recent doctoral students, V. Ramaswami and D. M. Lucantoni have already obtained research results in this area which go far beyond their theses; their work demonstrates that an algorithmic concern is a genuine source of mathematical inspiration and not, as is still occasionally heard, the first refuge of the incompetent analyst. Given the needs of the applications and the potential of the computer, the graduate education of numerate mathematicians is a major responsibility of the mathematical community and I view my supervision of a number of fine theses in this area as a contribution which is at least as important as my publications.

4. Education in Algorithmic Probability

Education in mathematics may be viewed as an expanding development of structural insights and manipulatory skills. The first are essential to understanding, the second lead to the simplest forms of formulas, the most transparent proofs of theorems or the most incisive solutions to equations. For many problems or theorems, it is useful to know different proofs or varying calculational approaches. These enable one to recognize those with greater versatility or potential for generalization.

In recent decades, the educational emphasis has been toward generality, simple elegant proofs and (hoped-for) insight at the expense of even purely analytic skills. It is now rare, for example, to find students of analysis able to solve even the moderately difficult exercises in, say, Whittaker and Watson's classical text [2]. One may succinctly say that the ability to handle generality and the search for elegance have developed at the expense of the ability to deal with complexity.

That shift in emphasis is unfortunate because the analytic skills are needed to identify those features of a complex model that lead to the simplification and ultimate tractability of its analysis. That same need, but greatly magnified, underlies algorithmic methodology.

Some years ago, in developing a course on computational probability, I realized that much contemporary didactic material has an emphasis inappropriate to the education of numerical mathematicians. In searching through the classical texts, I found that even those exercises requiring a specific analytic answer were too simple to present an algorithmic challenge. Students are conditioned to view a solution as an explicit formula in terms of elementary functions. The successful solution of complex problems is, at best, expressed in recursive schemes or differential or integral equations which call for a qualitative discussion and a numerical solution. In order to obtain challenging algorithmic problems, I had to draw on specialized texts in statistics and combinatorics in addition to inventing quite a number of simple stochastic models myself.

The primary purpose of these problems was not to develop situations complex enough to require recourse to the computer. What were sought were probability structures with enough features that a careful analysis of events was needed to obtain several alternate schemes of computation with varying storage requirements, qualities of accuracy and operational effort. Such problems require an excellent understanding of probability and develop algorithmic judgement. Actual computer programs are a necessary, but minor part of the course, although their correctness is carefully scrutinized and the significance of numerical results receives thorough interpretation.

With the exception of discrete event simulation, few algorithmic aspects of probability have yet entered the educational curriculum. When they do, I would prefer to see them integrated with the structural and analytic aspects of the subject. Therefore, I hope that future non-measure-theoretic texts on probability and stochastic models will include algorithms alongside theorems and analytic derivations, as well as substantial exercises to enhance the student's algorithmic skills.

5. Algorithmic Analysis and the Future of Stochastic Modelling

The difficulty of predicting the future is well known. Extrapolation from the present is hazardous because computational probability is a very young subdiscipline. Research to date has primarily focused on the implementation of existing theoretical results, such as the product-form solution for large networks of queues, and on some analytic procedures that can be naturally expressed in corresponding computer methods. The matrix-analytic results reviewed in [1] belong to that second category.

The future of algorithmic probability will depend, I believe, on how the mathematical community resolves two issues of principle. The first issue is germane to applicable mathematics generally; the second is specific to probability. Both issues clearly deserve to be discussed at greater length than is possible here.

Algorithmic mathematics has an essential experimental component, but for sound historical reasons, mathematics has hitherto denied itself the experimental method. Its purely deductive, dialectic nature is even codified into most philosophical definitions of mathematics. As long as algorithmic work remained a minor, mostly marginal activity of mathematicians, such a strict self-definition rarely affected individual identities or the nature of "acceptable" results and methods. With the emergence of the computer and the explosion of algorithmic methods, the issue of sources of mathematical knowledge beyond theorems and of the importance of algorithmic and even computer-experimental results can no longer be avoided. Particularly to those working in a discipline with a real-world orientation, such as probability theory, the issue is a vital one. The importance of practical algorithmic problems is such that they *will* be tackled by persons who are steeped in the new technology. What is at stake is whether this task will be carried out by those formed in the rigorous crucible of advanced mathematical education or whether it will be left to others. I believe that only individuals with a strong mathematical identity can effectively integrate theory and practice, isolate and establish general principles and formulate precisely the new problems raised by every important algorithm.

Criteria of relevance and societal support alone need not be persuasive. The fact remains that the purely mathematical gratification of algorithmic involvement can be rich and profound. New problems and conjectures can evolve from, or be tested by, computer experimentation. The algorithmic use of limit theorems requires the determination of normalization constants with a precision that is often not essential in proving the existence of the limit. The merits of asymptotic results as approximations can be numerically investigated, and there are pervasive problems of approximation in applying theoretical results involving infinite-dimensional operators or state spaces. To resolve such aspects of an algorithmic problem constitutes significant mathematical research, and deserves appreciation.

The second issue is related to the historical role played by the probabilistic model. Starting with the theory of errors, through regression and time series analysis to the classical models of epidemic and storage theory, there is the leitmotif of randomness as *deviation* from an ideal, deterministic norm. This narrow perception of the chance model, a legacy from analogies with the theory of mechanics, has broadened considerably during the past decades, but survives in most of our terminology and, more importantly, in the emphasis that is placed on the description of the *average behavior* of the stochastic model. The importance of average behavior and of the mechanical analogy in understanding the life history of a queue, an epidemic or a branching process is undeniable. It is often also the only aspect of the model that can be expressed in an informative analytic form. There is generally a major increase in difficulty in the derivation of distributions rather than moments, in large excursion results and in conditional (limit) theorems.

It is, however, undeniable that the average behavior of stochastic models is rarely surprising and that the most interesting phenomena are related to exceptional, rather than typical realization. As a few examples, we mention the early identification of large epidemics, the mechanisms causing overload conditions in queues, and the planning of inventories for situations with large emergency demands. In all cases of such problems which I have encountered, the simpler descriptors of average behavior were inadequate and sometimes misleading.

In order to obtain the necessary information for a thorough understanding of such models, one often needs to apply a laborious and hybrid methodology. Next to such theorems as can be proved, one needs to compute, not one, but several probability distributions and this for a variety of parameter values. The numerical results require cogent interpretation—which is often difficult—and, in some cases, confirmation by experimentation with realistic data.

There is often great reluctance to undertake such studies. Among the reasons commonly given are the time and effort involved, the commitment to a single methodological approach and the perceived difficulty of publishing such extensive studies and reaping appropriate professional reward for them. There is undoubtedly much validity to these reasons, which have all been heard in other fields when they were in a period of innovative flux. After long hesitancy, major fields of technological and medical endeavor are now viewing stochastic modelling as highly significant and useful. The problems they bring are difficult and nontraditional, and I have no doubt that algorithmic analysis will play a major role in their solution. As so often in science, ultimate rewards will go to those who dare to seize the day and help shape the future.

In starting this essay, I have described some impressions of the world of applied mathematics of the 1950s, when it was still shackled by computational impotence. Computer technology and algorithmic methodology have already brought us far along, yet in so short a time. One can only dream about what the contents of an essay such as this might be, if it were to be written 30 years hence.

References

[1] Neuts, M. F. (1984) Matrix-analytic methods in queuing theory. *European J. Operat Res.* 15, 2–12.

[2] Whittaker, E. T. and Watson, G. N. (1927) *A Course of Modern Analysis.* Cambridge University Press, London.

D. Vere-Jones

David Vere-Jones was born in London, England, on 17 April 1936. During the war his parents lived near Manchester, but later they moved to Wellington, New Zealand. He was educated at Cheadle Hulme School (England), Hutt Valley High School (New Zealand), Victoria University of Wellington, New Zealand, and Oxford University, England, where he was a Rhodes Scholar for 1958–61. After a year as an exchange student in Moscow, he returned to New Zealand at the end of 1962 to work with the Department of Scientific and Industrial Research. In 1965 he married Mary To Kei Chung, and moved to the Institute of Advanced Studies, Australian National University, Canberra. In 1970 he was appointed Professor of Mathematics at Victoria University, Wellington, where he is currently the head of the statistics group.

His early research interests centred round denumerable Markov chains, following his Ph.D. work with D. G. Kendall. As part of this work he developed the so-called "R-theory" for geometrically convergent Markov chains. In New Zealand he has been mainly concerned with point-process theory and seismology, and has written several research and review papers round these themes. He is currently completing a monograph on point-process theory with his colleague Daryl Daley from the Australian National University.

In recent years he has been a council member of the Bernoulli Society, and chairman of its East Asian and Pacific Regional Committee. He is currently an ISI council member. In 1982 he was elected a fellow of the New Zealand Royal Society, and is currently a member of its council as well as a member of the New Zealand Universities Entrance Board.

He has three children, Helen, Andrew and Martin. His interests include the theatre (his brother is a well-known New Zealand actor), tennis, walking, reading and languages.

Probability, Earthquakes and Travel Abroad

1. Youth and Early Training

When I was 12 years old, my parents emigrated from England to New Zealand, and this, perhaps, was the single most important event in my early life. The more relaxed atmosphere in the New Zealand schools, and the opportunity this gave for reading and study outside the pressure of competition, led, I am sure, to a greater success than I could have achieved in the tenser, more status-conscious environment of England. It also seemed to pave the way for the travel opportunities which are one of the great privileges of being an academic. From that time on, without any tremendous effort on my part, opportunities for travel have arisen which have allowed me to remain based in New Zealand, while living and working for periods in Australia, Britain, Russia, India, Japan and elsewhere. These opportunities have come about not through being an explorer, a journalist, an interpreter even, but through being a mathematician, and that at a modest level. Whatever dreams of adventure I may have had as a child, I never thought that such opportunities could come, of all things, from a career in mathematics. Be a mathematician and see the world? It may sound an unusual slogan, but in fact there can be few disciplines, if any, where mutual understanding is so independent of race, politics, culture or religion, and (with a few exceptions, perhaps) recognition of important work is so freely acknowledged on an international basis. Be this as it may, from those early days onwards, while other interests have come and gone, mathematics and travel have remained the dominant concerns of my working life.

At school I found mathematics a relatively easy subject, and enjoyed the parts of it I understood well—I can remember, at the age of 11 or so, covering pages with quadratic equations when I first realized that I could both write down an equation with given roots, and then recover the roots I had first thought of by solving the equation. A few years later I was filled with curiosity by the successive mysteries revealed in the early chapters of Hogben's *Mathematics for the Million*. However, it never occurred to me that I might follow mathematics as a career. Mathematics was handed down on tablets of stone. Only Newton, Euclid and similar heroes actually created mathematics. The rest of us had our work cut out to understand what they had done, let alone create anything remotely comparable ourselves. This view of things still seems to me substantially correct. What was not made clear at school was that there is a whole spectrum of creative opportunity in mathematics at a much more modest level, and that even at this modest level creative work in mathematics can be both enjoyable and useful.

I therefore drifted into the professional world of mathematics through what I suppose is a common mixture of inertia and chance opportunities.

Because success came more easily in mathematics, mathematics remained the centre of my undergraduate programme. Not of my life as a student, however, which became dominated by a consuming enthusiasm for the theatre. Two persons, in particular, helped to steady my somewhat faltering footsteps along the road to mathematics. One was the professor at Wellington, J. T. Campbell, a student of A. C. Aitken's, an ardent mathematical missionary, and a well-informed, perceptive teacher. It was through his intervention at the critical times that I first came to work with Peter Whittle, and later with David Kendall. As a third-year student, I was recommended by Campbell to the Applied Mathematics Laboratory of the New Zealand Department of Scientific and Industrial Research (DSIR), where I was able to work for two summers as a vacation student with Peter Whittle. A year later, when I was a fourth-year student, and nothing was really further from my mind (still much taken up with acting), he nudged me in the direction of a preselection committee for the Rhodes Scholarships and thereby opened up a path which ended with my studying for a D.Phil. at Oxford under the supervision of David Kendall.

It was with Peter Whittle that I first saw what being a mathematician meant. Peter at that time was easily mistaken for the errand boy from the sweet shop below the offices at Courtenay Place, close to the vegetable markets, where the Applied Mathematics Laboratory then had its premises. Riding in from a distant suburb on a battered bicycle, wearing shorts or sou'wester, he would carry his bicycle up the stairs, past the glorious piece of Meccano (an early integrating engine, which had somehow found its way to New Zealand) which graced the entry to the laboratory, and into a back room, where he changed into only slightly less disreputable garb to begin a working day as a mathematician. From the first encounter with Peter onward, I had no doubt that being a mathematician was just as worthy an enterprise as being an actor. It might still be a hard medium to work in, and less glamorous than the theatre, but no less worthy of the dedication and painful effort needed to bring any creative aspirations to fulfilment. There was a transparent simplicity about Peter's explanations of a mathematical point, or suggestions for tackling a problem, which allowed one a personal view of his mathematical intuition at grips with the world. I have never met anyone since who possessed this quality to quite the same extent.

Aitken's interest in matrices had been passed on to Campbell, who gave special lectures in this topic, which encouraged me to try some problems from Peter and others on Markov chains. It then became inevitable that I should choose a topic in this field to work on with David Kendall. He suggested a more detailed study of the phenomenon he had discovered recently of "geometric ergodicity" in denumerable Markov chains—i.e., a geometric rate of convergence of the transition probabilities to their equilibrium values. This became the main topic of my thesis, and of my first significant publications [3], [5].

My years at Oxford, however, were somewhat more stressful than my undergraduate years in New Zealand. For one thing, I was never quite at ease with my role as a Rhodes Scholar, despite the interest of meeting fellow scholars from all parts of the Commonwealth, and the unfailing patience and sense of humour of the Warden of Rhodes House, Bill Williams. But I was determined to experience the Oxford acting scene, and soon found myself caught up in tensions between several different worlds: the acting world; the world of the expatriate colonial; and the world of the mathematics postgraduate student. Undoubtedly the last was the most melancholy of the three; not a joyous state of life, in my experience. The mathematics student, facing his problem alone, has none of the psychological reassurance provided by the day-to-day practicalities of experimental work, and little of the social stimulus occasioned by the ardent discussion of literary or historical themes. Even fellow students with analogous problems have limited power to assist, or even to provide comfort. Socially, the mathematics student is often an outcast; social chit-chat on mathematical themes is hardly conceivable, and anathema to many mathematicians. The student may also be gauche and caught up with personal problems he is ill-placed to resolve. A year after I arrived in Oxford, one of my fellow mathematics students from New Zealand committed suicide; my own stock being none too high at the time, the insight into his circumstances was unsettlingly vivid. It was too close for comfort.

Under such circumstances it is not surprising that my relationship with David Kendall was also somewhat variable. I am a tremendous admirer of David's writing, and continue to take some pride in belonging, on no matter how eccentric an orbit, to an English "school" of probability distinguished for the excellence of its mathematical writing. But to acquire that skill myself, let alone the mathematical insights which it enhances and by which it is impelled, seemed hopelessly beyond my power, and still does (though one continues to struggle). At times David was hospitality and sympathy itself, but at others the perfection of his own writing and lecturing, by comparison with my own blotted copybook, seemed to form an impassable barrier. I must admit, however, that my relations with David relaxed considerably when I finally produced something half-way decent for my thesis, and we were both of us spared the continuing tension over whether passable results would emerge.

2. A Year in the USSR and a Trip to India

As the end of the D.Phil. grew near, another problem loomed. Where next? By then, the Applied Mathematics Division (as it had become) had enrolled me as a permanent staff member on leave without pay, so a return to New Zealand was inevitable unless I resigned, which I did not wish to do. Yet the thought of returning permanently to an office job in New Zealand was something I found hard to face. An opportunity for procrastinating was provided by that great benefactor of would-be escapists, the British Council. They advertised an exchange programme with the Soviet Union for post-graduate students, and further enquiries elicited the information that they positively encouraged the rather rare applications from nonlanguage students. Since I had, for my own interest, and stemming from an early enthusiasm for Chekhov's plays and short stories, already embarked on a programme of private tuition in Russian, this seemed just the chance to visit the homeland not only of Chekhov, but also of such giants of the probability world as Chebyshev, Markov, Kolmogorov, Gnedenko and Dynkin. I applied, and was accepted. Our departure was enlivened by a ponderous Foreign Office briefing on how to avoid being blackmailed by the Russian Secret Service (basically, by avoiding contacts with Russian persons of the opposite or similar sex; the advice, I am happy to say, was almost universally ignored). Shortly afterwards I found myself embarked on the good Soviet ship *Baltika*, chugging up the Baltic Sea to Leningrad.

There followed one of the most arduous but also most enriching years of my life. At last, I think, I grew up. I found an easy emotional affinity with many Russians, who, like me, were disorganized, enthusiastic, fond of and accepting eccentricity. Mathematically also I fell on my feet. I chose to work on queueing theory with B. V. (Boris Vladimirovich) Gnedenko at Moscow University. Boris Vladimirovich may not have been so highly regarded as a mathematician as some of his colleagues, but as a supervisor, as a source of support and assistance to foreign postgraduate students, as a representative of Russian culture and hospitality, he had no equal. His little flat in one of the forbidding towers of the Moscow University Building was then, and has now been for decades, a refuge for aspiring probability students from all parts of Russia and indeed from all quarters of the globe. He particularly delighted in the progress of his little nests of students from the third-world countries. He also played a major role in the postwar development of probability theory, particularly applied probability, in the countries of eastern Europe, and had many visitors from those countries. Entry to his circle provided me with a new world view, a form of Russian/European culture quite different from the colonial and the English varieties, and owing more, I think, to prerevolutionary traditions than to communist innovations. I also met a new view of mathematics and its creators, in the form of the Russian version of a "school" of mathematics.

In Oxford and Cambridge, though one can talk loosely of a school, the emphasis is so much on individual brilliance that the school is at best a loosely bound assemblage of discrete, light-emitting particles. There is little joy for the also-rans, the neutrinos of the mathematical world, who flit in twilight ignominy between the major particles. The Russian "school", on the other hand, is really the "collective" emerging in a mathematical context. Mathematical progress is achieved by the group as a whole. Even the humbler members can feel that they have a contribution to make, if only fractional, in moving the subject forward—but then the total progress is the sum of such fractions. Some may be larger, some smaller, but all are needed and of value.

I have often pondered over which view of mathematical progress is more valid. It is an uncomfortable truth (for most of us) that one really brilliant mathematician can make more progress in 10 days than 10 less gifted comrades could make in as many years. But that is not quite the whole story.

Major developments must, if they are to be properly recognized and assimilated, even initiated, take place within a matrix of mathematical culture far more extensive than could possibly be maintained by the major innovators themselves. Those forming part of this matrix have the lesser tasks of receiving, analysing, modifying and applying the major contributions. This work is also essential; it is worthy in its own right and should not be distorted in a vain attempt to make it an emulation of works of outstanding originality. The Russian experience allowed me to see a role for my own activities which was neither beyond my personal abilities, nor yet ignominious. Some further impressions of my year in Russia have appeared in [4].

Our exchange visit ended with a tour for the whole group through Soviet Central Asia and the Caucasus—a fine climax, but one which did nothing to lessen the feeling of relief, of pressure removed, which I think every Westerner experiences after a prolonged stay behind the Iron Curtain. It is not so much the difference in political and social styles, as the lurking knowledge that one is not really welcomed by the system. One is conscious of living in a state of unstable equilibrium; one has to be constantly on guard against minor misadventures, lest they suddenly explode into major difficulties.

Leaving Russia, I almost immediately met and joined a group of Oxford acquaintances travelling to India by jeep. This again is another story; let me mention only that through the contacts gained on this trip I was able some years later to spend a period as visiting lecturer at the Indian Institute of Science in Bangalore, so ending up well and truly infected with "Indian disease"—that fascination for the country, with all its myriad problems and contradictions, which, once contracted, is very hard to shake off.

3. Stochastic Models of Earthquakes

Finally, however, I was back in New Zealand, back in the office over the sweet shop and past the Meccano; only Peter Whittle, several years previously established in Cambridge, was no longer there to carry his venerable bicycle up the stairs. Almost immediately, I was set to work on a descriptive study of New Zealand earthquakes, designed in part to settle a dispute between the geologists and the seismologists over earthquake zoning. Resolve the dispute I never did (it turned into a truce eventually, as time passed and personalities changed) but the study introduced me to a field of application of statistics and probability which has provided a rich source of problems and ideas ever since. At the same time the work with the DSIR offered a much-needed opportunity to take stock of ideas picked up in the preceding years, and to gain experience in applications. I was able to lecture from time to time at the university, to read, and to work with some talented colleagues and vacation students. The work on earthquakes stimulated my interest in a topic Gnedenko had lectured about in Moscow, the theory of input streams (for queueing systems), or point processes. The earthquake analysis required the treatment of new aspects—second-order properties and spectral theory—in order to test for clustering and periodic effects. Working during one vacation with Robert Davies (now already the director of the division!) on testing for periodicities, we were getting close to inventing the right point-process concepts when Bartlett's classic paper on the spectral theory of point processes appeared, and at once the whole subject fell into perspective. It was hard to know whether to feel pleased or sorry, but the seismologists were looking for an answer, so we pressed on, incorporating Bartlett's analysis into our own discussion of the New Zealand earthquake data [11].

While the contact with the seismologists has been very valuable for my own work, I am less confident that it has been of much value to them. This impression prompted me, in a symposium on the role of probability organized by the Australian Academy of Science a few years ago, to take a rather pessimistic view of the value of stochastic models in the earth sciences [9]. It is only in rare instances, it seems to me, that a stochastic process lies at the heart of a geophysical phenomenon, in the sense that it governs its basic laws. A stochastic process is a process which we only partially understand, so that the probability aspect models our ignorance. This is a weak starting point for a research enterprise, and not one which appeals to the average scientist, at least in such a practical field as seismology. There may be some geophysical laws that are essentially a consequence of the mass interaction of chance events; indeed I have tried to explain the distribution of earthquake magnitudes in this way, using a branching-process model [7], and suspect there is more to this type of explanation than the seismologist is willing to admit. However, this may be a rare exception. If the heart of the phenomenon is not the mass action of chance events, then the introduction of probabilistic aspects may inhibit rather than advance understanding. Another problem is that the scientist is unlikely to rest content with this kind of explanation, not only because it yields priority to an alien discipline, but because it seems to negate the possibility of further analysis and prediction. In the final count, he is willing to challenge the verdict of ignorance, and to attempt to analyse the detailed configuration which causes the die to fall the way it does on a particular toss—or, in the seismologist's case, to study the genesis of each particular earthquake—thereby returning to a deterministic framework. It is only when one gets right down to the scale of fundamental particles that such an attempt may seem futile—and even here the position is controversial; is the randomness really a property of the particles themselves, or of our ignorance about them, and is there a difference between these views?

On the engineering side of seismology it is a different story. The engineer gathers information as best he may about the occurrence of earthquakes, and designs his structures accordingly, leaving to the philosopher and the scientist the fundamental questions over the nature of the phenomenon. In doing so he inevitably has recourse to statistical methods, including the use of simplified probability models for simulation and prediction purposes. This is a much more promising context for the application of statistical ideas; the goal of the investigation is clearly defined; there is a natural interpretation of optimality in terms of minimal cost or risk; and the engineer himself is much more receptive to ideas of approximate inference and simulation. There is lacking, perhaps, the intellectual challenge of seeking an ultimate explanation, but it is worthy of note that engineering has provided a tremendous stimulus to both statistics and probability modelling, not least in recent years.

4. Australian Interlude

In the meanwhile, other events were looming which were to have a decisive influence on my life. Out of the blue, after I had been back in New Zealand for less than three years, a letter arrived from Pat Moran offering me a job as a fellow at the Institute of Advanced Studies in the Australian National University. This was an opportunity too tempting to refuse—prestige, good pay, stimulating and eminent colleagues, no drudgery—how could I refuse? I did not, and compounded the felony by taking with me one of the Applied Mathematics Division's most loyal and valued employees, Mary To Kei Chung.

Mary had come to New Zealand from Hong Kong at much the same time that I had from England, and in fact we had taken several mathematics courses together (she claims that even at that time I was always stealing pencils from her, but I do not recall any such episodes). But where I had enjoyed all kinds of opportunities, she had worked away from scratch, overcoming financial and other problems, not least the differences in language and culture. She had finally obtained a master's degree, embarked on a successful career with the Applied Mathematics Division, and even, before we married, taken the decisive step of becoming an independent householder. To throw up all of this in favour of an absentminded ex-actor with an addiction to travel must have been a painful decision; I hope it has not been too much regretted. On my side also, the decision to settle down irrevocably was not taken without some qualms. But we have been married now for over 20 years, with three children and many trials and tribulations behind us. With any luck the hazard function is monotonic decreasing, so I live in hope that we shall continue to survive together happily.

The years in Canberra were particularly enjoyable and profitable. Personally, there were the joys as well as the stresses of adjusting to married life, the wonder of the arrival of our first two children, and the fascination of watching their early development. Mathematically, also, this was perhaps the most creative period of my life to date. The institute provided an ideal environment for research—a small group of stimulating colleagues; excellent library facilities; generous opportunities for travel to local and overseas meetings; Pat Moran's wry but penetrating comments on all kinds of matters; the many good students passing by—C. K. Cheong, Eugene Seneta, Mark Westcott, Robin Milne, Richard Tweedie, to mention only those with whom I worked most closely; contacts with colleagues in the other Australian universities, especially with Oliver Lancaster and Geoff Eagleson at Sydney; all these made for a rich context of mathematical ideas and opportunities; in return, perhaps, we all played some role in helping to create a particular Australian genre of applied probability and statistics.

5. Return to New Zealand

In 1970, after a year's sabbatical spent as a visiting professor in Manchester, I returned to New Zealand in the guise of Professor of Mathematics at the Victoria University of Wellington. The chair to which I was appointed happened to be the original chair of mathematics; earlier incumbents included T. MacLaurin, the foundation professor, a man of formidable talents who was also responsible for law and classics. After a short period in New Zealand he moved to the United States where he played a distinguished role in elevating the Massachusetts Institute of Technology to its present position of eminence. Another was D. M. Y. Somerville, whose texts on analytical and non-Euclidean geometry are still current; he was also a water-colour painter and some of his gentle landscapes still survive on the departmental walls. The last was my retiring teacher, J. T. Campbell. One cannot but wonder at the high calibre of these early appointments, and how they were able to survive the incredible workloads placed on them. Teaching a full range of courses in pure and applied mathematics, first year to third year, with even some additional courses, and at best one assistant, was standard practice for the times.

Fortunately, I was not asked to emulate these feats of antiquity, nor even to teach across the broad range of mathematics that the title of my position might seem to imply; it was envisaged rather that I would take responsibility for the work in statistics and probability, with the understanding that a new chair in pure mathematics would be (and was) established.

The mathematical scene in New Zealand is small and isolated even by comparison with Australia. The universities remain largely teaching institutions, and there is little community interest in such intellectual matters as mathematical research.

Despite or perhaps because of such difficulties, my colleagues have developed a "do-it-yourself" attitude to research, which no doubt would please the ghost of Rutherford should it choose to haunt his home universities. However, it is achieved at a high cost. Participation in fields of major current effort is rendered very difficult by the shortage of graduate students, the slow passage to New Zealand of critical papers, and the limited possibility for discussions with scientists sharing similar concerns. The best work is generally on applied topics which arise from local problems, and excite the combination of ingenuity and practical competence which are somewhat characteristic of the New Zealand mind. Even here much time is wasted because of lack of support staff. Local competence at the practical level is too often denied at the political and administrative level, where the tradition is still to call in overseas experts to solve local problems; this traditional pattern illustrates the gap that still exists between professional and academic staff, and the perceptions of the community at large.

Probability and statistics are potentially well-placed to bridge this gap. There is a New Zealand tradition, which extends from Aitken through his New Zealand students such as Campbell and Silverstone to more recent scholars such as Peter Whittle. The subject has a practical, agricultural bias, and offers plenty of scope for ingenuity. This has generally been borne out by my teaching experience: relatively speaking, and increasingly with time, the practical aspects have seemed easier to teach to local students than the theoretical aspects, statistics easier than probability, though my personal bias has them the other way round.

I do not wish to write in detail about my teaching and research in New Zealand, partly because it is still too close to hand. Most of my teaching has been routine; my research interests have continued to lie in the areas of point processes [2], Markov chains [1], applications in geophysics [8], [12], with occasional throwbacks to the realms of algebra and analysis [6], [10].

Two continuing concerns may be worth recording. One relates to the development of statistics (including applied probability and operations research) in the wider university context. We are currently faced in most universities with a confusion of first-year service courses, and a proliferation of "methods" courses offered at second- and third-year level by a large number of departments with an amazing range of both technical and teaching competence. There is in addition a continuing stream of honours and master's students with statistical themes for their projects or theses, but little idea of the subtlety or complexity of the questions they ask. Faced with all these on the one hand, and the necessarily limited, sometimes unduly academic courses offered within the framework of the mathematical disciplines on the other, one is conscious of a task of daunting magnitude. The aim should be to imbue all university staff with the awareness that there is such a thing as good statistical practice, then add the desire to incorporate such practice into their own teaching and research. Finally one would wish the staff to have the confidence that within the university they had access to the quality of advice and computational support that would allow them to put such desires into action.

As a first step in this direction, disregarding the cynics and scoffers, especially numerous in the traditional sciences, we set up an umbrella-type organization (Institute of Statistics and Operations Research—I have a gift for unfortunate acronyms) with the modest aims of bringing together the scattered practitioners of statistics and operations research across the university, for seminars, discussions, and mutual support in pressing for better facilities. As time has passed, this body has acquired an increasing range of responsibilities—first a postgraduate diploma in (applied) statistics and operations research; then a consulting function; most recently a range of new honours and master's courses in statistics and operations research.

The most important of these developments relates to the work in statistical consulting. Shortly after the institute was set up, a biometrician's post became available, which was converted into a university-wide consulting post, without departmental or faculty bonds, but attached to the institute, itself a cross-disciplinary body. Thanks to the zeal of the present occupant of this post, and some excellent support from the Computing Services Centre, this activity has developed into a major university service; it now involves two full-time consultant statisticians, strongly supported by the faculties, and directly engaged in a number of major research projects, mostly inside but some outside the university itself. This focus of activity also helps to increase the morale of the other statistics staff (most of whom continue with some consulting with time-honoured clients of their own) and highlights for students the opportunities as well as the interest of consulting work.

Another preoccupation, particularly in a small university, is the proper role for probability and statistics in relation to the other mathematical disciplines. Here it is necessary to point out that in New Zealand, as in Australia, the perception of the mathematical disciplines follows very much the British line of vision—that is to say, there is no clear distinction between fundamental and technical (*lycée* vs. *école polytechnique*, *hochschule* vs. *realschule*) aspects of mathematics, but rather an empirical tradition which, if anything, tends to emphasize applied aspects and to undervalue the role of pure mathematics. Statistics itself (in contrast to probability theory) is very much a product of this tradition, despite its long battle against the conservatives to get itself accepted as a mathematical discipline in its own right. This battle has, in some universities, led to rather bitter feelings and a determination to achieve independence at all costs.

Personally, I see such difficulties as birth pains, and I do not take the view that full independence is desirable, however necessary it may be for the statisticians to win the right to act on their own assessment of their needs, priorities and standards of excellence. The fact is that the rest of mathematics is changing also, and experiencing its own pressures to link up with applications and computing. Data engineering, if such a term exists, is by no means the prerogative of the statistician, insofar as one means by that an expert assessor of the role of chance. Chance may be important in modelling the world, but it is not clear that its role should be predominant. New developments in other branches of mathematics and computing may reduce its importance. I see some danger of statistics and probability themselves becoming a little rigid in their outlook and preoccupations, and being left stranded at their high-tide mark while other species start to populate the oceans. The more they become separated from the other mathematical disciplines the greater the danger.

Turning finally to more general activities, I have in recent years derived particularly great pleasure from the visits to New Zealand of younger colleagues from Japan, Italy and elsewhere. I have also enjoyed serving as chairman of the East Asian and Pacific Regional Committee of the Bernoulli Society. Trying to create links between statisticians across the immense distances and disparities of circumstance within this far-flung territory has been a quixotic enterprise which suited my taste entirely, and provided further exciting opportunities to travel and work among my Asian colleagues.

In New Zealand, among many other matters, I helped to set up the New Zealand Mathematical Society, and later served for six years as Subject Convenor in Mathematics for the Universities Entrance Board. Here I tried to secure a more general acceptance of the importance of statistics within the school programme; more generally I helped as best I could to sort out the problems of the mathematics programme in the upper secondary schools—an exceptionally difficult task, in which progress is hard to define, let alone achieve. But the effort brought further rewards in terms of the contacts with school mathematics teachers and their skills and aspirations.

To conclude, while I am honoured to have been included in this volume, the justification is not as transparent as I would have liked it to be. "Vere-Jones' theorem", that mathematical revelation with which I might once have hoped to stir the world, has never materialized, and I suspect that the chances of its doing so are monotonically decreasing. If this article has any justification, it may be as an encouragement to the gifted but not altogether brilliant to persevere. One does not have to be a Kolmogorov, a Fisher, or a Feller, to make useful contributions to probability and statistics, or to enjoy a rewarding life within the local and international mathematical and statistical communities.

References

[1] Athreya, K., Tweedie, R. and Vere-Jones, D. (1980) Asymptotic behaviour of a point process with Markov-dependent intervals. *Math. Nachrichten* 99, 301–313.

[2] Daley, D. J. and Vere-Jones, D. (1972) A summary of the theory of point processes. In *Stochastic Point Processes*, ed. P. A. W. Lewis, Wiley, New York, 299–383.

[3] Vere-Jones, D. (1962) Geometric ergodicity in denumerable Markov chains. *Quart. J. Math.* (2) 13, 7–28.

[4] Vere-Jones, D. (1964) The mathematician's tale. *Survey* 52 (Report on Soviet Science), 52–60.

[5] Vere-Jones, D. (1967, 1968) Ergodic properties of non-negative matrices, Parts I and II. *Pacific J. Math.* 22, 361–386; 26, 601–620.

[6] Vere-Jones, D. (1971) Finite bivariate distributions and semi-groups of non-negative matrices. *Quart. J. Math.* (2) 22, 247–270.

[7] Vere-Jones, D. (1976) A branching model for crack propagation. *J. Pure Appl. Phys.* 114, 711–725.

[8] Vere-Jones, D. (1978) Earthquake prediction—a statistician's view. *J. Phys. Earth.* 26, 129–146.

[9] Vere-Jones, D. (1979) Distribution of earthquakes in space and time. In *Chance in Nature*, ed. P. A. P. Moran, Australian Academy of Sciences, Canberra, 72–90.

[10] Vere-Jones, D. (1985) An identity involving permanents. *Linear Algebra Appl.* To appear.

[11] Vere-Jones, D. and Davies, R. B. (1966) A statistical survey of earthquakes in the main seismic region of New Zealand—Part II: Time series analysis. *N.Z.J. Geol. Geophys.* 9, 251–284.

[12] Vere-Jones, D. and Smith, E. G. (1981) Statistics in seismology. *Commun. Statist. Theor. Math.* A 10, 1559–1585.

K. R. Parthasarathy

Kalyanapuram Rangachari Parthasarathy was born on 25 June 1936 in Madras, into a modest but deeply religious Brahmin family. After completing his secondary schooling at the P.S. High School, Madras, in 1951, he studied mathematics at the Vivekananda College of Madras University, and became fascinated with mechanics. In 1956 he obtained a scholarship to study statistics at the Indian Statistical Institute (ISI), Calcutta, and later completed his Ph.D. there in 1962.

His first trip away from India took him to the Steklov Institute, Moscow, in 1962–3, where he came under the influence of Professor A. N. Kolmogorov and his school. He lectured at the ISI in Calcutta for two years before joining the University of Sheffield as a lecturer in probability and statistics in 1965. His outstanding research led to rapid promotion to a senior lectureship in 1966 and a readership in 1967; in 1968, he was appointed to a professorship in the Statistical Laboratory, Department of Mathematics, of the University of Manchester, a post which he held for two years.

In 1970 he decided to return to India with his wife and two children, and was in turn professor at the University of Bombay (1970–3), at the Indian Institute of Technology, Delhi (1973–6), and at the ISI, Delhi Centre, where he is currently working.

He is a fellow of the Indian National Science Academy, Delhi, and of the Indian Academy of Sciences, Bangalore, and was awarded the Bhatnagar Memorial Prize in 1976. He has written two books and numerous papers on probability theory and its applications in quantum mechanics. He was married in 1965; he and his wife have two children. His interests are literature, philosophy and music, both Indian and Western.

From Information Theory To Quantum Mechanics

1. My Early Education

The first 20 years of my life were spent in the intensely religious and ritualistic atmosphere of a south Indian Brahmin family confined to the towns of

Thanjavur and Madras and the paddy fields of the village of Thalanayar in Thanjavur district. Emphasis was laid at home on the philosophy and teachings of the Vaishnavite saints Ramanuja and Desika, the Hindu epics *Ramayana* and *Mahabharata*, the devotional songs of Tamil saints known as the *Alwars* and finally the *Bhagavad Gita*. To this day, I believe in their basic idea that personal and global peace are possible only through renunciation and action with detachment.

At school I was exposed to the geometry of triangles and circles, as far as rigorous mathematics was concerned. Even though the axiomatic approach to geometry was given extensive coverage, a similar effort was not made in arithmetic and algebra. My first contact with modern mathematics took place at Vivekananda College, Madras, in 1953–6 while I was reading for the BA Hons. in mathematics. The topics which I enjoyed most were classical mechanics and mathematical analysis. It was my good fortune to have had the most enthusiastic and dedicated teachers in T. R. Raghava Sastri, S. Krishnamurthy Rao and K. Subramaniyan. They taught me the fundamentals of projective geometry, analysis and classical mechanics. The result which made the greatest impression on me during this period was Lagrange's variational principle for writing down the equations of motion in generalized coordinates. Hamilton's principle of least action (which was outside the syllabus of our examination) was just being introduced in a somewhat vague manner by Krishnamurthy Rao; he died suddenly owing to a neglected carbuncle on his back.

The term "probability" had not reached my ears. Had I been given an opportunity to select a theme for further study at that time, my first choice would have been classical mechanics. During these college years, we never got the idea that we could formulate problems and prove our own theorems. I at least was under the impression that God transmitted mathematical ideas through a few geniuses. The sole aim of our classroom teaching was to solve problems from prescribed textbooks in order to obtain first-class honours in the examination. Any deviation from this path was due to the enthusiasm of a very small number of teachers.

2. Student Days in the Indian Statistical Institute

On the completion of my honours course it was suggested to me by the elders in the family that I might apply for the position of "upper division clerk" in the office of the Accountant General, or start a new career in Ayurveda, an Indian system of medicine based on native herbs. Thanks to the suggestion of my friend M. K. Mani, I wrote an examination conducted by the Indian Statistical Institute (ISI), Calcutta and was admitted into a programme with the rather bizarre title, "Three Year Advanced Professional Statisticians' Training Course" in 1956, which also happened to be the silver jubilee year of the institute.

Those were the days of five-year plans, through which India's beloved and most dynamic prime minister Pandit Jawaharlal Nehru unleashed tremendous forces on our sleeping continent, and created a large number of scientific institutions and economic programmes. Professor P.C. Mahalanobis, the Director of ISI known simply as "Professor" to its workers, students and devotees, enjoyed the confidence of Nehru and played a dominating role in the formulation of five-year plans as a member of the Planning Commission. In view of his position, phrases like "planning for national development", "statistics, the key technology of the twentieth century", "perspective planning", "operational research approach to planning", "statistical quality control in industry" etc., reverberated in the long narrow corridors of the Research and Training School of the Institute.

The need for a large army of well-trained statisticians to monitor and formulate new economic programmes was being constantly emphasized. As a consequence we, the students, were bombarded with courses and examinations on diverse topics like Indian economics, national income, time series and index numbers, psychometry, biometry, sample surveys, numerical analysis, descriptive and official statistics and so on. At the same time a substantial dosage of instruction on estimation and inference, design of experiments, multivariate analysis, sequential analysis, computational linear algebra and elementary probability theory was also included.

At ISI, statistics appeared to me a new religion rather than a technology. In probability theory emphasis was laid on the combinatorial results arising from coin tossing, dice throwing and card shuffling. Of course, the normal distribution and the distributions arising from simple transformations of independent and identically distributed normal random variables were introduced. Liapunov's central limit theorem was stated and used, but not proved. The goal was action rather than contemplation. The manually operated Facit desk calculator was a constant companion of every student. I still remember the day when Chou En Lai, the Chinese prime minister, visited our class room and all the students whipped out their calculators on a signal from our teacher Dr S. K. Mitra.

As a result of the silver jubilee celebrations ISI had many distinguished visitors, among whom I may mention Professor R. A. Fisher, J. Neyman and the Soviet physicist N. N. Bogoliubov. We had unlimited amounts of lectures, tea and delicious Bengali sweets. A big controversy on fiducial inference between the giants of statistics was once terminated by C. R. Rao reminding the speakers and the audience that it was time for tea.

In the jungle of courses imposed on me at breakneck speed there was one flower. I refer to the well-organized and delicately presented course on inference and Wald's sequential analysis by Professor R. R. Bahadur. For the first time, theorems were stated and proved in a sequence. The Neyman–Pearson lemma, the likelihood ratio as a Radon–Nykodym derivative, Wald's sequential probability ratio test, and several interesting examples rolled down in succession, making me sense the order amidst chaos. I cannot forget the characteristic phrase with which Bahadur conveyed at one stroke the spirit of sequential analysis: when best of five is played and one of the players wins the first three games it is futile to continue. It was also during this course that I felt how the definition of a sample space for a problem could become delicate.

After nearly two and a half years of intense instruction with stress on practice (including crop cutting in the agricultural fields of Giridih) the students were asked to do a project. After spending a few days in our well-equipped library my choice was made along a theoretical line. I would present a seminar on the moment problem: if $\mu_0 = 1, \mu_1, \mu_2, \ldots$ is a sequence of real numbers such that $\sum_{0 \leq i,j \leq n} a_i a_j \mu_{i+j} \geq 0$ for all real numbers $a_1, a_2, \ldots$ and nonnegative integers n, then there exists a probability distribution F on the real line such that $\mu_j = \int x^j \, dF(x)$ for all $j = 0, 1, 2, \ldots$.

In order to understand its solution I sought the assistance of my friend V. S. Varadarajan, who came to ISI to do his Ph.D. after completing his honours in statistics in the same year as I did. He had already done important research work on the interplay between topology and measure [51] by obtaining new proofs of the Riesz representation theorem, existence of regular conditional probability measures and Prokhorov's compactness theorem. Time and again he emphasized the importance and basic nature of Kolmogorov's consistency theorem for defining measures on infinite-dimensional spaces. Much later this influence of Varadarajan enabled me to deduce Bochner's theorem for positive definite functions and the Pontryagin–Van Kampen duality theorem for locally compact abelian groups from Kolmogorov's consistency theorem in [27] and [35]. The same idea is exploited in [38] to obtain the Gelfand–Neumark–Segal construction.

During the same period I came into close contact with R. Ranga Rao who had initiated research in the delicate problem of improving the Berry–Esseen estimates for the rate of convergence in multidimensional central limit theorems and also introduced the novel idea of uniformity of weak convergence over measurable convex sets in $\mathbb{R}^n$ (see [46], [5]).

The ISI, in addition to the meagre scholarship of 150 rupees ($15) per month, provided each student with a generous book grant. This enabled me to purchase *Measure Theory* by P. R. Halmos, *General Topology* by J. L. Kelley and *Mathematical Methods of Statistics* by H. Cramér. Through innumerable conversations with Varadarajan and Ranga Rao and the study of these texts I improved my background of basic knowledge and finally prepared my seminar on the moments problem as presented in the book of Shohat and Tamarkin [49]. The net result was that I obtained my certificate for the successful completion of the course and a scholarship to do research for the Ph.D. degree under the supervision of C. R. Rao.

At the suggestion of C. R. Rao and with the assistance of D. D. Joshi I started learning the subject of information theory through the relaxed and masterly presentation of A. I. Khinchine in his little book [18] which had just been acquired by our library. For the first time I had seen that the ideas of measure and integral were not just lifeless theorems, but meant to be used in building models for solving problems in day-to-day life. I understood the role of notions like shift-invariant probability measure, ergodic theorems, entropy rate of a source, information rate of a channel etc., in Shannon's formulation of coding theorems. To this day I am amazed at how Shannon could achieve so much with so little, starting from his scientific poem on the derivation of $-\sum p_i \log p_i$ as a measure of information.

Much greater is our admiration when we look at the impact of his concepts on the developments in the theory of dynamical systems from the hands of A. N. Kolmogorov and Ya. G. Sinai [47]. During my study of information theory I came across *Lectures on Ergodic Theory* by P. R. Halmos [10]. One look at the proof of the result that there exists a large class of ergodic measure-preserving transformations for the Lebesgue measure in a simplex and the combination of this result with my experience in information theory suggested the following problem: does there exist a large class of ergodic probability measures for the bilateral shift transformation? This led me to the following result: in the convex set of all probability measures invariant under the bilateral shift acting on a countable product of copies of a complete and separable metric space the set of extreme points, namely, the ergodic probability measures, is a dense G_δ in the weak topology.

I sent this result for publication to the *Illinois Journal of Mathematics* [22] and very soon I received an appreciative report from the referee. With great delight I ran to C. R. Rao's office and showed him the report. Instantly, I was granted a substantial increase in my scholarship. As the head of the Research and Training School, C. R. Rao enjoyed supreme power and commanded the confidence of all his colleagues. In the present-day democratic atmosphere of ISI such a recommendation needs the approval of a properly constituted committee. Soon I was happy to learn that my little piece of research was presented by J. C. Oxtoby [21] at an international symposium.

When I learnt how an invariant measure could be expressed as a direct integral of its ergodic components I began to investigate whether the entropy rate of a stationary source could be expressed as an integral of the entropy rates of its ergodic components. This led me to write [23] and also prove in [24] a coding theorem for channels whose transition probabilities do not transform an ergodic source into another ergodic source. Much later, I learnt that V. A. Rohlin of Leningrad [47] and K. Jacobs of Erlangen (private communication) had obtained more general results in the context of dynamical systems. Even though I stumbled on these results in my lonely path without interaction and experience of participation in a regular seminar programme, I was delighted to note that nearly 15 years after their publication my results were reprinted in the collection [9].

The results of [22] and [23] enabled me to obtain my Ph.D. degree. In fact, ISI was declared an institute of national importance by an act of parliament in 1959, and authorized to confer its own degrees. The first convocation of ISI was held on a grand scale on 12 February 1962 and the degree of Doctor of Science (*honoris causa*) was conferred upon S. N. Bose (of Bose–Einstein statistics), R. A. Fisher, Jawaharlal Nehru, A. N. Kolmogorov and W. A.

Shewhart. Nehru and Kolmogorov could not be present on this occasion. Along with these eminent men J. Sethuraman and I received the Ph.D. degree for our theses. I had a cutting from the Indian newspaper *The Hindu* where my face was visible (to me at least) in the large crowd. My grandfather Raghunathachari, who spared no effort in educating me since childhood, kept this cutting at his bedside and showed this picture of the first Ph.D. in the family to every visitor at our residence in Madras as well as Thalanayar.

We had lost Varadarajan to Princeton, but the ISI had acquired a new and fertile mind in S. R. S. Varadhan. Ranga Rao, Sethuraman, Varadhan and I lectured to each other on diverse topics like measure theory, general topology, harmonic analysis in locally compact abelian groups, Peter–Weyl theory of representations of compact groups, functional analysis and weak convergence of probability measures. After Varadarajan's departure Ranga Rao was the natural leader of our group, and at his suggestion the problem of construction of a large class of indecomposable probability distributions on topological groups was investigated. R. A. Fisher's example of an indecomposable density function and Paul Lévy's work [19] gave the necessary clues. The outcome of our effort was the paper [43] where it was shown that in the convolution semigroup of all probability measures on an infinite complete and separable metric group with weak topology the set of indecomposable measures is an everywhere dense G_δ.

It is interesting to note that the spirit of this result is similar to the earlier result on ergodic measures. In this context we could not answer the following question: in a compact infinite group how rich is the collection of absolutely continuous (with respect to Haar measure) indecomposable probability measures? Again Ranga Rao came out with another important suggestion that we should develop a theory of infinitely divisible distributions and central limit theorems on locally compact abelian groups along the lines of the famous book [8] of B. V. Gnedenko and A. N. Kolmogorov. So dominating was the pace and influence of Ranga Rao and Varadhan that my pet information theory was all but forgotten. The result of our effort was the paper [44] which was followed by Varadhan's famous Ph.D. thesis [52] containing the Lévy–Khinchine representations for infinitely divisible distributions and central limit theorems on a Hilbert space.

Meanwhile another important event took place; the visit of Academician A. N. Kolmogorov to ISI in April 1962. He was undoubtedly the most admired probabilist in the world, and we just could not believe that he was in our midst in Calcutta. Both Professor Mahalanobis and C. R. Rao took the greatest pains to keep the students and faculty members of ISI in contact with the best scientific minds from abroad and we continue to maintain this tradition to this day, against great odds. At the suggestion of C. R. Rao and A. N. Kolmogorov arrangements were made for my visit to the Steklov Institute of the USSR Academy of Sciences, and I flew to Moscow at the end of October 1962.

3. Visit to Moscow and Return to ISI

In Moscow I came into contact with Yu. V. Prokhorov, Ya. G. Sinai, A. N. Shiryaev, V. N. Tutubalin, V. V. Sazonov and the visiting Czech probabilist P. Mandl. After overcoming the difficulties of language and food with a mighty effort, I began attending the seminars of E. B. Dynkin, Ya. G. Sinai and the famous Monday evening programme of I. M. Gelfand. I felt completely overwhelmed by the pace of activities in Moscow, where mathematics appeared to me a mighty industry. I began reading the famous papers of Gelfand and Neumark on the representations of classical groups. Even though I did a little bit of work in probability theory and wrote [25] and [36], I was slowly leaning towards the theory of representations of matrix groups.

I returned to Calcutta in November 1963. On reaching Howrah station I was pleasantly surprised to see Ranga Rao greeting me with warmth and affection. Varadarajan had returned from the USA and Varadhan had left for the Courant Institute in New York. Varadarajan lost no time, and began teaching me the fundamentals of the theory of Lie groups and Lie algebras. He gave me a brief outline of the progress that he had made with Varadhan and Ranga Rao earlier in the theory of infinite-dimensional representations of complex semisimple Lie algebras. The strategy was to construct them as algebraic closures of finite-dimensional representations. We had a long battle with the maximal ideals of the universal enveloping algebra of a Lie algebra. There were no regular hours of work. After office hours our discussions continued at Varadarajan's residence in a narrow lane not far from ISI where Mrs Veda Varadarajan showered on us limitless hospitality and pleasant distractions in music, the Sanskrit language, Russian culture and so on. The results of our massive effort during this period were written up by Varadarajan in his characteristically clear style from the USA and published as [45].

In spite of our furious activity in the representation theory of semisimple Lie algebras, it was the course that Varadarajan gave on the mathematical foundations of quantum mechanics, following G. W. Mackey [20], during his last year at ISI which made the greatest impression on me and revived my interest in the mechanics that I had learnt with enthusiasm in my college years. Here was a subject with its roots in nature where ideas like probability, group representation and the spectrum of an operator came to life.

The conventional triple $(X, \mathscr{F}, P)$ of classical probability theory consisting of a sample space X, a σ-algebra $\mathscr{F}$ of subsets of X and a probability measure P is replaced by the new triple $(H, \mathscr{P}(H), \rho)$ where H is a Hilbert space, $\mathscr{P}(H)$ is the lattice of orthogonal projection operators on H and ρ is a nonnegative self-adjoint operator of unit trace. The probability of an *event* $Q \in \mathscr{P}(H)$ under the *state* ρ is just tr(ρQ), where tr denotes trace. Random variables are replaced by self-adjoint operators and called *observables*. *Expectation* of the observable A in the state ρ is tr ρA. Variance of the observable A in the state ρ is tr $\rho A^2 - ($tr $\rho A)^2$.

In order to obtain expressions for physically relevant observables and their expectations for a single particle or a system of particles, one appealed to the underlying symmetry groups and their representations. With the results of Von Neumann, Wigner, Bargmann, Gleason and Mackey following in quick succession and culminating in the Dirac equation for the electron in the masterly exposition of Varadarajan, a whole new world was opened to me. Rather abruptly in 1965, the curtain was drawn on the stage of our little group of mathematicians in Calcutta, with the departure of Varadarajan to Los Angeles, Ranga Rao to Illinois and myself to Sheffield. Calcutta was a difficult city, and most of our life was spent in temporary shacks with asbestos roofs and bamboo partitions. All along, we knew that our existence here was fragile, and with age and marriage, the inevitable departure from Calcutta to greener pastures in scientific as well as material terms took place.

4. Sheffield and Manchester

At the suggestion of Joe Gani, then at the University of Sheffield, I started a seminar course on probability theory along the lines of research pursued by Varadarajan, Ranga Rao, Varadhan and me during the Calcutta period. This led me to write a comprehensive set of notes on probability measures on metric spaces with emphasis on the convolution operation in general groups. Professor Eugene Lukacs, who visited the department at Sheffield, suggested the publication of these notes in a series published under his editorship by Academic Press, and I readily agreed. Thus [26] was published in 1967, and I subsequently left to take up a chair in the University of Manchester Statistical Laboratory, within the Sheffield–Manchester Joint School of Probability and Statistics.

With the help of Roger Plymen of the mathematics department at Manchester and his graduate students B. Falkowski and Miss P. Adamson, I started a seminar on the mathematical foundations of quantum mechanics. During this period I was invited by Ray Streater of Bedford College, London to give a lecture on infinitely divisible distributions on topological groups. Ray revealed to me the remarkable connections that exist between stochastic processes with independent increments, continuous tensor products of Hilbert spaces and first-order cocycles of group representations. He gave a brief outline of his work [50] and that of Araki and Woods [3] and Araki [2]. Subsequently a brilliant young mathematician, Klaus Schmidt, came to Manchester from Vienna as a British Council scholar seeking a change from the lines of his research for his Ph.D. in equidistribution theory. I presented to him a brief outline of the Araki–Streater theory of continuous tensor products and indicated that, in this context, it might be profitable to look at positive-definite kernels invariant under a group action.

The Russian novelist A. I. Solzhenitsyn has written that, "When a great opportunity comes and knocks at the door, its first tap may be no louder than the beating of one's heart" and it is necessary to seize it. Ray's remark and my subsequent contact with Klaus turned out to be such an opportunity; it has led to a long-standing collaboration between us which resulted in the publication of [37], [38], [39]–[41]. (See also [6], [48].) The seminar programme at Manchester enabled me to come into contact with John Lewis of Oxford and Robin Hudson of Nottingham. Their lectures, followed by several conversations, helped me immensely in obtaining a clearer perspective of the notion of stochastic processes from the quantum mechanical point of view.

5. Return to India and ISI

During my visit to The Australian National University in the summer of 1968, at the invitation of Ted Hannan, I stayed at University House and I had the pleasure to be in the company of Professor and Mrs M. H. Stone and Professor K. Mahler at the breakfast table almost every day. One day, rather casually, Professor Stone asked me, "Why don't you go back to India and work for the good of Indian mathematics?" I remarked rather cynically that there was very little chance of my getting a job outside ISI, Calcutta. Stone's reply was that he would "arrange" a position for me outside Calcutta!

In 1969 Professor S. S. Shrikhande of the University of Bombay wrote to me that the University Grants Commission of India had "created" a chair for me at his centre, and I should accept it. With the encouragement of my wife Shyama and against the wishes and warnings of my dear friends, especially Varadarajan, I came to Bombay with my family at the beginning of the great monsoon season in 1970. The university provided me with comfortable accommodation, and very soon I settled down to work in the congenial atmosphere of Shrikhande's department. I started a seminar on quantum theory at the luxurious premises of the Tata Institute of Fundamental Research (TIFR). The institute was generous enough to provide me with office space in their excellent air-conditioned building with the best-equipped library in India. The Mathematics Institute at the University of Warwick in Britain, and the personal interest that Joe Gani, then of CSIRO, Australia, showed in my welfare enabled me to continue my international contacts; but for this help I wonder whether I could have continued my research work.

In 1973 the Indian Institute of Technology (IIT), Delhi, made a generous offer (by Indian standards), and, being very much impressed by their spacious and pleasant surroundings, I accepted it. My work with Klaus Schmidt and J. L. Doob's comments in the appendix to the book [8] of B. V. Gnedenko and A. N. Kolmogorov constantly reminded me that there should be a way of generalizing the classical central limit theorems of P. Lévy and A. I. Khinchine to products of uniformly infinitesimal arrays of positive definite functions on Lie groups, and this would provide a link with the limit theorems of quantum statistical mechanics. Even though I succeeded in proving such limit theorems in [28], [30] without invoking "probability measure", I have come nowhere near understanding this connection with quantum theory.

I began to devote myself to the teaching responsibilities at the IIT. The IIT's in India have a national character, and the selection of students for the B. Tech. courses is done on the basis of an all-India entrance examination which is fiercely competitive. The top 25 among the students who graduate from any IIT go abroad and work for the multinational companies. Students are inevitably better than the teachers, and I enjoyed teaching them. However, their examination system discouraged me completely. Hardly anyone can pass the entrance examination unless he or she has been grilled for a year in one of the numerous tutorial colleges run solely for this purpose. Seldom did I meet here a student from rural India.

Furthermore, the technologists who control the IITs do not understand the nature of mathematical research and teaching, and make the budgeting of mathematics departments a persistent headache for their chairmen. They would grant limitless amounts of finance for hurriedly prepared projects, but not an extra rupee for enriching the library by a single new journal. I often wonder how one can prepare a "project" honestly in mathematical research.

On 31 December 1974 Prime Minister Indira Gandhi inaugurated the new campus of the Indian Statistical Institute in New Delhi, and soon after this C. R. Rao suggested to me that when the buildings were completed I should think of moving over to ISI. Already, in my spare hours, I was teaching a course on probability theory, Brownian motion and the Ito calculus for the M. Stat. students of ISI at their old premises in Yojana Bhavan (the centre of economic planning in India). Towards the end of 1976 I gladly joined ISI once again after a lapse of nearly 12 years and freed myself from the perpetual examination fever of the IIT. My pedagogical zeal during the IIT years led me to write the textbook [29] in the hope that this would make probability theory a part of the undergraduate and graduate curriculum in the Indian universities. By this time the Indian scientific community had generously showered on me awards like the Bhatnagar prize, fellowship of the Indian National Science Academy, Delhi and the Indian Academy of Sciences, Bangalore.

With the assistance of my students I organized a seminar on the perturbation theory of linear operators along the lines of the great book [17] of T. Kato. Our goal was to understand the properties of Schrödinger operators of quantum mechanics. The publication of the lecture notes [31], [34] enabled me to learn a new subject, and two of our students at ISI to begin their research work. The second volume of these notes received a bad review in the *Bulletin of the London Mathematical Society* which among other things, including [26], was later quoted as evidence of scientific dishonesty by some of my colleagues.

I would only stress the usefulness and importance of writing lecture notes for many advanced courses with the help of students, and making them available to the student community at a price we can afford in India, without being discouraged by the unexpected hazards and criticisms that may follow.

This is particularly relevant when Western publishers are escalating the prices of their books and preventing the easy spread of scientific knowledge in economically disadvantaged countries. Added to this is the lurking fear that in the near future scientific knowledge may be stored only in computers and communicated through telephone networks; without the expensive imported equipment (which may frequently break down in our environment) the pleasure of self-study by conventional methods may be denied to us.

In the summer of 1981 I visited the universities of Warwick and Nottingham and continued my discussions with Robin Hudson and Ray Streater; we exchanged ideas on the results of [11], [12], [32] and [42]. We were led to the conclusion that it should be possible to develop a noncommutative or quantum version of Itô's stochastic calculus. Whereas Ray and his group proceeded along the algebraic direction depending on the theory of Von Neumann algebras [4], Robin and I proceeded along the probabilistic route with our feet entrenched in the classical Itô calculus.

If you look at a classical probability space $(\Omega, \mathscr{F}, \mu)$ a real-valued random variable on Ω can be interpreted as the self-adjoint operator of multiplication by that random variable. If $(\Omega, \mathscr{F}, \mu)$ is the Brownian motion probability space so that $w \in \Omega$ is a typical Brownian path, denote by $Q(t)$ the self-adjoint operator of multiplication by $w(t)$. Then $\{Q(t), t \geq 0\}$ is an operator-valued function, but we call it an "operator-valued process". All the operators $Q(t)$, $t \geq 0$ commute. If $f(t, w)$ is a nonanticipating Brownian functional the process $\{\int_0^t f(s, w) \, dw(s), t \geq 0\}$ can also be interpreted as the operator process $\{\int_0^t f(s, Q) \, dQ(s), t \geq 0\}$. Since the probability measure μ of Brownian motion is quasi-invariant under a large class of translations by smooth (deterministic) functions, it suggests that there should be a parallel self-adjoint operator process $\{P(t), t \geq 0\}$ such that $[P(t), P(s)] = 0$ for all s, t and $[Q(t), P(s)] = Q(t)P(s) - P(s)Q(t) = 2i \min(t, s)$.

All the information about the pair $\{Q, P\}$ is contained in $A(t) = \frac{1}{2}(Q + iP)(t)$, $t \geq 0$. Instead of dealing with Q, P, one may as well deal with $\{A(t), A^\dagger(t), t \geq 0\}$ where $A^\dagger(t) = \frac{1}{2}(Q - iP)(t)$ is adjoint to $A(t)$. A and $A^\dagger$ are called the *annihilation* and *creation* processes respectively. The Hilbert space $L_2(\mu)$ admits the tensor product decomposition $L_2(\mu) = H_t \otimes H^t$ where H_t and H^t are respectively the spaces of square-integrable functionals of $\{w(s), s \leq t\}$ and $\{w(s) - w(t), s \geq t\}$. An operator-valued map F on $[0, \infty)$ where $F(t) = F_0(t) \otimes 1^t$, $F_0(t)$ an operator in H_t and 1^t the identity in H^t, is called an adapted process. If E, F, G are three such adapted processes, one proceeds to define the process

$$X(t) = \int_0^t (E \, dA + F \, dA^\dagger + G \, dt), \qquad t \geq 0.$$

If $E = F$ and G are multiplications by nonanticipating Brownian functionals then $X(t)$ turns out to be multiplication by the random variable $\int_0^t (E \, dw + G \, dt)$. One can couple $L_2(\mu)$ to an initial Hilbert space H_0 and consider the same problem in $H_0 \otimes L_2(\mu)$. If $X(t) = X(0) + \int_0^t (E \, dA + F \, dA^\dagger +$

$G\,dt$) we write $dX = E\,dA + F\,dA^\dagger + G\,dt$. If

$$dX_i = E_i\,dA + F_i\,dA^\dagger + G_i\,dt, \qquad i = 1, 2$$

it turns out that

$$d(X_1 X_2) - (dX_1)X_2 - X_1(dX_2) = E_1 F_2\,dt. \tag{1}$$

We first established (1) in special cases using the classical Itô calculus. A preliminary version of this idea [16] and quantum diffusion process [12] was presented at an international symposium held in January 1982 at Bangalore as a part of the golden jubilee celebrations of ISI. Robin Hudson visited ISI in order to attend this symposium and also to work with me in Delhi as an exchange visitor of the Royal Society and the Indian National Science Academy. The final version of (1) in the boson case was obtained during my subsequent visit to Warwick and Nottingham in the summer of 1983, relying entirely on the commutation rules of quantum fields which have their origin in the celebrated Heisenberg uncertainty principle.

The classical Itô formula for Brownian motion (as well as the Poisson process) turned out to be an offspring of Heisenberg commutation relations. Furthermore, one considers quantum stochastic differential equations of the form

$$U(0) = 1, \qquad dU = U\{E\,dA - E^\dagger dA^\dagger + (iH - \tfrac{1}{2}EE^\dagger)\,dt\}, \tag{2}$$

where $\{U(t), t \geq 0\}$ is an unknown adapted process, E and H are known adapted processes and $H(t)$ is self-adjoint for each t. If $E \equiv 0$ and $H(t)$ is independent of t, (2) is nothing but the abstract Schrödinger equation of quantum theory. Under fairly general conditions, equation (2) was solved in [13].

However, the most interesting question is the unitarity of the operators $U(t)$. The unitarity of $U(t)$ was established in [15] when E and H are constant bounded operators and the unbounded case is still elusive except in some special examples. Corresponding to the "noisy" Schrödinger equation (2) one can also formulate a "noisy" Heisenberg equation. For all the gory details and some applications of these ideas we refer to [7], [13], [14], [42]. For the subsequent and natural developments in the fermionic stochastic calculus the reader may refer to the recent thesis [1] of D. Applebaum. The discussions which ensued between me and Luigi Accardi of Rome, who visited ISI during the winter of 1983, led us to an algebraic version of Itô's formula (1) encompassing the boson and fermion cases.

Of course, it will take several more years to comprehend the intricacies of a quantum diffusion theory and assess the merit or usefulness of this circle of ideas. Nevertheless it was a great excitement for me personally to participate in this endeavour, cultivate and strengthen close friendship with many colleagues in India and abroad in the process, and also demonstrate the possibility of doing some research work in India when supported by adequate contact with the international community.

6. Concluding Remarks

It has been my particular good fortune to have been provided with a job of mathematical research which is probably a luxury in an economically backward nation like India, and I feel that I am well paid and looked after for all the pleasures that I am allowed to indulge in. It definitely concerns me and several of my distinguished colleagues in ISI when we see in our own campus the comfortable homes that we have established for ourselves and our assistants with the sweat of labourers in the neighbouring jhuggis, who toil in the cold winter as well as the hot summer of Delhi, and whose children run around the campus barefoot in rags. Right opposite our campus is the old village of Katwaria Sarai, epitomizing all the economic, demographic and social problems of India, defeating the effectiveness of all economic theories and programmes and making us constantly wonder whether scientific research can in any way alleviate the misery of our fellow brethren. However, the intellectual battle has to go on, for "knowledge that is not used for the winning of further knowledge does not even remain, it decays and disappears". If there is no hope in science, where else does it lie?

I would like to conclude this essay with sincere apologies for any factual errors in the text that was written mostly from memory, and for any offence that I might have caused to any one.

Publications and References

[1] Applebaum, D. (1984) Fermion Stochastic Calculus. Ph.D. Thesis, University of Nottingham.

[2] Araki, H. (1970) Factorisable representations of current algebra. *Publ. Res. Inst. Math. Sci. Kyoto Univ.* A 5, 361–422.

[3] Araki, H. and Woods, E. J. (1966) Complete boolean algebras of type I factors. *Publ. Res. Inst. Math. Sci. Kyoto Univ.* 2, 157–242.

[4] Barnett, C., Streater, R. F. and Wilde, I. F. (1982) The Itô–Clifford integral. *J. Functional Anal.* 48, 172–212.

[5] Bhattacharya, R. N. and Ranga Rao, R. (1976) *Normal Approximation and Asymptotic Expansions.* Wiley, New York.

[6] Erven, J. and Falkowski, B. J. (1981) *Low Order Cohomology and Applications.* Lecture Notes in Mathematics 877, Springer-Verlag, Berlin.

[7] Frigerio, A. (1984) Covariant Markov dilations of quantum dynamical semigroups. Preprint.

[8] Gnedenko, B. V. and Kolmogorov, A. N. (1954) *Limit Distributions for Sums of Independent Random Variables.* Addison-Wesley, Cambridge, MA.

[9] Gray, R. M. and Davisson, L. D. (EDS.) (1977) *Ergodic and Information Theory.* Dowden, Hutchinson and Ross, Stroudsburg, PA.

[10] Halmos, P. R. (1953) *Lectures on Ergodic Theory.* Chelsea, New York.

[11] Hudson, R. L. and Ion, P. D. F. (1981) The Feynman-Kac formula for a canonical Wiener process. *Proc. Colloq. Random Fields: Rigorous Results in Statistical Mechanics and Quantum Field Theory.* North-Holland, Amsterdam.

[12] Hudson, R. L. and Parthasarathy, K. R. (1983) Quantum diffusions. In *Theory and Applications of Random Fields*, ed. G. Kallianpur *et al.*, Lecture Notes in

Control and Information Sciences 49, Springer-Verlag, Berlin, 111–121.

[13] Hudson, R. L. and Parthasarathy, K. R. (1984) Quantum Itô's formula and stochastic evolutions. *Commun. Math. Phys.* 93, 301–323.

[14] Hudson, R. L. and Parthasarathy, K. R. (1984) Stochastic dilations of uniformly continuous, completely positive semigroups. *Acta Applicandae Math.*

[15] Hudson, R. L. and Streater, R. F. (1981) Itô's formula is the chain rule with Wick ordering. *Phys. Letters* 86 A, 277–279.

[16] Hudson, R. L., Karandikar, R. L. and Parthasarathy, K. R. (1983) Towards a theory of non-commutative semi-martingales adapted to Brownian motion and a quantum Itô's formula. In *Theory and Applications of Random Fields*, ed. G. Kallianpur et al., Lecture Notes in Control and Information Sciences 49, Springer-Verlag, Berlin, 96–110.

[17] Kato, T. (1976) *Perturbation of Linear Operators*. Springer-Verlag, Berlin.

[18] Khinchine, A. I. (1957) *Mathematical Foundations of Information Theory*. Dover, New York.

[19] Lévy, P. (1952) Sur une classe de lois de probabilité indécomposables. *C. R. Acad. Sci. Paris* 235, 489–492.

[20] Mackey, G. W. (1963) *The Mathematical Foundations of Quantum Mechanics*. Benjamin, New York.

[21] Oxtoby, J. C. (1961) On two theorems of Parthasarathy and Kakutani concerning the shift transformation. In *Ergodic Theory, Proc. Internat. Symp. Tulane University*, Academic Press, New York, 203–215.

[22] Parthasarathy, K. R. (1961) On the category of ergodic measures. *Illinois J. Math.* 5, 648–656.

[23] Parthasarathy, K. R. (1961) On the integral representation of the rate of transmission of a stationary channel. *Illinois J. Math.* 5, 299–305.

[24] Parthasarathy, K. R. (1963) Effective entropy rate and transmission of information through channels with additive random noise. *Sankhyā* A 25, 75–84.

[25] Parthasarathy, K. R. (1964) The central limit theorem for rotation groups. *Theory Prob. Appl.* 9, 248–257.

[26] Parthasarathy, K. R. (1967) *Probability Measures on Metric Spaces*. Academic Press, New York.

[27] Parthasarathy, K. R. (1972) A probabilistic approach to the Pontrjagin duality theorem. *Period. Math. Hungar.* 2, 21–26.

[28] Parthasarathy, K. R. (1974) The central limit theorem for positive definite functions on a locally compact group. *J. Multivariate Anal.* 4, 123–149.

[29] Parthasarathy, K. R. (1977) *Introduction to Probability and Measure*. Macmillan India, New Delhi (Russian translation: MIR, Moscow, 1983).

[30] Parthasarathy, K. R. (1977) The central limit theorem for positive definite functions on Lie groups. *Symposia Mathematica, Pub. Ist. Naz. Alt. Mat. XXI*, Academic Press, New York, 245–256.

[31] Parthasarathy, K. R. (1979) *Lectures on Functional Analysis II*. ISI Lecture Notes 6, Macmillan India, New Delhi.

[32] Parthasarathy, K. R. (1982) On a class of time inhomogeneous nonsingular flows and Schrödinger operators. *Math. Z.* 179, 123–133.

[33] Parthasarathy, K. R. (1984) A remark on the integration of Schrödinger equation using quantum Itô's formula. *Lett. Math. Phys.*

[34] Parthasarathy, K. R. and Bhatia, R. (1978) *Lectures on Functional Analysis I*. ISI Lecture Notes 3, Macmillan India, New Delhi.

[35] Parthasarathy, K. R. and Bingham, M. S. (1968) A probabilistic proof of Bochner's theorem on positive definite functions. *J. London Math. Soc.* 43, 626–632.

[36] Parthasarathy, K. R. and Sazonov, V. V. (1964) On the representation of infinitely divisible distribution on a locally compact abelian group. *Theory Prob. Appl.* 9, 118–122.

[37] Parthasarathy, K. R. and Schmidt, K. (1972) Factorisable representations of current groups and the Araki–Woods imbedding theorem. *Acta Math.* 128, 53–71.

[38] Parthasarathy, K. R. and Schmidt, K. (1972) *Positive Definite Kernels, Continuous Tensor Products and Central Limit Theorems of Probability Theory.* Lecture Notes in Mathematics 272, Springer-Verlag, Berlin.

[39] Parthasarathy, K. R. and Schmidt, K. (1975) Stable positive definite functions. *Trans. Amer. Math. Soc.* 203, 161–174.

[40] Parthasarathy, K. R. and Schmidt, K. (1976) A new method for constructing factorisable representations of current groups and current algebras. *Commun. Math. Phys.* 50, 167–175.

[41] Parthasarathy, K. R. and Schmidt, K. (1977) On the cohomology of a hyperfinite action. *Monatsh. Math.* 84, 37–48.

[42] Parthasarathy, K. R. and Sinha, K. B. (1982) A random Trotter–Kato product formula. In *Statistics and Probability, Essays in Honour of C. R. Rao,* ed. G. Kallianpur *et al.,* North-Holland, Amsterdam, 553–565.

[43] Parthasarathy, K. R., Ranga Rao, R. and Varadhan, S. R. S. (1962) On the category of indecomposable distributions on topological groups. *Trans. Amer. Math. Soc.* 102, 200–217.

[44] Parthasarathy, K. R., Ranga Rao, R. and Varadhan, S. R. S. (1963) Probability distributions on locally compact abelian groups. *Illinois J. Math.* 7, 337–369.

[45] Parthasarathy, K. R., Ranga Rao, R. and Varadarajan, V. S. (1967) Representations of complex semi-simple Lie groups and Lie algebras. *Ann. Math.* 25, 383–420.

[46] Ranga Rao, R. (1962) Relations between weak and uniform convergence of measures with applications. *Ann. Math. Statist.* 33, 659–680.

[47] Rohlin, V. A. (1967) Lectures on the entropy theory of transformations with invariant measure. *Uspehi Mat. Nauk* 22, 3–26 (*Russian Math. Surveys* 22, 1–52).

[48] Schmidt, K. (1971) Limits of uniformly infinitesimal families of projective representations of locally compact groups. *Math. Ann.* 192, 107–118.

[49] Shohat, J. A. and Tamarkin, J. D. (1943) *The Problem of Moments.* Math. Surveys 1, American Mathematical Society, New York.

[50] Streater, R. F. (1969) Current commutation relations, continuous tensor products and infinitely divisible group representations. *Rend. Soc. Int. Fisica E. Fermi* XI, 247–263.

[51] Varadarajan, V. S. (1961) Measures on topological spaces. *Mat. Sb.* 55, 35–100 (*Amer. Math. Soc. Transl.* (2) 48, 161–228).

[52] Varadhan, S. R. S. (1962) Convolution Properties of Distributions on Topological Groups. Ph.D. Thesis, ISI, Calcutta.

Marius Iosifescu

Marius Vicenţiu Viorel Iosifescu was born on 12 August 1936 in Piteşti, Romania, the son of Victor and Ecaterina Iosifescu. After completing his secondary education at the Nicolae Bălcescu gymnasium in Piteşti, he studied mathematics at the University of Bucharest and was awarded the degrees of Doctor of Mathematics in 1963, and Doctor of Sciences six years later.

From 1959 to 1962 he worked as a consultant at the Central Board of Statistics in Bucharest. After a short period from 1961 to 1963 as assistant professor of mathematics at the Bucharest Polytechnic Institute, he joined the Institute of Mathematics of the Romanian Academy as a research worker in the probability division. When in 1964 the latter was expanded into the Centre of Mathematical Statistics he moved to it and rose to be its head in 1976; this is the position he currently holds.

In 1971 Dr Iosifescu was elected an overseas fellow of Churchill College, Cambridge, England. Subsequently, he has been visiting professor at the universities of Paris, Mainz, Frankfurt-am-Main and Bonn.

Dr Iosifescu has published a large number of research papers in functions of real variables, operations research, and probability, and is the author of nine books. He has twice been awarded prizes by the Romanian Academy, in 1965 and 1972, for his work in stochastic processes and their applications.

Dr Iosifescu is a member of the International Statistical Institute, the Biometric Society and the Bernoulli Society for Mathematical Statistics and Probability. He is also a member of the editorial boards of several international and Romanian mathematical journals.

Dr Iosifescu married Ştefania Eugenia Zamfirescu on 21 July 1973; they have one son. His hobbies are music and playing the violin.

From Real Analysis to Probability: Autobiographical Notes

To the memory of Gheorghe Mihoc and Octav Onicescu

1. Early Years and Education

I was born on 12 August 1936 in Pitești, then a small town of some 20,000 inhabitants situated on the river Argeș, about 120 km north-west of Bucharest. It has now become an important industrial centre with more than 150,000 inhabitants. During my childhood, Pitești enjoyed the reputation of being a clean and attractive town with very good secondary schools modelled after the French lycées. My parents were teachers there, and as a child I thought I would also become a teacher.

Except during the Second World War, I can honestly say that my parents managed to give my sister and me a happy childhood. The terrible realities of war took the form of frequent American air raids, one of which completely destroyed our home; they were later followed by the unpleasant consequences of Romania's defeat. My aversion to brute force, intolerance, violence and arbitrariness doubtless dates from that time. All in all, I would not exchange the landscape of my childhood, with its light and shade, for anything in the world.

During the period 1947–54 at the Nicolae Bălcescu (formerly Ion C. Brătianu) gymnasium I was fortunate to have excellent teachers in almost all subjects. I also indulged in sports, and it is from that time that my love for good music, especially classical music, originated. This was based on private study of the violin encouraged by my father. On leaving the gymnasium, I had a good knowledge of French and a passable one of Russian, a fact for which, as a mathematician, I had many opportunities to congratulate myself later. I learned English long after, in the late 1960s; as to German, I regret that my knowledge is negligible. I have never found the time to fill this gap in my education. And curious though it may seem, I never learned Latin at school, in spite of the fact that Romanian is a romance language. I have highlighted my knowledge of widely spoken languages, because for those whose mother tongue is not among them, a good education cannot be conceived without their study.

My interest in mathematics was aroused when I was 14, and I started tackling problems proposed for solution in the Romanian journal for young people called *Revista Matematică și Fizică*. This was a successor to the celebrated *Gazeta Matematică*, which during the period 1895–1948 served as an ideal apprentices' magazine for the independent work of many Romanian mathematicians of the past. I remember well how a few successful attempts at solving problems gave me the impetus to continue on this path. If my interest in mathematics became a passion, this was largely due to Dumitru Mihalașcu

(1908–84) my only mathematics teacher at the gymnasium. I shall always revere his memory for his tactful guidance, patience, and understanding.

In 1953 and again in 1954 I was able to win first prize in the National Mathematical Olympiads organized by the Society for Mathematical and Physical Sciences of Romania. This body was also the initiator of the International Mathematical Olympiads, of which the first two sessions were held in 1959 and 1960 in Romania. These successes led me naturally to the idea of becoming a mathematician. Apart from the subjective motivation, there also existed a favourable objective context to a mathematical career. In 1949, under the auspices of the Academy of the Romanian People's Republic, which was the reorganized Romanian Academy founded in 1866, a research mathematical institute was set up in Bucharest. Patterned after the corresponding Soviet institutions, the Institute of Mathematics of the Academy of (from 1965) the Socialist Republic of Romania played an overwhelming role in the flourishing of mathematical research in this country between 1949 and 1975. The best mathematics students entered this institute almost automatically after finishing their studies. Thus, partly for subjective and partly for objective reasons, my dream was to study mathematics and enter the Institute of Mathematics. Of course, at the time, my idea of what mathematical research meant was quite imprecise. I spontaneously held the Halmos-like view that "problems are the heart of mathematics" [16]. Later on, I realized that what really counts is the relevance of the problems.

2. The University of Bucharest, 1954–59

After graduating from the gymnasium I enrolled in the Faculty of Mathematics and Physics (now Mathematics, since Physics separated in 1962) at the University of Bucharest. This tells only half the story, so perhaps I should give some details. My parents were teachers who wished to save me from the difficulties of their profession, fearing that I would also become a teacher. Consequently, they insisted that I should choose the more practical profession of engineer. After much debate I agreed to study both engineering and mathematics. So, I first enrolled in the Faculty of Electrotechnics of the Bucharest Polytechnic. But three weeks were quite enough to prove my complete incompatibility with the training of engineers. I went to consult Professor Gheorghe Mihoc (1906–81), then the dean of the Faculty of Mathematics and Physics, who knew me from the Mathematical Olympiads. He agreed at once to enrol me in his faculty, and my excursion into engineering came to an end, except for a short period seven years later when I returned to the polytechnic as an assistant professor of mathematics.

My university years (1954–59) fell in a period when all the great names of interwar Romanian mathematics were still active. Thus, among my professors were Dan Barbilian (1895–1961), Alexandru Froda (1894–1973), Alexandru Ghika (1902–64), Grigore C. Moisil (1906–73), Miron Nicolescu (1903–75),

Octav Onicescu (1892–1983), Nicolae Teodorescu, Victor Vâlcovici (1885–1970), and Gheorghe Vrânceanu (1900–79). To these should be added Cabiria Andreian Cazacu, Tudor Ganea (1922–71), C. T. Ionescu Tulcea, Solomon Marcus and Gheorghe Marinescu, who made their names as mathematicians after the Second World War.

I should mention that the gymnasium syllabus for mathematics did not then go beyond elementary mathematics; the situation is radically different now. On my own initiative, I had learned something of the algorithmic aspects of single- and multivariable calculus and analytical geometry. Consequently, the course in ε–δ analysis given by Ionescu Tulcea in my first university year, which maintained a high level of mathematical rigour, was something of a shock. From that course, I learned what a mathematical proof is; I should confess that the course was decisive in my formation as a mathematician. In the second year, the calculus course was given by Marcus who, following his research interests, oriented it towards real functions. From Marcus, I learnt to ask "How do we know in mathematical analysis?" and, consequently, how to ask and answer mathematical questions. As a result I was able to do my first piece of original mathematical work (see the next section for details); Professor Marcus was my first and, unfortunately, last supervisor.

During my university years I also took courses in algebra (with Froda and Moisil), in foundations of geometry and differential geometry (with Vrânceanu and his assistants), in calculus of variations (with Marinescu), in complex functions (with Cabiria Andreian Cazacu), in differential equations (with Teodorescu), in functional analysis (with Ghika) in mechanics (with Vâlcovici), in algebraic number theory (with Barbilian) in probability (with Onicescu), in real functions (with Froda and Nicolescu) and in general topology (with Ganea).

The course in number theory given by Barbilian was quite unusual, as was Barbilian himself as a man and mathematician. Though his lectures had been prepared well in advance, spontaneous ideas for new proofs and results flashed through his mind during the class, thus offering us the living spectacle of creative mathematics. I have no doubt this happened because Barbilian is one of the four or five poets of genius in Romanian literature; as a poet he is known under the pen-name of Ion Barbu. Barbilian's double career as a scientist and an artist offers us support for Karl Weierstrass' well-known opinion (see [2], p. 432) that a mathematician who is not also something of a poet will never be a perfect mathematician. Marcus [61], p. 142, remarks that while the mathematician Barbilian can be easily identified in Barbu's poetry, we still do not have a serious analysis of the poet Barbu in Barbilian's mathematical texts. I am convinced that a thorough investigation of the Barbilian phenomenon could shed light on whether mathematics is an art or a science, or both (see, e.g. [5]).

3. Work in Real Analysis

My papers in real analysis are concerned with two subjects: differential properties of real functions of a real variable; and properties of derivatives of real functions of a real variable. These papers were motivated by questions raised by Marcus.

3.1. Differential Properties of Real Functions of a Real Variable

The theory of differentiation of real functions of a real variable depends on Lebesgue's theorem asserting the almost everywhere differentiability of any monotonic function of a real variable. Using this theorem in 1924 the Romanian mathematcan Simion Stoilow (1887–1961) proved the following theorem (see [69], [70] and [71]):

Let f be a continuous real function defined on an interval $I \subset R$. There exists a set $A \subset f(I) = \{y = f(x): x \in I\}$ of Lebesgue measure 0 such that whatever $y \in f(I) - A$:

(i) *At any point x of the first kind of the level set $L_y = \{x: f(x) = y\}$ (i.e. such that x is a unilateral limit of points in L_y) there exists a Dini derivative[1] (finite or infinite) which equals the opposite Dini derivative. At any isolated point of L_y there exists a unique derivative (finite or infinite).*

(ii) *If the set of values taken by the two left or right Dini derivatives at the points of L_y, where the side may vary with the point, is bounded (unbounded but not containing the values $-\infty$ and ∞) then L_y is finite (countable).*

Stoilow's theorem allows us to obtain the Denjoy 1915 theorem (see below) concerning the Dini derivatives of a continuous function. Moreover, it contains a lot of results (sometimes in stronger versions), which have subsequently been derived by authors who ignored Stoilow's papers. A penetrating analysis of the matter can be found in [61], pp. 69–91 (see also [60] and [1], pp. 147–72).

In my papers [24], [25] and [27][2] I took up the problem of the role played by the hypothesis of continuity in part (i) of Stoilow's theorem. Thus I was led to define what I have called property (m). A real function defined on an interval I is said to have the one-sided (two-sided) property (m) at $x_0 \in I$ if on one side (on both sides) of x_0 there is (are) one (two) interval(s) with an extremity in x_0 on which either $f(x) > f(x_0)$ or $f(x) < f(x_0)$. Clearly, the one-sided property (m) can hold only at points of the first kind of the level sets, while the two-sided property (m) can hold only at isolated points of the level sets. The basic result,

---

[1] For the definition of the Dini derivatives see, e.g. [6], p. 52.
[2] Papers [25] and [27] are in German. An explanation is necessary in view of my earlier remark (see page 251). At the time of their publication there were strict quotas for languages in our *Revue (Roumaine) de Mathématiques Pures et Appliquées*. The same situation occurred again with my papers [26] and [28] which were published in Russian.

which I have established via some constructions involving the association of four monotonic functions with any rational point in I, reads as follows.

Basic lemma. *Given an arbitrary real function f defined on an interval I there are two sets B and $C \subset I$ of Lebesgue measure 0 such that:*

(i) *At any point in $I - B$ at which f has the one-sided property (m) there exists a Dini derivative (finite or infinite) which equals the opposite Dini derivative; at any point in $I - B$ at which f has the two-sided property (m) there exists a unique derivative (finite or infinite).*

(ii) *At any point in $I - C$ at which f has the one-sided property (m) there exists a finite Dini derivative which equals the opposite Dini derivative; at any point in $I - C$ at which f has the two-sided property (m) there exists a unique finite derivative.*

Now, a real function f defined on an interval I is said to be a Darboux function or to have the intermediate value property if, whatever $x_1, x_2 \in I$, for y any number between $f(x_1)$ and $f(x_2)$, there is a number x_3 between x_1 and x_2 such that $f(x_3) = y$. It is easy to see that for a Darboux function the one-sided property (m) holds exactly at the points of the first kind of the level sets, while the two-sided property (m) holds exactly at the isolated points of the level sets. Accordingly, the basic lemma can be stated as a property of Darboux functions, which finally leads to the conclusion that part (i) of Stoilow's theorem holds for a Darboux (rather than a continuous) function f.[3]

It is worth mentioning that the basic lemma above has a definite independent methodological value. Indeed, it allows an almost immediate proof, perhaps the simplest one presently known (see [6], p. 65 for an account of the standard proof) of the celebrated Denjoy–Young–Saks theorem.[4] According to this, for an *arbitrary* real function f defined on an interval I, there exists a set $D \subset I$ of Lebesgue measure 0 such that at any point in $I - D$ two opposite Dini derivatives are simultaneously either finite and equal or infinite and unequal, the upper one being $+\infty$. (It then follows that at any point in $I - D$ two associated Dini derivatives are either unequal, at least one of them being infinite, or equal and finite.)

Let me conclude these considerations about Stoilow's theorem by discussing an important special case. It refers to the class of continuous real functions f defined on an interval I for which all level sets $L_y, y \in f(I)$, are finite. It follows that any function in this class is differentiable almost everywhere.[5] This result, obtained by André Marchaud in [58], is in my opinion a very beautiful one; my first published paper [21] was intended to give a "simple" proof of Marchaud's

---

[3] It is still an open problem whether part (ii) can be extended to functions more general than the continuous functions f.

[4] The Denjoy 1915 theorem I referred to above is historically the first version (for continuous functions) of this theorem.

[5] Clearly, this conclusion holds more generally for any Darboux function for which all level sets are discrete (for a direct proof see [59]). Such a function can have at most a denumerable set of points of discontinuity [25].

theorem. I soon discovered a slip in my reasoning. See [23], where I also showed that "finite" cannot be replaced by "at most denumerable". Nevertheless, I now think that "continuous" can be replaced by "almost everywhere continuous", which would constitute a considerable improvement of Lebesgue's fundamental theorem. It is perhaps interesting to note that Marchaud's name has recently been quoted (see [55]) in an applied probability context; it appears that Marchaud [57] defined the concept of a fractional derivative. This is nothing but a confirmation of the constant interplay between mathematical analysis and probability theory.

3.2. Properties of Derivatives of Real Functions of a Real Variable

My work on derivatives started from a result of the Czechoslovak mathematicians V. Hruška and V. Jarník (see [19]) asserting that the ratio of two finite derivatives, though not always a derivative, is a Darboux function.[6] The proof essentially makes use of the Cauchy intermediate value formula

$$\frac{F(b) - F(a)}{G(b) - G(a)} = \frac{F'(\theta)}{G'(\theta)} \qquad (1)$$

for some $\theta \in [a, b]$. It is then quite natural to ask whether the product of two derivatives (which is not always a derivative) is a Darboux function. From the very beginning I looked for a counterexample (a wrong proof had been given and no formula analogous to (1) exists for a product). I still remember my enthusiasm when at the end of my second university year I succeeded in finding the desired counterexamples [22]. One of them is as follows. The functions

$$f_1(x) = \begin{cases} \sin(\frac{1}{x}) & \text{if } x \neq 0 \\ 0 & \text{if } x = 0 \end{cases} \quad \text{and} \quad f_2(x) = \begin{cases} \sin^2(\frac{1}{x}) & \text{if } x \neq 0 \\ \frac{1}{2} & \text{if } x = 0 \end{cases}$$

are derivatives. Then the product of the derivatives $f_2 - f_1$ and $f_2 + f_1$ is the function

$$f(x) = \begin{cases} -\frac{1}{4}\sin^2(\frac{2}{x}) & \text{if } x \neq 0 \\ \frac{1}{4} & \text{if } x = 0 \end{cases}$$

which, clearly, is not a Darboux function. It also appeared that the difference or the sum of squares of two derivatives may not be a Darboux function.

A subsequent paper [26] took up the problem of finding conditions for the product of two derivatives to be a derivative. Let me quote just the following two results.

---

[6] As is well known, a derivative (finite or not) is a Darboux function. (This result, proved by G. Darboux in 1875, gave the name to the concept.) It is certainly less well known that a heuristic proof of the Darboux theorem is contained in Galileo Galilei's 1638 work *Discorsi e dimonstrazioni matematiche intorno à due nuove scienze*, where one can read that "Plato perhaps had the idea that a body cannot go from rest to a positive velocity ... without passing through all smaller velocities" (see R. P. Boas, *Amer. Math. Monthly* 88 (1981), p. 156).

(i) *Let $f \in L^2(I)$. A necessary and sufficient condition that f and f^2 both be derivatives is that all points $x \in I$ be Lebesgue points of f, i.e.*

$$\lim_{\substack{y \to x \\ y \in I}} \frac{1}{y-x} \int_x^y [f(x) - f(t)]^2 \, dt = 0, \qquad x \in I.$$

(ii) *Let f and g be bounded derivatives on an interval I. A sufficient condition for fg to be a derivative is that at any point in I at least one of the functions f and g be approximately continuous.*[7]

Finally, in [43] conditions are given for the sum of squares of two derivatives (think of the arc length formula) to be a Darboux function. One of them is as follows:

If f is a continuous function and g is a derivative on an interval then $f^2 + g^2$ is a Darboux function.

I have published no more papers on real functions since 1962, the year of publication of [43]. I am flattered that these papers, published more than 20 years ago, are still frequently quoted in research papers and monographs (see e.g. [6] and [73]). I have also noted rediscoveries of weaker versions of my results. I should confess that it is with melancholy but without regret that I look back to my early research years. From time to time my attention is captured by some open problem in real analysis, but lack of time forces me to give it up soon. Perhaps I shall be able to work in real analysis again after my retirement.

4. Postuniversity Years. Doctoral Degrees

I graduated with honours in July 1959. My diploma paper was based on part of the work outlined above. Contrary to my expectations, as a result of unfavourable (political) circumstances I was only offered a position as mathematics teacher at a gymnasium in Călăraşi, a small harbour on the Danube. The fact that Barbilian's career had also started there did not make the post any more attractive. I therefore refused the offer and began looking for other possibilities. In the end, help came once again from Professor Mihoc, on whose recommendation I got a position as consultant at the Central Board of Statistics in Bucharest. My main duty there was to improve the mathematical education of the personnel. Of course, this was not terribly exciting but, living in Bucharest, it was at least possible for me to keep up some contact with mathematics.

Actually, this was also the beginning of a new phase in my mathematical career. On the one hand, I started studying some mathematical statistics as required by my job. On the other hand, Professor Mihoc suggested that I should try to carry out research in probability. He proposed that I first

---

[7] For the definition of this notion see, e.g. [6], p. 18.

become acquainted with dependence with complete connections, a concept which was his own and Onicescu's joint creation. This suggestion was quite timely, as a book [8] by George Ciucu and Radu Theodorescu had just appeared, gathering together all that was then known in the field.

My conversion from real analysis to probability was by no means difficult. The former is an indispensable basis for the latter and, moreover, as I have already mentioned, I had taken a solid course in probability with Onicescu.

In the period 1960–8, both independently and jointly with Radu Theodorescu, I was able to complete some 25 papers on dependence with complete connections. In July 1963 I got my Ph.D. degree (in the Soviet nomenclature of the time, this was called a "candidate of science" degree) with a thesis on chains with complete connections with an arbitrary set of states, the revised version of which was published in Russian as [28]. (This paper was awarded a prize of the Romanian Academy in 1965.) The jury was composed of Onicescu, Mihoc, Moisil, N. Teodorescu and Alexandru Climescu, a very distinguished professor of mathematics at the Jassy Polytechnic. A second doctoral degree, which is roughly equivalent to the English D.Sc., was granted to me in December 1969 for the whole of my mathematical work. This time the jury was composed of Onicescu, Mihoc, Nicolescu, Marinescu, Marcus, Romulus Cristescu and Ciprian Foiaş, the last two members being leading representatives of newer generations of Romanian mathematicians.

Meanwhile, I changed jobs several times. First, at the end of 1961 I obtained the position of assistant professor of mathematics at the Bucharest Polytechnic Institute, giving up my job at the Central Board of Statistics a few months later. In June 1963 I succeeded at last in becoming a research worker in the probability division of the Institute of Mathematics of the Academy (my old dream!). This was also the end of my regular teaching activities. When in 1964 the Centre of Mathematical Statistics was set up, I moved to it and have remained there ever since.

5. The Centre of Mathematical Statistics

The Centre of Mathematical Statistics developed as an independent entity from the probability division of the Mathematical Institute in April 1964. The promoter of this development was Professor Mihoc, who became the centre's first director (until 1976). It should be emphasized that Mihoc's motivation was a visionary understanding of the part probability and statistics would play in the not very distant future, rather than the considerable expansion of these fields at the time. The appearance of the centre also constituted a revival, at a different level, of the Institute of Statistics, Actuarial Science and Computation of Bucharest University, which developed in 1941 from the School of Statistics founded in 1931 by Professor Onicescu. This institute, dismantled in 1947, was one of the first centres of postgraduate studies in statistics in the world.

The centre was intended to carry out both basic research and applications, the latter consisting mainly of *genuine* applied mathematics, which is to be understood as setting up realistic mathematical models. This idea was implemented by bringing together mathematicians with strong backgrounds in probability, statistics, and operations research, and economists, physicians, biologists and engineers. After years of work these people finally succeeded in learning the art of talking to one another in order to solve various problems in economics, industry, medicine, and other fields of application.

Special mention should be made here of the activity of the biomathematics group, which had been led since the founding of the centre by Dr Petre Tăutu, a physician by training, a man of encyclopaedic culture (including some mathematics) and firm artistic tastes. Our encounter had been fruitful for both of us; in fact, this group did the first genuine biomathematics[8] work in Romania. Regrettably, in 1976 the group was moved away from the centre, with all the easily foreseeable consequences. This move followed the transfer in 1974 of the centre from the Academy to the Ministry of Education, and in 1976 to the National Institute of Metrology. It is since this last transfer that I have become the head of the centre, which ceased to be an autonomous institution and became a division (laboratory) of the National Institute of Metrology. Fortunately, there has been much understanding for our work in this institute. Moreover, especially in the last four years, we have received generous financial support for basic research from the National Council for Science and Technology.

My collaboration with Tăutu dates back to the centre's pre-1974 existence. It started spontaneously when, a few weeks after the centre's birth, Tăutu suggested to me a detailed plan for reviewing all the existing stochastic models describing biological or medical phenomena. I was seduced by his ideas, the more so as I thought this enterprise would benefit my theoretical work. We embarked on the project and very soon became aware of the plethora of "models" which, far from describing biological laws, were actually simple exercises in calculus and/or stochastic processes. This was an extreme but not infrequent case; there were, however, a few models having the value of genuine natural laws such as, for example, Mendel's laws. After some three years of hard work, the finished result of our collaboration appeared: this was a 350-page book in Romanian entitled *Stochastic Processes and Applications in Biology and Medicine*, which was published in 1968. The book was favourably received in Romania, a prize of the Academy being awarded to it. Moreover, most important, as indisputable a pioneer of serious applied probability as David G. Kendall recommended its translation into English. The English

---

[8] By biomathematics I understand the setting up of (realistic) mathematical models in biology. A more comprehensive meaning has been proposed to refer to "any interaction arising between biology and mathematics" (see [7]) and comprises both mathematical biology and biological mathematics; the former refers to the activity of using mathematics for biological research while the latter includes the developments of special branches, methods or techniques in mathematics itself, particularly suitable for biological research.

version, almost twice as long, was published in two volumes in 1973 [45] as Volumes 3 and 4 of the Springer Biomathematics Series. We tried to eliminate various shortcomings in the original Romanian version, but many of them remained. As remarked in some reviews of the book, the basic shortcoming seemed to derive from the heterogeneity of the intended readership (mathematicians, biologists, physicians, etc.) and the extreme width of the coverage. The book should have been followed by a series of more detailed studies of some topics (e.g. models in physiology and pathology); we had projected such extensions but they did not materialize, mainly because Tăutu left the centre in 1973 (when he joined the Deutsches Krebsforschungszentrum in Heidelberg). Even so, I believe our book has played some part in the development of biomathematics, if only as a reference text. It is amusing to note that quite recently I have found the book useful for myself, when completing with some colleagues a book on stochastic modelling [48].

Returning to the activities of the Centre of Mathematical Statistics, mention should be made of its contribution to the scientific cooperation between Romanian and foreign scientists. The centre was entrusted with the organization of the Anglo-Romanian Conference on Mathematics in the Archaeological and Historical Sciences in Mamaia in 1970, under the auspices of both the Royal Society and the Romanian Academy, and of the Eighth International Biometric Conference in Constanţa in 1974. The centre has also been organizing (in principle, every third year) the Braşov conferences on probability theory in 1968, 1971, 1974, 1979, and 1982. The founders of these conferences were Professors Onicescu and Mihoc. Starting from a small national colloquium in 1955 and continuing with a larger international one in 1962, the Braşov conferences found in the centre a most appropriate planner and organizer; they have become a notable event in the calendar of international periodic scientific conferences. Through the years we have welcomed in Braşov many of the most distinguished probabilists, statisticians, biomathematicians and operations researchers from all over the world (see [64]). Since 1968 I have become increasingly involved in the planning and running of the Braşov conferences, and have had direct responsibility for the last two. The work was hard but rewarding, as the *Proceedings* volumes can witness. We intend to go on with the Braşov conferences in spite of the recent loss of their founders, which marks the passing of an era.

In retrospect, 20 years after its foundation, one can say that the Centre of Mathematical Statistics has greatly contributed to the development of probability, statistics and operations research in Romania. In fact, through those who started their careers at the centre and who have now settled abroad, this contribution has a definite international character. The creation of the centre was the crowning achievement of the lifelong scientific and organizational activities of Professors Onicescu and Mihoc, whose existence was clearly a historical accident. Other countries have not had such a chance; fully conscious of the fact that scientific centres are fragile constructions, highly sensitive to the loss of their creators, we look to the future with circumspection.

6. Work in Probability

Let me begin this section with a trivial remark. One speaks about probabilistic modelling as if this were a separate activity, or more precisely, an applied development of probability theory. It should be clear, however, that probability theory itself consists of the study of more or less intricate mathematical models of real random phenomena. The underlying sample space is a mathematical model of a real random experiment; a stochastic process is a mathematical model of a real process, the evolution of which obeys some probability laws. A Markov process is a special type of stochastic process, namely a model of a real process, whose evolution retains its most recent recollection of the past: once in a given state at a specific time, the way in which that state was reached does not affect the future evolution of the process. My point is that even the "purest" probabilists are in fact dealing with models in their work. The applied probabilist is concerned with devising special models for particular real-life situations, by making use of general concepts as raw material. The craft of probabilistic modelling is much more widespread than it might seem at first sight.

6.1. Dependence with Complete Connections

My first contribution to probability pertains to the point made above; it involves the setting up of a general definition of the concept of dependence with complete connections. A discussion of the original motivation for this concept in a special case can be found in [62].

Consider a collection

$$((W, \mathscr{W}), (X, \mathscr{X}), u, P), \tag{2}$$

where $(W, \mathscr{W})$ and $(X, \mathscr{X})$ are measurable spaces, u is a mapping of $W \times X$ into W which is $(\mathscr{W} \times \mathscr{X}, \mathscr{W})$-measurable, P is a transition probability function from $(W, \mathscr{W})$ to $(X, \mathscr{X})$ so that $P(w, \cdot)$ is a probability measure on $\mathscr{X}$ for any $w \in W$ and $P(\cdot, A)$ is a real-valued $\mathscr{W}$-measurable function for any $A \in \mathscr{X}$. I called (2) a (homogeneous) random system with complete connections (r.s.c.c. for short) by analogy with the special case of a Markovian system (see below). Let us denote by $(\Omega, \mathscr{K})$ the countable-fold product of $(X, \mathscr{X})$ with itself. Write $\xi_n(\omega)$ for the nth coordinate of $\omega \in \Omega, n \in N = \{1, 2, \ldots\}$, and for any (fixed) $w \in W$ define $\zeta_0(\omega) = w$, $\zeta_{n+1} = u(\zeta_n, \xi_{n+1})$, $n \in N_0 = \{0, 1, 2, \ldots\}$. On account of the well-known Ionescu Tulcea theorem there exists a probability P_w on $\mathscr{K}$ such that P_w-a.s.

$$\mathsf{P}_w(\xi_1 \in A) = P(w, A), \quad A \in \mathscr{X},$$
$$\mathsf{P}_w(\xi_{n+1} \in A | \zeta_n, \xi_n, \ldots, \xi_1, \zeta_0) = P(\zeta_n, A), \quad A \in \mathscr{X}, n \in N.$$

The sequence $\xi = (\xi_n)_{n \in N}$ might therefore be called an infinite-order chain, while, as is not difficult to see, the sequence $\zeta = (\zeta_n)_{n \in N}$ is a Markov chain. In the very special case where $(W, \mathscr{W}) = (X, \mathscr{X})$, $u(w, x) = x$, and P is a transition probability function on $(X, \mathscr{X})$ (i.e. from $(X, \mathscr{X})$ to itself), the two random chains ξ and $(\zeta_n)_{n \in N}$ do coincide and reduce to a Markov chain with state space $(X, \mathscr{X})$ and transition probability function P.

Since the publication of my paper [28], practically all of the significant work in dependence with complete connections has been given in the framework of (2) and the family of probability spaces $(\Omega, \mathscr{K}, \mathsf{P}_w)$, $w \in W$.

Let me stress here the difference between my approach and that in [20], a basic paper on dependence with complete connections, where the collection (2) is the starting point, too. The nomenclature "infinite-order chain" for the sequence ξ is abusive. Actually, the name infinite-order chain is reserved in the strict sense for an r.s.c.c. (2) with $(W, \mathscr{W}) = (X^{-N_0}, \mathscr{X}^{-N_0})$ and $u(w, x) = (\ldots, x'_{-n}, \ldots, x'_{-1}, x'_0)$, where $x'_{-n} = x_{-n+1}$, $n \in N$, $x'_0 = x$, if $w = (\ldots, x_{-n}, \ldots, x_{-1}, x_0)$. In this case $w = (\ldots, x_{-n}, \ldots, x_{-1}, x_0) \in X^{-N_0}$ can be interpreted as a "path", to mean the left-infinite sequence of successive states of a system (with state space X): x_0 is the state at the last observation time, x_{-1} the state a time unit before, and so on going back endlessly in time. Next, $P(w, x)$ can be thought of as the probability of being in state $x \in X$ at time $t + 1$ conditional on the path $w \in X^{-N_0}$ up to time t for any $t \in (-N_0) \cup N$. Clearly, the elements ξ and P_w, $w \in W$, constructed above do not agree with this interpretation of a strict-sense infinite-order chain. The latter implies the existence of a *doubly* infinite sequence of random variables. The approach in [20] is directed to answering this requirement by imposing suitable topological assumptions on $(X, \mathscr{X})$. My approach—which does not require any topological assumption on $(W, \mathscr{W})$ and $(X, \mathscr{X})$—is complementary, covering all the cases different from the strict-sense infinite-order chain case. A list of examples of the former can be found, for example, in [40]. One of the most important is undoubtedly that of the general stochastic model for learning, which is nothing but a paraphrase of the concept of an r.s.c.c. (see Norman [63], pp. 12 and 24). Another important example will be discussed below (Section 6.3).

6.2. Strictly Stationary φ-Mixing Sequences

An r.s.c.c. is said to be uniformly ergodic if for any $r \in N$ there exists a probability P_r^∞ on $\mathscr{X}^r = r$-fold product σ-algebra of $\mathscr{X}$ with itself such that

$$\varepsilon_n = \sup |\mathsf{P}_w((\xi_n, \ldots, \xi_{n+r-1}) \in A) - \mathsf{P}_r^\infty(A)| \to 0$$

as $n \to \infty$, where the upper bound is taken over all $w \in W$, $A \in \mathscr{X}^r$, $r \in N$. The uniformly ergodic r.s.c.c.'s enjoy many interesting properties. Thus, according to a result of Cohn (see [47], pp. 136–7) the sequence ξ is φ-mixing under P_w and, moreover, $\varphi_w(n) \leq \varepsilon_n + \varepsilon_{n+1}$ whatever $w \in W$, $n \in N$. Remember that this means that

$$\varphi_w(n) = \sup |\mathsf{P}_w(A_2 | A_1) - \mathsf{P}_w(A_2)| \to 0$$

as $n \to \infty$, where the upper bound is taken over all $A_1 \in \mathscr{K}_{(1,t)}$ such that

$P_w(A_1) \neq 0$, $A_2 \in \mathcal{K}_{(t+n, t+n+r-1)}$, $t, r \in N$. Here $\mathcal{K}_{(u,v)}$ denotes the σ-algebra generated by the random variables $\xi_u, \ldots, \xi_v$, $u \leq v$. Next, under uniform ergodicity there exists a probability P_∞ on $\mathcal{K}$ such that ξ is a strictly stationary sequence on $(\Omega, \mathcal{K}, P_\infty)$. This was stated by G. Theiler, but the complete proof was given in [34] and [66]. Moreover, ξ is φ-mixing under P_∞, too, and $\varphi_\infty(n) \leq \varepsilon_n$, $n \in N$.

I was thus naturally involved in researches on strictly stationary φ-mixing sequences. My paper [29] develops a methodology for proving the (first) central limit theorem with remainder and the (first) law of the iterated logarithm for such sequences. The basic auxiliary results in [29] were frequently used by many authors (including myself in order to obtain the functional law of the iterated logarithm [30]).

In 1974 in my address at the Fifth Braşov Conference (see [32]) I conjectured that if $(\eta_n)_{n \in N}$ is a strictly stationary φ-mixing sequence with $E\eta_1^2 < \infty$ and $\sigma_n^2 = \text{var}(\eta_1 + \cdots + \eta_n) \to \infty$ as $n \to \infty$ then both (a) the functional central limit theorem and (b) the functional law of the iterated logarithm hold. This conjecture is not yet settled. Two important steps towards settling (a) have been recently made. Herrndorf [18] has shown that part (a) of the conjecture is not true if there exists a strictly stationary φ-mixing sequence with $\lim_{n \to \infty} \sigma_n^2 = \infty$ and $\liminf_{n \to \infty} \sigma_n^2/n = 0$ (no example of such a sequence is known). On the other hand, Magda Peligrad [65] has proved that under the additional assumption $\liminf_{n \to \infty} \sigma_n^2/n \neq 0$, part (a) of the conjecture is true. Therefore to get a complete solution one has to see whether a strictly stationary φ-mixing sequence with $\sigma_n^2 \to \infty$ and $\liminf_{n \to \infty} \sigma_n^2/n = 0$ may exist. As to part (b) of the conjecture, I know of no attempt to confirm or invalidate it.

6.3. An r.s.c.c. in Probabilistic Number Theory

Let me report now on an r.s.c.c. pertaining to probabilistic number theory. I am at a loss to explain this unexpected community of structure of mathematical models in psychology and the Euclidean algorithm underlying the continued-fraction expansion.

It is well known that each irrational number y in the unit interval has a unique infinite continued-fraction expansion of the form

$$y = \cfrac{1}{a_1(y) + \cfrac{1}{a_2(y) + \cdots}},$$

where the $a_n(y)$ are natural numbers determined as follows. Put $Ty = 1/y$ (mod 1). Then $a_1(y) = 1/y - Ty$, and $a_{n+1}(y) = a_n(Ty) = a_1(T^n y)$, $n \in N$. Endowing the unit interval with the σ-algebra of Lebesgue measurable sets, the a_n become random variables defined almost everywhere with respect to any probability measure that assigns probability 0 to the set of rational numbers (in particular with respect to Lebesgue measure λ).

The metric theory of continued fractions is concerned with the study of the random sequence $(a_n)_{n \in N}$. The first problem of this theory was raised in 1812 by Gauss who in a letter to Laplace stated that

$$\lim_{n \to \infty} \lambda(r_n^{-1} < u) = \frac{\log(1+u)}{\log 2} \tag{3}$$

for each u in the unit interval, and asked for an estimate of the error

$$E_n(u) = \lambda(r_n^{-1} < u) - \frac{\log(1+u)}{\log 2}.$$

Here $r_n(y) = 1/(T^n y)$, $n \in N$, i.e.

$$r_n = a_n + \cfrac{1}{a_{n+1} + \cfrac{1}{a_{n+2} + \cdots}}.$$

R. Kuzmin first proved (3) in 1928, giving an error estimate $E_n(u) = O(q^{\sqrt{n}})$ as $n \to \infty$ with $0 < q < 1$. One year later, Paul Lévy gave a different proof, improving the error estimate to $E_n(u) = O(q^n)$ with $q < 0.7$. In 1974, E. Wirsing showed that the optimal value of q is $0.30366300289873265860\ldots$ (For the exact references to these results see my paper [35].)

It is not difficult to prove that

$$\lambda(r_1 > t) = 1/t,$$

$$\lambda(r_{n+1} > t | a_1, \ldots, a_n) = \frac{s_n + 1}{s_n + t}, \qquad t \geq 1, \quad n \in N,$$

where

$$s_n = \cfrac{1}{a_n + \cfrac{\cdot}{\cdot \cdot + \cfrac{1}{a_1}}}.$$

Hence

$$\lambda(a_1 = k) = \frac{1}{k(k+1)},$$

$$\lambda(a_{n+1} = k | a_1, \ldots, a_n) = \frac{s_n + 1}{(s_n + k)(s_n + k + 1)}, \qquad k, n \in N.$$

Noting that $s_{n+1} = 1/(a_{n+1} + s_n)$, we are led to consider the r.s.c.c. $((W, \mathcal{W}), (X, \mathcal{X}), u, P)$ for which W is the unit interval $[0, 1]$, $\mathcal{W}$ is the σ-algebra of Borel sets in $[0, 1]$, $X = N$ and $\mathcal{X}$ = the σ-algebra of all subsets of N;

$$u(w, x) = \frac{1}{w + x}, \qquad P(w, x) = \frac{w + 1}{(w + x)(w + x + 1)}, \qquad w \in [0, 1], \quad x \in N.$$

Clearly, starting with $w = 0$, the sequences ξ and ζ associated with the above r.s.c.c. are just $(a_n)_{n \in N}$ and $(s_n)_{n \in N_0}$, $s_0 = 0$, under $P_0 = \lambda$ as probability measure. In this way, the r.s.c.c. theory allows us to obtain as simple corollaries all the basic results of the metric theory of continued fractions. The details are to be found in my paper [31], essentially written in Cambridge (see Section 7) and motivated by Doeblin's celebrated paper [11], which was the first to suggest the part that dependence with complete connections might play in the metric theory of continued fractions. Doeblin's paper contains an astonishing number of results, some of them still waiting for complete proofs. My paper [33] provides some of these.

The study of the continued-fraction expansion suggested a new approach to the existence theorem for strict-sense infinite-order chains with a denumerable state space taken to be N. (See Section 6.1.) The basic idea is to consider a path $(\ldots, x_{-1}, x_0)$ as the continued-fraction expansion (read inversely) of a $y \in [0, 1]$. This has been developed in my paper [44] written jointly with Aurel Spătaru.

Recently, Sofia Kalpazidou [50] has also been able to obtain the main results of the metric theory of the continued-fraction-to-the-nearest-integer expansion as corollaries of r.s.c.c. theory. Recall that the continued-fraction-to-the-nearest-integer expansion of an irrational y in the interval $[-1/2, 1/2]$ has the form

$$y = \cfrac{\varepsilon_1(y)}{a_1(y) + \cfrac{\varepsilon_2(y)}{a_2(y) + \cdot\,\cdot\,\cdot}},$$

where $\varepsilon_n(y) \in \{-1, 1\}$, $2 \leq a_n(y) \in N$, $a_n(y) + \varepsilon_{n+1}(y) \geq 2$, $n \in N$. Clearly, the definitions of u and P are now much more complicated than in the case of the continued-fraction expansion, and the computations are more difficult.

The moral to be drawn from the above is that a lot of effort and time would have been saved if, instead of resorting to *ad hoc* arguments, the general r.s.c.c. theory had been applied. May we hope that in future, the situation will not be repeated for other special expansions of real numbers.

6.4. More Probabilistic Number Theory

Another paper of mine on probabilistic number theory is [41], which is the result of some work I did in Bonn during the winter semester 1981–2 (see Section 7). In it a probabilistic framework is set up for the celebrated Riemann hypothesis concerning the complex zeros of the zeta function. It appears that the conclusion "the Riemann hypothesis is true with probability 1" is not at all absurd, but, moreover, gives good reasons for believing the hypothesis. Actually, such a conclusion is the only one we can presently afford. A similar situation arises as to the Borel normality of, e.g., the number π: the probability is 1 that π is a normal number, but we do not yet know (will we ever know?)

whether π is normal or not. My motivation for becoming involved in this matter has been Denjoy's paper [10].

I cannot help mentioning that as to Fermat's last theorem, it is possible to prove that the Fermat equation for an exponent greater than 3 has at most finitely many solutions with probability 1 (see [13] and [39]). What Gerd Faltings has, among other things, recently proved in settling Mordell's conjecture (see, e.g. [3]) is just the above statement without the probability connotation!

6.5. Books on Dependence with Complete Connections

The 1960s were a very active period of research on dependence with complete connections both in Romania and abroad. Jointly with Radu Theodorescu I co-authored the book [47], which sums up the work done in the field up to 1969. I remember with pleasure the letter I received from Frederick Mosteller, a pioneer of stochastic models for learning, telling me he had found remarkable the appearance of a book on mathematical learning in Springer's "Yellow Peril" series. I should concede that the book does not make easy reading and that several parts of it could have been better produced. In spite of these shortcomings, I think the book played a useful part in disseminating interest in mixing random variables, dependence with complete connections and stochastic models for learning.

The work done since 1969 clearly justified a new synthesis. This is the subject of the book [15], written jointly with my colleague Șerban Grigorescu. I think the new book is quite the opposite of the old one as to readability and clarity. We hope that it will find a foreign publisher willing to edit an English translation of it.

We also hope that our new book will be able to rule out the rooted prejudice according to which dependence with complete connections does not go much beyond Markovian dependence. In part the blame for this misunderstanding lies with the associated Markov chain (see Section 6.1). Even this chain is of a special type and cannot be brought under established theories for Markov chains. I am thoroughly convinced that dependence with complete connections is a field where much interesting work is being done and has yet to be done. Noticeable recent developments are to be found in [42] and [51] with reference to iterated maps as dynamical systems (crude models for turbulence). In a series of seminars started in 1984 at the Centre of Mathematical Statistics, the relations between random evolutionary processes (see, e.g. [68]) and dependence with complete connections are being examined. Also, an essential step towards establishing a theory of continuous parameter dependence with complete connections has recently been made by Pruscha [67].

6.6. Markov Chains

In 1975 I was approached by the Technical Publishing House of Bucharest with a proposal to write something on finite Markov chains. The move followed unsuccessful attempts to arrange a Romanian translation of J. G. Kemeny and J. L. Snell's 1960 classic *Finite Markov Chains*. The result was a book first published in Romanian in 1977 followed in 1980 by a revised and expanded English translation [38]. I was quite surprised by the unanimous acclaim for the latter. The success of this book may be explained by the happy blending of the classical material with new results under a vision which benefited from the experience I have gained during 20 years of research work in *much more general* topics.

6.7. Game Theory and Stochastic Programming

Last but not least, I wish to mention my paper [46] written jointly with Radu Theodorescu, where a game-theoretic definition of the solution of a stochastic programming problem is proposed. This definition has been adopted by a series of writers who, in every case, fail to mention its origin, and has been discussed in a few monographs (see [14], [49], [53], [72]).

7. Travels and Visits

The best opportunities for scientific work, especially in recent years, have been provided for me during extended visits abroad. These visits are also occasions for becoming acquainted with new people, renewing old acquaintances and, of course, seeing the world. For us in Romania, they are becoming more and more vital as a result of increasing difficulties in obtaining new books and journals, with their ever-rising costs and our shortage of hard currency. I am happy that it has not been excessively difficult for me to get permission to travel (unfortunately, on all but one occasion, without my family).

First of all, I have greatly benefited from the exchange agreements between the Romanian Academy and foreign learned societies. Thus, within this framework, I paid two- or three-week visits to the USSR (in 1964 and 1970), Czechoslovakia (in 1965 and 1967), Poland (in 1966, 1975, and 1976), Hungary (in 1972), France (in 1979), Austria (in 1983), and Great Britain (in 1984). On these occasions, besides visiting research institutes and universities, I sometimes attended scientific gatherings, too. Under the exchange agreement between the Royal Society of London and the Romanian Academy I spent the 1971 Michaelmas Term and the 1972 Lent Term (October 1971–March 1972) as an overseas fellow of Churchill College at the Statistical Laboratory of Cambridge University. My stay in Cambridge, with its unique scientific atmosphere, was one of my happiest and most productive periods. The architect of this visit in all its details was Professor David Kendall who honoured me with his friendship. I shall always be grateful to him for his invaluable help, the more so as he has further been instrumental in helping all of us at a difficult time for the Centre of Mathematical Statistics (immediately after 1975). During my Cambridge stay I gave seminars in London (at Imperial College and University College), Oxford, Sheffield and Swansea.

The publication of my books must certainly have stimulated invitations from abroad. Thus I held visiting professorships at the Université René Descartes in Paris (February–September 1974), the Johannes Gutenberg-Universität in Mainz (December 1977–February 1978), the Johann Wolfgang Goethe-Universität Frankfurt-am-Main (October 1979–March 1980), and the Rheinische Friedrich Wilhelms-Universität Bonn (November 1981–February 1982). Except for this last case, which was a research appointment (under the auspices of the Sonderforschungsbereich 72), I taught, lecturing on dependence with complete connections in Paris and Mainz and on Markov chains in Frankfurt-am-Main. For making these arrangements possible, I am grateful to Professors Bui Trong Lieu, Wolfgang Bühler, Hermann Dinges and Walter Vogel, and to priv. doz. dr Ulrich Herkenrath. During my stays in West Germany I also gave seminars in Braunschweig, Duisburg, Düsseldorf, Eichstätt, Heidelberg and München, and was able, at the invitation of Professor Paul Deheuvels, to visit in 1980 and 1982 the Institut de Statistique des Universités de Paris, lecturing on my research work. (My first talks in Paris, at the invitation of Professors Bui Trong Lieu and Albert Tortrat, were given in April 1972 on my way back from Cambridge to Bucharest.)

I was invited to lecture in 1976 and 1981 at the Stefan Banach International Mathematical Center for Raising Research Qualifications in Warsaw. While in 1976 I arrived a little late and my stay in Poland turned into an exchange-type visit (I nevertheless published a paper [36] in the corresponding *Proceedings* volume), in September 1981 I was able to speak, as arranged, about stochastic approximation and dependence with complete connections. I returned to Poland in August 1983 to attend the International Congress of Mathematicians. Thus I have followed almost on the spot the recent developments in Poland, and have been quite impressed by the fact that, in spite of unfavourable circumstances, Polish mathematics has been able to continue its progress without much damage.

In May 1978 I lectured on mixing sequences of random variables at the Third International Summer School on Probability Theory and Mathematical Statistics in Varna (Bulgaria). (See [37].)

Although I was an invited speaker at the fifth (1968), sixth (1971), seventh (1974), eighth (1978), and ninth (1982) Prague Conferences on Information Theory, Statistical Decision Functions and Random Processes, for various reasons (the first of the above conferences was simply cancelled) I was able to go to Prague to attend such a conference only in 1978. Instead, I attended as an invited speaker the three Vilnius International Conferences on Probability Theory and Mathematical Statistics organized in 1973, 1977 and 1981. I was also present in Vilnius in 1985, too, for the fourth conference. I first went there in 1970 and cannot forget the warm welcome and the excellent hospitality extended to me by my very active Lithuanian colleagues headed by Professors J. Kubilius, V. Statulevičius and B. Grigelionis.

I attended the London (1969) and Warsaw (1975) sessions of the International Statistical Institute, to which I was elected as a titulary member in 1970.

My personal preference is for smaller specialized gatherings. (The only merit I can find for mammoth conferences like the International Congress of Mathematicians or ISI sessions is the chance to listen to a large number of important scientists at a time.) The last two conferences I have attended are of this type and I found them exceptionally well organized and interesting: they were the Mathematical Learning Models—Theory and Algorithms Conference held in May 1982 in Bad Honnef (West Germany) and the Fourth Pannonian Symposium on Mathematical Statistics held in September 1983 in Bad Tatzmannsdorf (Austria).

Finally, not very frequently, I have received invitations from different universities to give one or several lectures. I was able to respond favourably to such invitations in May 1971, when I was invited to Vienna by Professor Leopold Schmetterer, and in May 1984, when I was invited to Salonika by Professor P.-C. G. Vassiliou. My first visit to Greece was a quite unforgettable experience which I should not hesitate to repeat.

My travels abroad have not taken me beyond the borders of Europe. In particular, I have not yet been to the United States, despite the fact that I have had several invitations to go there. Unable so far to solve the financial problems of travelling costs, I retain some hopes for the future.

8. Other Activities

I should like to mention some activities which I have found time-consuming, and infringing upon my scientific productivity, especially in the last 10 years when I have also become involved in administrative duties. These I accepted solely from a sense of social responsibility.

8.1. Doctoral Students

While I have not been involved in regular teaching for over 20 years, I have been granted since 1971 the right to supervise doctoral students under the auspices of the Centre of Mathematical Statistics. Their number varies from year to year but it has never been below 10, a figure which would certainly startle any western professor. Recently, following the deaths of Professors Onicescu and Mihoc, most of their doctoral students have joined my quota, so that I dare not even mention the present figure. Our system does not impose a time limitation on the preparation of a Ph.D. thesis, and many of these students will never succeed in completing their dissertations. The only advantage is that the candidates, most of them teaching mathematics in higher technical institutes, are able to keep in contact with recent mathematical developments. Among them are some genuinely gifted candidates. It is quite rewarding to have doctoral students such as—to mention just a few—Mioara

Buiculescu, Radu Gologan, Aurel Spătaru, I. M. Stancu-Minasian, and Sorin Rădulescu (who has recently defended a thesis on global inversion theorems, making me return for a while to my old subject).

In addition, frequently enough, I am asked—both from Romania and abroad—to give my opinion on Ph.D. theses, on candidates for vacant teaching positions or even on candidates for fellowships in learned societies.

8.2. Refereeing and Editorial Work

I have been on the editorial boards of the *Zeitschrift für Wahrscheinlichkeitstheorie und verwandte Gebiete* since 1972 and the *Journal of Multivariate Analysis* since 1976. This of course implies a lot of refereeing work. Frequently also, I am asked to referee papers for other journals.

Also, since 1972 I have been a member of the editorial boards of the Romanian mathematical journals *Revue Roumaine de Mathématiques Pures et Appliquées* and *Studii şi Cercetări Matematice* (*Mathematical Studies and Researches*), becoming in 1973 one of the six secretaries of both editorial boards. For fortuitous reasons, the sextet gradually lost its members and in 1979 I remained alone to carry the responsibility for *Revue Roumaine*. In the period 1973–83, besides refereeing I did a lot of editorial work (ranging from preparation of manuscripts for printing, to proof-reading). I learned many things about the printing of journals and books, at the expense of much of my time. In 1983 I was promoted deputy chief editor of the *Revue Roumaine*. In this way, much of the load has been taken from my shoulders but, on the other hand, I have other difficult problems to face (e.g. how to reduce the backlog).

I must also mention the co-editing (this means again almost all the editorial duties ending with proof-reading) of the proceedings of the fourth, fifth, sixth, and seventh Braşov Conferences on Probability Theory[9] as well as a Festschrift for Professor Onicescu, which appeared in 1983 under the title *Studies in Probability and Related Topics: Papers in Honour of Octav Onicescu on His 90th Birthday* published by Nagard, a Romanian-born Italian publisher. The last has been a particularly difficult enterprise which I would not dare to repeat.

While writing these lines I cannot help remembering G. H. Hardy's words "Exposition, criticism, appreciation is work for second-rate minds" ([17], p. 61). Is this assertion still true, and has it ever been true? My own experience indicates that a great many papers submitted for publication are far from reaching the quality or originality to be desired. The same is also true of some published material; this is an obvious result of the publish-or-perish pressures which lead in turn to the overdevelopment of some originally attractive ideas,

---

[9] There were no published proceedings of the first three Braşov conferences. The initiative of publishing the proceedings of the Fourth Braşov Conference was taken by Professor Miron Nicolescu, then President of the Romanian Academy and Director of the Institute of Mathematics.

and the keeping alive of artificial side problems with a perseverance worthy of a better cause. Still worse, there frequently occurs duplication of already known results (as illustrated by the admirable Fejér–Erdös dictum "everybody writes, nobody reads") and even plain plagiarism. I have recently discovered six plagiarisms by the same "author." The present state of things might be somewhat counteracted by better refereeing.[10] We need, above all, an increasing number of authoritative survey papers, the ideal form of which would be Hilbert-like, i.e. pointing out really important research directions and/or important and difficult open problems. This cannot be work for second-rate minds! I suspect that even Hardy did not take his assertion literally. In line with the ideas above, the number of invited survey papers at the Braşov conferences has been steadily increasing.

9. Concluding Remarks

I am grateful to Professor Gani for providing me with the opportunity to write about my life and work. While I became interested in probability somewhat under compulsion, a fact I shall never regret, the work I have done might be characterized as unfashionable. In fact, this may not be too bad a thing, as has recently been argued in [12].

I have not discussed subjects on which there is much debate, such as good or bad mathematics, mathematics and computers, mathematics and society, or the health of mathematics. I have nothing interesting to add to the debate and, after all, the situation will surely become clearer in the not too distant future.

It is my strong conviction that practical efficiency can be based only on theoretical depth. Thus any applied mathematician should maintain an interest in areas of pure mathematics.

In quite another direction, I think that to maintain one's spiritual health any mathematician (pure or applied) and, more generally, any scientist should become interested in some artistic or humanistic endeavour. My *violon d'Ingres* is precisely music and the violin, to the point that had I not been a mathematician I should have wanted to be a musician. As a practical consequence of my involvement in both mathematics and music, when I find myself bound to attend an essentially useless meeting or to listen to some boring people my defensive reaction is either to start thinking about some mathematical problem, or equally, to rememorize some musical score. For me it becomes increasingly clear that mathematics and music are among the very few oases of light in a world increasingly darkened by folly, intolerance and dishonesty.

I think it is indisputable that, so far, the deepest applications of mathemat-

---

[10] I was delighted to read in Lindley [56], a plea for a strong refereeing system. However, I doubt that the idea of (handsomely) paying the referees can be implemented.

ics have been made in physics. At the other end, in most of the social sciences, the part played by mathematics has essentially been in organizing information. It is not inconceivable that we shall see the setting up of some new mathematical structures adapted to the scientific needs of the less mathematized sciences in future. However, we cannot completely exclude the impossibility of any meaningful mathematization in specific concrete problems. If this statement appears too vague, I might say instead that there seem to be good prospects for research on multidimensional-parameter stochastic processes, for example.

Concerning the future, conditional on the nonextinction of mankind by nuclear or other weaponry, I should say I am not too optimistic. The international mathematical community is presently under obvious bureaucratic and financial pressures (see, e.g. [9], and [54]). In part these pressures may be due to some unfulfilled promises made at a time when, to obtain funds, it was maintained that mathematics was good for everything, which surely is not true. Of course, our subject is strongly placed within science, but will it survive and develop if the trend against basic research continues?

Let me end these notes by making explicit my view of mathematics. Some have maintained that mathematics is, a way of looking at life [52], or one of the most important of human activities [4]. For me mathematics is simply a way of bearing life.

Acknowledgement

Beginning with my family, I should like to thank the very many people of various professions and social positions, in Romania and abroad, too numerous to be named here, who, at one time or another, in one way or another, have helped me to overcome difficulties of every kind, thus allowing me to be able to tell my story.

Publications and References

[1] Andreian Cazacu, Cabiria and Marcus, S. (1983) *Simion Stoilow* (in Romanian with a French summary). Ed. ştiinţifică şi enciclopedică, Bucureşti.

[2] Bell, E. T. (1937) *Men of Mathematics*. Simon and Schuster, New York.

[3] Bloch, S. (1984) The proof of the Mordell conjecture. *Math. Intelligencer* 6 (2), 41–47.

[4] Bonsall, F. F. (1982) A down-to-earth view of mathematics. *Amer. Math. Monthly* 89, 8–15.

[5] Borel, A. (1983) Mathematics: Art and science. *Math. Intelligencer* 5 (4), 9–17.

[6] Bruckner, A. M. (1978) *Differentiation of Real Functions*. Lecture Notes in Mathematics 659, Springer-Verlag, Berlin.

[7] Chytil, M. K. (1982) The elucidation of the concept of biomathematics. In *Progress in Cybernetics and System Research* VI, Hemisphere, Washington, DC, 35–40.

[8] Ciucu, G. and Theodorescu, R. (1960) *Processes with Complete Connections* (in Romanian). Ed. Acad. R.P. Romîne, Bucureşti.

[9] David, E. E., Jr. (1984) Toward renewing a threatened resource: Findings and

recommendations of the *ad hoc* committee on resources for the mathematical sciences. *Notices Amer. Math. Soc.* 31, 141–145.

[10] Denjoy, A. (1964) Probabilités confirmant l'hypothèse de Riemann sur les zéros de $\zeta(s)$. *C.R. Acad. Sci. Paris* 259, 3143–3145.

[11] Doeblin, W. (1940) Remarques sur la théorie métrique des fractions continues. *Compositio Math.* 7, 353–371.

[12] Dyson, J. F. (1983) Unfashionable pursuits. *Math. Intelligencer* 5 (3), 47–54.

[13] Erdös, P. and Ulam, S. (1971) Some probabilistic remarks on Fermat's last theorem. *Rocky Mountain J. Math.* 1, 613–616.

[14] Golstein, E. and Youdine, D. (1973) *Problèmes particuliers de la programmation linéaire*. Mir, Moscow.

[15] Grigorescu, Ş. and Iosifescu, M. (1982) *Dependence with Complete Connections and Its Applications* (in Romanian). Ed. ştiinţifică şi enciclopedică, Bucureşti.

[16] Halmos, P. R. (1980) The heart of mathematics. *Amer. Math. Monthly* 87, 519–524.

[17] Hardy, G. H. (1969) *A Mathematician's Apology* (with a Foreword by C. P. Snow). Cambridge University Press, London.

[18] Herrndorf, N. (1983) The invariance principle for φ-mixing sequences. *Z. Wahrscheinlichkeitsth.* 63, 97–109.

[19] Hruška, V. (1946) Une note sur les fonctions aux valeurs intermédiaires. *Časopis Pěst. Mat. Fys.* 71, 67–69.

[20] Ionescu Tulcea, C. T. (1959) On a class of operators occurring in the theory of chains of infinite order. *Canad. J. Math.* 11, 112–121.

[21] Iosifescu, M. (1956) On a theorem of A. Marchaud (in Romanian). *Com. Acad. R.P.R.* 6, 1169–1171.

[22] Iosifescu, M. (1957) On the product of two derivatives (in Romanian). *Com. Acad. R.P.R.* 7, 319–321.

[23] Iosifescu, M. (1958) Sur les fonctions continues dont les ensembles de niveau sont au plus dénombrables. *Rev. Math. Pures Appl.* 3, 439–441.

[24] Iosifescu, M. (1959) Propriétés différentielles des fonctions jouissant de la propriété de Darboux. *C.R. Acad. Sci. Paris* 248, 1918–1919.

[25] Iosifescu, M. (1959) Ueber die differentialen Eigenschaften der reellen Funktionen einer reellen Veraenderlichen. *Rev. Math. Pures Appl.* 4, 457–466.

[26] Iosifescu, M. (1959) Conditions that the product of two derivatives be a derivative (in Russian). *Rev. Math. Pures Appl.* 4, 641–649.

[27] Iosifescu, M. (1959) Ueber eine Erweiterung eines Satzes von S. Stoilow. *Rev. Math. Pures Appl.* 4, 725–729.

[28] Iosifescu, M. (1963) Random systems with complete connections with an arbitrary set of states (in Russian). *Rev. Math. Pures Appl.* 8, 611–645; Addenda, *ibid.* 9, 91–92. (English translation obtainable from Addis Translations, Menlo Park, Calif.)

[29] Iosifescu, M. (1968) La loi du logarithme itéré pour une classe de variables aléatoires dépendantes. *Teor. Verojatnost. i. Primenen.* 13, 315–325; adendum *ibid.* 15 (1970), 170–171. (English translation: *Theory Prob. Appl.* 13 (1968), 304–313; 15 (1970), 160.)

[30] Iosifescu, M. (1972) On Strassen's version of the loglog law for some classes of dependent random variables. *Z. Wahrscheinlichkeitsth.* 24, 155–158.

[31] Iosifescu, M. (1974) On the application of random systems with complete connections to the theory of f-expansions. *Progress in Statistics I (European Meeting of Statisticians, Budapest, 1972)*, Colloq. Math. Soc. J. Bolyai 9, ed. J. Gani *et al.*, North-Holland, Amsterdam, 335–365.

[32] Iosifescu, M. (1977) Limit theorems for φ-mixing sequences. A survey. *Proc. 5th Conf. Probability Theory (Braşov, 1974)*, Ed. Acad. R.S. România, Bucureşti, 51–57.

[33] Iosifescu, M. (1977) A Poisson law for ψ-mixing sequences establishing the truth of a Doeblin's statement. *Rev. Roumaine Math. Pures Appl.* 22, 1441–1447.

[34] Iosifescu, M. (1977/78) Dependence with complete connections. Lecture notes, Fachbereich Mathematik, Johannes Gutenberg-Universität in Mainz. [Revised version in *Sequential Methods in Statistics* (Papers, XVIIIth Semester, Stefan Banach Internat. Math. Center, Warsaw, 1981), Banach Center Publ. 16, PWN, Warsaw, 245–262.

[35] Iosifescu, M. (1978) Recent advances in the metric theory of continued fractions. *Trans. 8th Prague Conf. Inform. Theory, Statist. Decision Functions, Random Processes (Prague, 1978)*, A, Reidel, Dordrecht, 27–40.

[36] Iosifescu, M. (1979) The tail structure of nonhomogeneous finite state Markov chains: A survey. In *Probability Theory* (Papers, VIIth Semester, Stefan Banach Internat. Math. Center, Warsaw, 1976), Banach Center Publ. 5, PWN, Warsaw, 125–132.

[37] Iosifescu, M. (1980) Recent advances in mixing sequences of random variables. In *3rd Internat. Summer School Probability Theory and Math. Statistics (Varna, 1978)*, B"lgar Akad. Nauk, Sofia, 111–138.

[38] Iosifescu, M. (1980) *Finite Markov Processes and Their Applications*. Wiley, Chichester and Ed. tehnică, Bucharest.

[39] Iosifescu, M. (1982) On the random Fermat's equation. *An. Univ. Craiova Ser. Mat. Fiz.-Chim.* 10, 61–65.

[40] Iosifescu, M. (1983) Notes on the general stochastic model for learning. In *Studies in Probability and Related Topics: Papers in Honour of Octav Onicescu on His 90th Birthday*, ed. M. C. Demetrescu and M. Iosifescu, Nagard, Roma, 287–300.

[41] Iosifescu, M. (1984) On the random Riemann hypothesis. *Proc. 7th Conf. Probability Theory (Brașov, 1982)*, Ed. Acad. R.S. România, București, 435–450.

[42] Iosifescu, M. (1985) Ergodic piecewise monotonic transformations and dependence with complete connections. *Proc. 4th Pannonian Symp. Math. Statist.* (Bad Tatzmannsdorf, Austria, 1983), A, ed. F. Konecny et al., Akadémiai Kiadó, Budapest and Reidel, Dordrecht, 167–173.

[43] Iosifescu, M. and Marcus, S. (1962) Sur un problème de P. Scherk concernant la somme des carrés de deux dérivées. *Canad. Math. Bull.* 5, 129–132.

[44] Iosifescu, M. and Spătaru, A. (1973) On denumerable chains of infinite order. *Z. Wahrscheinlichkeitsth.* 27, 195–214.

[45] Iosifescu, M. and Tăutu, P. (1973) *Stochastic Processes and Applications in Biology and Medicine: I. Theory; II. Models*. Biomathematics, 3 and 4, Ed. Academiei București and Springer-Verlag, Berlin.

[46] Iosifescu, M. and Theodorescu, R. (1963) Sur la programmation linéaire. *C.R. Acad. Sci. Paris* 256, 4831–4833.

[47] Iosifescu, M. and Theodorescu, R. (1969) *Random Processes and Learning*. Grundlehren der math. Wiessenschaften 150, Springer-Verlag, Berlin.

[48] Iosifescu, M., Grigorescu, Ș., Oprișan, G. and Popescu, G. (1984) *Elements of Stochastic Modelling* (in Romanian). Ed. științifică și enciclopedică, București.

[49] Judin, D. B. (1974) *Mathematical Methods for Control under Conditions of Incomplete Information (Problems and Methods of Stochastic Programming)* (in Russian). Sovetskoe Radio, Moscow.

[50] Kalpazidou, Sofia (1985) On a random system with complete connections associated with the continued fraction to the nearer integer expansion. *Rev. Roumaine Math. Pures Appl.* 30, 527–537.

[51] Keller, G. (1982) Stochastic stability in some chaotic dynamical systems. *Monatsh. Math.* 94, 313–333.

[52] Kendall, D. G. (1969) Mathematics—a way of looking at life (in Romanian). *Gaz. Mat. Ser.* A 74, 4–10.

[53] Kolbin, V. V. (1977) *Stochastic Programming*. Reidel, Dordrecht.

[54] Lang, S. (1984) Circular A-21: A history of bureaucratic encroachment (excerpts). *Math. Intelligencer* 6 (1), 33–46.

[55] Laue, G. (1980) Remarks on the relation between fractional moments and fractional derivatives of characteristic functions. *J. Appl. Prob.* 17, 456–466.

[56] Lindley, D. V. (1984) Refereeing. *Math. Intelligencer* 6 (2), 56–60.

[57] Marchaud, A. (1927) Sur les dérivées et sur les différences des fonctions de variables réelles. *J. Math. Pures Appl.* 6, 337–425.

[58] Marchaud, A. (1933) Sur une condition de quasi-rectificabilité. *Fund. Math.* 20, 105–116.

[59] Marcus, S. (1957) Sur un théorème de M. A. Marchaud et sur les fonctions dérivables presque partout. *C.R. Acad. Sci. Paris* 244, 2345–2347.

[60] Marcus, S. (1966) Quelques aspects des travaux roumains dans la théorie des fonctions des variables réelles. *Rev. Roumaine Math. Pures Appl.* 11, 1123–1138.

[61] Marcus, S. (1975) *From the Romanian Mathematical Thinking* (in Romanian). Ed. ştiinţifică şi enciclopedică, Bucureşti.

[62] Mihoc, G. (1982) A life for probability. In *The Making of Statisticians*, ed. J. Gani, Springer-Verlag, New York, 22–37.

[63] Norman, M. F. (1972) *Markov Processes and Learning Models*. Academic Press, New York.

[64] Onicescu, O. (1981) Quelques réflexions sur les conférences de Braşov sur les probabilités. *Proc. 6th Conf. Probability Theory (Braşov, 1979)*, Ed. Acad. R.S. România, Bucureşti, 11–14.

[65] Peligrad, M. (1985) An invariance principle for φ-mixing sequences. *Ann. Prob.* 13, 1304–1313.

[66] Popescu, G. (1978) Asymptotic behaviour for random systems with complete connections. I (in Romanian). *Stud. Cerc. Mat.* 30, 37–68.

[67] Pruscha, H. (1983) Learning models with continuous time parameter and multivariate point processes. *J. Appl. Prob.* 20, 884–890.

[68] Siegrist, K. (1981) Random evolution processes with feedback. *Trans. Amer. Math. Soc.* 265, 375–392.

[69] Stoilow, S. (1924) Sur les transformations continues d'une variable. *C.R. Acad. Sci. Paris* 179, 807–810.

[70] Stoilow, S. (1924) Sur l'ensemble où une fonction continue a une valeur constante. *C.R. Acad. Sci. Paris* 179, 1585–1586.

[71] Stoilow, S. (1925) Sur l'inversion des fonctions continues. *Bull. Soc. Math. France* 53, 135–148.

[72] Vajda, S. (1972) *Probabilistic Programming*. Academic Press, New York.

[73] Ver Eecke, P. (1983) *Fondements du calcul différentiel*. Presses Universitaires de France, Paris.

W. J. Ewens

Warren Ewens was born on 23 January 1937 in Canberra, Australia. He studied mathematics and statistics at the University of Melbourne and graduated BA in 1958 and MA in 1960. He moved to The Australian National University in 1961 and completed his Ph.D. there in 1963, working under the direction of P. A. P. Moran. At the same time he was a lecturer in mathematical statistics in the undergraduate section of the same university. He first visited the USA in 1964, doing postdoctoral work with Sam Karlin at Stanford. In 1967 he was appointed Professor of Mathematics at the newly-formed La Trobe University in Melbourne. A period as visiting professor at the University of Texas Department of Zoology in 1971 reinforced his preference for studying genetical questions in close contact with biologists, and was followed in 1972 by a move to the Department of Biology at the University of Pennsylvania. Since 1977 he has divided his time between Pennsylvania and the Department of Mathematics at Monash University, Australia.

Professor Ewens' research is centred in the mathematical and statistical theory of population genetics, particularly human genetics. He finds that the most rewarding approach to research problems is through his collaboration with his biological and medical colleagues at Pennsylvania, followed by a period devoted to the purely mathematical aspects of the work at Monash. This hedonistic existence is partly repaid by teaching statistics to large numbers of students in both places. Although his original interest was in the evolutionary side of population genetics, his present research is devoted almost entirely to the application of statistical methods to elucidate the genetic basis of various diseases in man, a field which he feels will grow enormously in the next few years with increasing knowledge of the genetic material of each individual at the DNA level and increasingly powerful computer techniques.

Professor Ewens is a member of the Australian Academy of Science and has been active in a number of bodies associated with statistics, most recently as chairman of the scientific advisory board of the Australian Twin Registry. He is married, has a son and a daughter, and plays too much bridge and too little tennis.

The Path to the Genetics Sampling Formula

1. Introduction

Since the highlights of our professional life are what makes it interesting and rewarding, I have decided to describe in this account the events surrounding the derivation of the so-called "Ewens sampling formula" of population genetics, which has formed a high point in my career as an applied probabilist. Before doing this I will briefly describe my background and early training in population genetics theory.

I enrolled as an undergraduate student at the University of Melbourne in 1955, intending to specialize in my two main interests at high school, mathematics and economics. However the economics department, like many at the time (and several now), did not encourage a mathematical approach to economic problems and I changed my enrollment to mathematics and statistics, this latter subject having only recently been introduced as an undergraduate major. Although I have maintained an interest in mathematical economics, I am confident that mathematical genetics, in which I subsequently became interested, is at least as relevant; I find it more challenging and interesting than mathematical economics, and in retrospect do not regret the change in the direction of my interests. This was the first in a series of steps that determined, more by luck than by judgement, what my future career would be.

The professors I had were all Cambridge-trained and, in the old Cambridge tradition, their courses emphasized solutions to practical problems, both in applied mathematics and statistics. Although we learned plenty of theory, it was largely directed towards specific practical applications. This approach provides possibly the best preparation for graduate and research work: one learned to train one's mind to recognize the most likely method of solving a problem. Except for the real mathematical high-flyers, skills of this kind are probably as useful in research, or at least the research I became interested in, as heavy mathematical machinery. Thus although I am completely aware of its deficiencies, I do not regret my "old-fashioned Cambridge" undergraduate training at all.

At that time it was the custom at Melbourne to complete a two-year master's degree before embarking on a Ph.D. Following an interest from my undergraduate work, I wrote my MA thesis on a problem in sequential analysis, at that time a comparatively new area of statistics. However, I became increasingly convinced that, while I wanted to continue my research in a mathematical and statistical framework, my interests lay in applying mathematics and statistics in some area of the natural sciences (biological or physical), and I spent part of my time, during my MA work, thinking about what might be a suitable field.

2. Graduate Work In Population Genetics

One person with whom I discussed this problem was Peter Finch. He had recently arrived in Melbourne from England, and although his own research interests lay elsewhere, suggested that recent developments in genetical knowledge would lead to an increased interest in, and need for, theoretical work in mathematical population genetics, particularly in problems associated with biological evolution. Thus when I secured a lectureship at The Australian National University (ANU) early in 1961, working in the undergraduate department headed by Ted Hannan (whose humanity, ability and concern for a young colleague I admired and appreciated, then and later), I decided to study at the same time for a Ph.D. under Pat Moran. Moran's legendary skills in applied stochastic processes, among other topics, were at that time directed towards precisely that area. I felt fortunate in more than one way; in those post-Sputnik days, with more academic posts opening up than there were staff available to fill them, one could secure a tenured job without even having started a Ph.D.

Moran had assembled a particularly strong group at the ANU; apart from himself and Ted Hannan, this included Jo Moyal, Joe Gani and, soon after, Peter Finch, with Evan Williams of CSIRO closely involved. I was helped and influenced by all of these people, in particular by Gani, who eased many of the problems encountered by a new research student and with whom I wrote my first paper in genetics; he also later facilitated my going to Stanford for postgraduate work. Thus the atmosphere in the department was one of friendliness, but it was also one of intellectual toughness. Moran's line with graduate students was to give them several unusual problems on arrival, to send them away to solve them and learn about research the hard way, and to persist with them only if they showed real signs of research ability. I only just scraped through this initiation, but then and now believe it preferable to the more comfortable approach so often found in other departments where almost unnatural attempts are made to get a mediocre student through his graduate work.

Before describing the two problems Moran gave me, it may be interesting to mention the relation of the ANU department with two outstanding figures also working in population genetics. At that time R. A. Fisher lived in Australia, but never visited our department and had no contact with us. This was probably just as well, as I would have raised with him several points in his work that I could not understand or felt were wrong, and his irascible temperament, which I had seen in full operation in Melbourne, would no doubt have caused him to respond strongly to my comments. On the other hand, we had a most amiable relationship, at a distance, with Sam Karlin and Jim McGregor of Stanford, who had recently switched their interests to genetics. Indeed this switch was largely the result of a talk given at Stanford by Joe Gani in which he described a genetical model introduced by Moran, essentially involving a birth-and-death process with quadratic birth and death coefficients, a process for which, fortunately and coincidentally, Karlin and

McGregor had recently developed much of the desired theory. Because of Gani's connection with Stanford, I was subsequently able to carry out postgraduate work there to my great advantage, as I will soon describe.

The first of the two problems I was given on arrival by Moran concerned "self-sterility" alleles. Self-sterility is an unusual genetic phenomenon, occurring in a few plant species, the characterizing feature of which is that a plant of genetic type S_iS_j at a "sterility" gene locus cannot be fertilized by S_i or S_j pollen. Its offspring will then be S_iS_k or S_jS_k, each with probability $\frac{1}{2}$, where $k \neq i, j$ and S_k is the pollen-derived gene. Observed self-sterility populations exhibit, as might be expected, some unusual genetic features and the last of the many bitter controversies between R. A. Fisher and S. Wright (Fisher [4], Wright [19]) concerned the correct mathematical description of the genetical evolution of a self-sterility species, with a view to explaining and quantifying these features. My first problem was to carry out my own analysis and thus to assess the models of Fisher and Wright.

Perhaps the main difficulty I had was to make sense of the theory used by Fisher and Wright for this problem, which in both cases relied on a diffusion process describing the random changes in the frequency x of a specified allele (say S_1). The standard forward Kolmogorov diffusion equation

$$\frac{\partial}{\partial t}f(x;t) = -\frac{\partial}{\partial x}\{m(x)f(x;t)\} + \frac{1}{2}\frac{\partial^2}{\partial x^2}\{v(x)f(x;t)\} \qquad (1)$$

describing the density $f(x;t)$ of x at time t requires an expression for the mean $m(x)\delta t$ and the variance $v(x)\delta t$ of the change in x in a small time interval δt, and the general properties of the process depend on the form of $m(x)$ and $v(x)$. If either $x = 0$ or $x = 1$ or both form, in the language of Feller, an "exit" boundary, there can be no stationary distribution for the frequency x. The expressions for $m(x)$ and $v(x)$ used by Fisher and Wright did imply that $x = 0$ was an exit boundary, yet their analysis was based on the stationary distribution formula

$$f(x) = C\{v(x)\}^{-1} \exp 2 \int^x m(y)/v(y)\,dy, \qquad (2)$$

where C is some constant, derived from (1) by the successive equations

$$0 = -\frac{d}{dx}\{m(x)f(x)\} + \frac{1}{2}\frac{d^2}{dx^2}\{v(x)f(x)\}, \qquad (3)$$

$$0 = -m(x) + \frac{1}{2}\frac{d}{dx}\{v(x)f(x)\}. \qquad (4)$$

If indeed there is no stationary distribution, it is not clear what interpretation can be given to (2), a point I shall return to later. I therefore decided to try a different method. The mathematical properties of the self-sterility process made the simplest approach one in which a Markov chain with an absorbing state describing the number X of S_1 genes is modified to a "return" process in which on reaching the value 0, the random variable X is returned immediately to its initial value. This return process has a stationary distribution which, because of the unusual genetical properties of the sterility model, was easily evaluated, and whose terms clearly have the interpretation of being proportional to the mean number of times in the original process that X takes each possible transient value before reaching 0. This was enough to give me the answers I wanted for the self-sterility question.

The second problem I was given initially had no connection with the sterility model, although as has been hinted, it turned out to be very closely associated with it. If the drift and diffusion coefficients $m(x)$ and $v(x)$ in (1) take the form

$$m(x) = x(1-x)p(x), \qquad v(x) = Kx(1-x), \tag{5}$$

where K is a constant, and the diffusion random variable x lies in $[0, 1]$, with $p(x)$ an arbitrary polynomial in x, then $x = 0$, $x = 1$ are both "exit" boundaries. At these points, no boundary conditions need be, or indeed can be, imposed, since with the form (5) for $m(x)$ and $v(x)$, the process automatically stops at $x = 0$ and $x = 1$. As mentioned above, in these circumstances no stationary distribution for x on $(0, 1)$ can exist.

Despite this, Feller [3] and Kolmogorov [15] asked what interpretation could be given to the function $f(x)$ obtained as a solution to the "stationarity" equation (3) when $m(x)$ and $v(x)$ have the form (5). In the simple case $m(x) = 0$, Feller gave the solution

$$f(x) = C\{x(1-x)\}^{-1} \qquad (C = \text{constant}), \tag{6}$$

which is actually a solution of the less general equation (4), while Kolmogorov obtained the more general expression

$$f(x) = Ax^{-1} + B(1-x)^{-1} \qquad (A, B = \text{constants}), \tag{7}$$

which is a general solution of (3). Neither (6) nor (7) is integrable on $(0, 1)$, and both are found by inappropriate "stationarity" methods not applying for the coefficients (5), and yet my second task was to find an interpretation for (6) and (7), as well as for the more complicated functions arising when $m(x) \neq 0$.

To make progress it seemed necessary to create a process admitting a stationary distribution, and an obvious candidate was the return process described above for self-sterility populations. The density function $f(x; t)$ in a return process does not satisfy (1), because of the discontinuous jumps from a boundary back to the initial point. Nevertheless its stationary distribution can be found and, for the case $m(x) = 0$, turns out to be

$$f(x) = \begin{cases} 2(1-p)/T(p)(1-x), & 0 \leq x \leq p, \\ 2p/T(p)x, & p \leq x \leq 1, \end{cases}$$

where $T(p) = -2\{p \log p + (1-p)\log(1-p)\}$ is the mean absorption time in the original process. This solution is of the form (7), provided the constants A and B are allocated different values for $0 \leq x \leq p$ and $p \leq x \leq 1$, thus providing an interpretation for (7). There is no interpretation for the less general expression (6). The reason for this is that the right-hand side in (4) can be regarded as a probability flux, which is 0 for a process admitting continuous sample paths, and a stationary distribution which is not 0 for the stationary distribution of the return process, which involves discrete jumps in the value of the random variable.

It is clear that the function $f(x)$ has an interpretation in the original absorbing process, since if we define $t(x)$ as $f(x)T(p)$, then $t(x)\delta x$ is the mean time spent by the random variable in $(x, x + \delta x)$ before reaching 0 or 1. This led me to consider the pre-absorption behavior of a number of genetic models. One particular model was the following. If two alleles A_1 and A_2 are selectively equivalent, and A_1 genes mutate to A_2 at rate u with no reverse mutation, then eventually the frequency x of A_1 is absorbed at 0. Suppose that the population is diploid and of fixed size N and that the original frequency of A_1 is $(2N)^{-1}$ (corresponding to a single initial A_1 gene). It is possible to obtain the functions $t(x)$ for this case, and for $x \geq (2N)^{-1}$, the only values of importance, we find

$$t(x) = 2x^{-1}(1 - x)^{\theta - 1}, \tag{8}$$

where $\theta = 4Nu$. Having calculated this and similar functions, I turned to other problems for the remainder of my Ph.D. thesis.

3. The Stanford Seminars

I finished my Ph.D. in late 1963, and in early 1964 spent six months as a postdoctoral student at Stanford working under Sam Karlin, whose connection with Moran's group, largely through Joe Gani, I have described earlier. Karlin had at that time organized a regular Monday evening seminar in population genetics, whose participants, besides himself, were (during my time at Stanford) Jim McGregor, Oscar Kempthorne, Walter Bodmer and myself. Karlin and McGregor were then at the height of their fame in stochastic process theory, and Kempthorne at his peak in statistics and mathematical genetics. Bodmer had just started his studies on the human histocompatibility system which subsequently led him to the directorship of the Imperial Cancer Research Laboratory in London. It is hardly necessary to emphasize the stimulus on me of this seminar series and the influence it had on my future research.

One matter often discussed was the new problems in population genetics that might follow from the double-helix gene model of Watson and Crick. One of the first visiting speakers in this seminar was Jim Crow, who described work recently completed with Kimura in which the implications of the double-helix model were explored. The molecular structure of a gene allows an effective infinity of allelic types (rather than the two or three considered in classical population genetics), and this led Kimura and Crow [10] to propose a model in which an infinite number of allelic types $A_1, A_2, A_3, \ldots$ is possible at a given locus, with the assumption that any new mutant gene is assumed to be of a novel allelic type not previously seen in the population. Assuming selective neutrality, one was therefore led to consider a model in which if, in generation

t, there were n_i genes of allelic type A_i ($i = 1, 2, 3, \ldots$), the probability that in generation $t + 1$ there would be m_i such genes, together with m_0 new mutant genes (all of novel and different types) is

$$\Pr\{(n_1, n_2, \ldots) \to (m_0, m_1, \ldots)\} = \frac{(2N)!}{\prod\limits_{i \geq 0} m_i!} u^{m_0} \prod_{i \geq 0} \left\{\frac{n_i(1-u)}{2N}\right\}^{m_i}. \quad (9)$$

While there can be no concept of stationarity for the frequency of any allelic type, the various partitions (of $2N$) describing the patterns of unlabelled allelic frequencies do have a stationary distribution, so there exists a stationary probability for any quantity defined solely by the partition pattern. One such quantity is the probability that two genes drawn at random are of the same allelic type, which Kimura and Crow found to be, approximately,

$$\Pr\{\text{homozygosity}\} = (1 + \theta)^{-1}. \quad (10)$$

It was clear to me that further information on this model could be found by using equation (8). Any particular allele must arise, in this model, with initial frequency $(2N)^{-1}$ and then, after a random time whose mean can be found approximately from (8) by integrating the function $t(x)$ from $(2N)^{-1}$ to 1, will leave the population, never to return. Since on average $2Nu$ new allelic types arise per generation by mutation, an ergodic argument shows that in any given stationary generation the mean number of allelic types existing in the population is

$$\bar{n} = 2Nu \int_{(2N)^{-1}}^{1} t(x)\, dx$$

$$= \theta \int_{(2N)^{-1}}^{1} x^{-1}(1-x)^{\theta-1}\, dx. \quad (11)$$

Further, the mean number of allelic types having frequency in $(x, x + \delta x)$ is

$$n(x)\delta x = \theta x^{-1}(1-x)^{\theta-1}\delta x, \quad (2N)^{-1} \leq x \leq 1. \quad (12)$$

It is clear that the Kimura–Crow formula (10) can be derived from this expression. If a given allele has frequency x, the probability that two genes drawn at random are of this allelic type is x^2. Since from (12) we can also say that the probability that there is an allele having frequency in $(x, x + \delta x)$ is $n(x)\delta x$, the required formula can be found by integrating $x^2 n(x)$ from 0 to 1.

I found several similar results for the infinitely-many-alleles model, but did not pursue them any further at the time. The reason for this was that, on Kempthorne's advice, I decided to read and assimilate all the literature then available on population genetics theory (advice one could follow then but not now), to get a historical perspective of the subject and to understand the evolutionary questions which the theory attempted to answer. I eventually spent several years on this undertaking, considering questions of the evolution of dominance, multilocus evolutionary behavior, rates of evolution, survival probabilities of new mutants in various circumstances, the so-called fundamental theorem of natural selection, and other topics, all of which had received considerable attention in the literature. I even attempted to contribute to this literature myself. I have always regarded this task as my real grounding in population genetics, and have never regretted a moment of the time spent on it.

Two developments during the late 1960s, however, revived my interest in the infinitely-many-alleles model. The first was the great amount of sample data then becoming available which described the extent and nature of genetic variation in natural populations. These data are in the form of a vector

$$\{k; n_1, n_2, \ldots, n_k\}, \quad (13)$$

indicating that in a sample of $\Sigma n = n_i$ genes taken from a population, k allelic types were observed, with n_1 genes of one allelic type, n_2 of another, and so on. The second development, partly prompted by the unexpectedly large degree of genetic variation revealed by these data, was the neutral, or non-Darwinian, theory of evolution of Kimura [9]. This theory suggested that the great bulk of the variation did not arise from selective forces acting on allelic types of different fitness, but reflected merely the random fluctuations in allelic frequencies expected from the model (9).

The formulation of this theory brought about a sharp debate in population genetics between "neutralists" and "selectionists", which is still unresolved, and which was particularly acrimonious when I next visited the USA in early 1971, this time at the Department of Zoology at the University of Texas. As it happened, Crow was visiting this department at the same time and we were asked, by those who had data of the type described by (13), to derive the theoretical stationary distribution of sample vectors of this form under the (neutral) model (9). Once this was done, the next aim was to devise statistical tests of the neutral theory, comparing any observed data vector with its neutral theory distribution.

Questions of this sort involved a change in direction in population genetics theory. Until recent times the theory had been largely deductive: well into the century, many scientists had been sceptical that evolution, as a Mendelian genetic process, could and would occur under the Darwinian paradigm. The great work of Fisher, Wright and Haldane was largely aimed at showing that, assuming reasonable values for mutation rates, selective differences and other genetic parameters, evolution would indeed take place. However, their theories could not be checked against data, since very little information on the genetic constitution of individuals in populations was then available. During the 1960s this situation changed: various laboratory techniques gave ever-increasing information on genetic variation both within and between populations. This gradually led to an emphasis, in theory and practice, on an inductive approach: using the data, one could hope to test statistically various theories about the evolutionary process. Devising tests of the neutral theory was one of the first aims of the inductive approach, and with my interest in statistics and the infinitely-many-alleles model, I felt a keen desire to find the probability distribution of the vector (13). Then, if possible, using this distribution I hoped to devise statistical tests of the neutral theory: indeed this seemed a challenge towards which much of my training and experience necessarily directed me.

After several false starts, I gradually formed the view that the distribution of the vector (13) was somewhere implicit in the function $n(x)$ defined in (12), which by this time I had called the "frequency spectrum". All the same, it was far from clear that the distribution of (13) could be found with any facility, and that even if it could, its form would lead to the statistical testing procedures that were being sought. The distribution of (13) would presumably depend on the two unknown parameters N and u, whose estimation might prove difficult; further, what was required was a distribution of partitions (of the number n), of which there is an enormous number. On the other hand, approximations were clearly necessary and allowable, with terms of order N^{-1} being safely ignorable, as in the derivation of (10).

4. The Sampling Formula

In retrospect, the derivation of the partition distribution seems straightforward. It was certainly not so at the time: even focusing on the frequency spectrum, I initially followed a number of lines of argument that led nowhere. Then, one evening, the entire distribution suddenly fell into place. If the frequency spectrum could be used to find the probability that two genes taken at random in any generation are of the same allelic type, it could also be used to find the corresponding probability for three, four, or in general i genes. These calculations had the advantage that any formula arising from them would not involve N and u separately, but rather jointly through the single parameter θ. The probability that a sample of i genes will yield the same allelic type is, to the degree of approximation allowed,

$$\theta \int_0^1 x^i \{x^{-1}(1-x)^{\theta-1}\}\,dx = (i-1)!/[(1+\theta)(2+\theta)\cdots(i-1+\theta)]. \quad (14)$$

Comparing samples of sizes $i-1$ and i, this implied that if the first $i-1$ genes drawn are of the same allelic type, the probability that the last one drawn is again of this type is $(i-1)/(i-1+\theta)$. In other words, given that the first $i-1$ genes are of the same allelic type, the probability that the next gene is of a new allelic type is

$$\theta/(i-1+\theta). \quad (15)$$

Equation (14) is sufficient to give the probability of data vectors of the form $\{1;n\}$. Further, the probability that an allele of frequency x occurs in a sample of i genes at least once is $1-(1-x)^i$, and from (12) this implies that the mean number of different allelic types in a sample of this size is

$$\theta \int_0^1 \{1 - (1-x)^i\}\{x^{-1}(1-x)^{\theta-1}\}\,dx$$

$$= \frac{\theta}{\theta} + \frac{\theta}{1+\theta} + \frac{\theta}{2+\theta} + \cdots + \frac{\theta}{i-1+\theta}. \quad (16)$$

By comparing the cases $i-1$ and i, the unconditional probability that the ith gene drawn is of a new allelic type unlike any in the first $i-1$ draws is seen to be $\theta/(i-1+\theta)$.

This probability is identical to the one given in (15), and suggested an unexpected "non-Bayesian" possibility, namely that if we imagine the genes in the sample drawn one by one, the probability that on draw i we observe a gene of an allelic type not seen on the previous $i-1$ draws is $\theta/(i-1+\theta)$, irrespective of the allelic composition of the genes on those first $i-1$ draws. Assuming this to be true, it was easy to set up a recurrence relation leading to

Pr{k allelic types in a sample of n genes}

$$= |S_n^k|\theta^k/\{\theta(1+\theta)(2+\theta)\cdots(n-1+\theta)\}, \quad (17)$$

where S_n^k is a Stirling number of the first kind. This gives the marginal distribution of the first element k in the vector (13).

There was no further time that evening to extend this argument to find the distribution of the entire vector (13), but on the bus to work next morning it became clear, continuing to assume the non-Bayesian hypothesis, what form this distribution would take. Again imagining the genes in the sample to be drawn sequentially, the probability that on draw i one observed a new type is $\theta/(i-1+\theta)$, and the probability that one observed some gene already observed h times in the previous $i-1$ draws is $h/(i-1+\theta)$. Thus the probability of any given ordered sequence containing n_1 genes of one type, n_2 of a second type, ..., n_k of a kth type, for $n_1 \neq n_2 \neq \cdots \neq n_k$, is

$$\theta^k(n_1-1)!\cdots(n_k-1)!/\{\theta(1+\theta)(2+\theta)\cdots(n-1+\theta)\}, \quad (18)$$

with a trivial amendment when some of the n_i are equal. Finally, multiplying by an obvious combinatorial coefficient to allow for all possible orders, the probability of the data vector (13) becomes

$$n!\theta^k/\{(\prod n_i)(\prod \alpha_j!)\theta(1+\theta)(2+\theta)\cdots(n-1+\theta)\}, \quad (19)$$

where α_j is the number of n_i's taking the value j.

Although it took a little time to confirm the exact form (19), it was immediately clear that a comparison of (17) and (19) yielded two completely unexpected conclusions. The first was that k is a sufficient statistic for θ, which is the only unknown parameter in the distribution (19). The reason why this was unexpected was that, using (10), θ had been previously estimated by the equation

$$(1+\hat{\theta})^{-1} = \sum n_i^2/n^2, \quad (20)$$

which experience had shown to have poor estimation properties. The sufficiency of k for θ shows why this is so, since the estimator $\hat{\theta}$ deriving from (20) uses the essentially uninformative part of the vector (13) to estimate θ.

I had lectured on sufficiency many times in statistics classes, and had always felt disappointed, despite the great importance of this concept, that most sufficient statistics (such as $\bar{x}$ for μ in a normal distribution) are not novel in the sense that one would tend to use them even without any knowledge of their sufficiency properties. Here, on the other hand, was an example of a totally unexpected sufficient statistic. The second conclusion followed from the first, that the neutral-theory distribution of $n_1 \cdots n_k$ given k was free of θ and thus of any unknown parameters, so that in principle a testing procedure of the neutrality hypothesis could be formulated despite our lack of knowledge of N and u. All these conclusions, reached within a few hours, left me in a state of great excitement, tempered only by the thought that they relied entirely on the as yet unproved non-Bayesian assumption.

By the generous intervention of Crow, I was able to present these tentative results a few days later at the Berkeley Symposium on non-Darwinian evolution. I could find no one there who was optimistic about the non-Bayesian assumption on which they were based; of the colleagues to whom I wrote only one, Sam Karlin, then in Israel, felt it might be true and, further, claimed that he would be able to verify its validity on his return to Stanford a few months later. I visited Karlin at Stanford on his return, and to cut a long story short he was able, with McGregor, to prove the correctness of the sampling distribution (19) by mathematical induction, thus implicitly confirming the validity of the "non-Bayesian" assumption. Incidentally, since in the paper [2] in which I stated all of these conclusions I did not write down the central distribution (19) (giving only the distributions (17) and (21) below which are equivalent to it), and since I did not prove the correctness of (19), it is somewhat ironical that in later years, under the impetus of John Kingman, the distribution (19) came to be named after me.

I next spent a good deal of time considering the statistical tests of the neutral theory, based on the parameter-free conditional distribution of $n_1, \ldots, n_k$, given k, namely

$$\Pr\{n_1, \ldots, n_k | k\} = n! / [|S_n^k|(\prod n_i)(\prod \alpha_j!)]. \tag{21}$$

There is little point in describing these here, other than to remark that (21) implies that under selective neutrality the least likely configuration for $(n_1, \ldots, n_k)$ is one in which $n_1 = n_2 = \cdots = n_k$. This conclusion ran counter to several expressed views and crude tests of neutrality in which it was felt that selective equality of alleles would imply rough equality of allele numbers. The reason why this does not occur is that the alleles arise by new mutations at different times in the past, and thus will tend to assume different frequencies in any one given generation.

It was of greater interest to me to explore the consequences of the sufficiency of k for θ, and in particular to compare the variances of the estimator for θ deriving from (20) and that deriving from k, which from (17) is easily shown to be the solution of the equation

$$k = \frac{\hat{\theta}}{\hat{\theta}} + \frac{\hat{\theta}}{1+\hat{\theta}} + \frac{\hat{\theta}}{2+\hat{\theta}} + \cdots + \frac{\hat{\theta}}{n-1+\hat{\theta}}, \quad (22)$$

an equation bearing an interesting relation to (16). I could make little theoretical progress on this comparison; fortunately, when in Stanford I had mentioned this problem to Harry Guess, then a student of McGregor, who was at that time running computer simulations of the model (9). By attaching a few extra orders to his program, Guess was able to compare these two variances. It transpired that the estimate deriving from (22) has a variance typically one quarter or one third of that of the estimator given by (20), as well as being almost unbiased, whereas the estimator (20) has considerable bias. These conclusions confirmed the predictions of the theory with unexpected rapidity.

I should also mention that John Gillespie, whom I had met at this time, had felt for some time that θ should be estimated by using k rather than by using $n_1, \ldots, n_k$. His argument was that θ is simply twice the mean number of new allelic types to arise in each generation. For the model (9), all that is useful in estimating this mean number is the fact that an allele occurs at all: the frequency with which it occurs relates purely to the random changes in frequency once this allele has arisen. The argument is of course more complicated than this since these random changes do depend on θ: nevertheless, the theory confirms the essential correctness of Gillespie's view.

Once the sampling distribution (19) had been confirmed, a number of further questions arose. An immediate one was to delineate the range of models (besides (9)) which lead to a distribution of the form (19), with possibly a more generalized definition of the parameter θ. This question was taken up by Geoff Watterson of Monash University, with whom I discussed this and other questions on my return to Australia at the end of 1971. Watterson's approach to these problems, developed over several years, started by assuming a model of K different possible alleles, whose frequencies $x_1 \cdots x_K$ under the neutral model have a joint density of the Dirichlet form

$$f(x_1, \ldots, x_K) = M \prod_{i=1}^{K} x_i^{\alpha-1}, \quad 0 < x_i < 1, \quad \sum x_i = 1, \quad (23)$$

with M a constant, and $\alpha = \theta/(K-1)$. One would then like to start with (23) and let $K \to \infty$, to derive results for the infinitely-many-alleles model (9).

Unfortunately there is no direct limit for (23), as K increases without bound. Watterson noted, however, that in a paper concerned with quite a different problem, Kingman [11] had shown that a limiting ($K \to \infty$) distribution for the first j order statistics $x_{(1)}, x_{(2)}, \ldots, x_{(j)}$ exists, for any j. Although it is difficult to write this limiting "Poisson–Dirichlet" distribution down explicitly, the fact that it exists justified Watterson in moving freely between a "finite K" model admitting distributions such as (23), and "infinite K" models such as (9). This was enough to attract Kingman to take up some of the outstanding problems concerning the distribution (19); Kingman and Watterson were then able to characterize rather precisely the class of evolutionary stochastic processes leading to sampling distributions of the form given by this distribution.

Following on from this, Watterson, Kingman and others subsequently developed quite general theory, often applying outside the domain of population genetics, and thus worth describing briefly. Perhaps the most interesting conclusions reached by Kingman concern the concept of random partitions. It had been remarked above that the distribution (19) is a distribution of the partitions of the number n. Now n is simply the total number of genes in the sample, and hence has no special significance: indeed, the sample could originally have been of size $n + 1$ genes, one of which taken at random had been lost, and thus a consistency condition must hold between the partition probabilities of a sample of n genes and one of $n + 1$ genes. If we define

$$\mathbf{a} = (a_1, a_2, \ldots, a_n) \tag{24}$$

as the partition observed in a sample of n genes (where the notation implies that a_j allelic types are observed exactly j times, with $\Sigma j a_j = n$), and if $P_n(\mathbf{a}) = P_n(a_1, \ldots, a_n)$ is the probability of such a partition, then P_n should satisfy the equation

$$P_n(a_1, a_2, \ldots, a_n) = \frac{a_1 + 1}{n + 1} P_{n+1}(a_1 + 1, a_2, \ldots, a_n, 0)$$
$$+ \sum_{r=2}^{n} \frac{r(a_r + 1)}{n + 1} P_{n+1}(a_1, a_2, \ldots, a_{r-1} - 1, a_r + 1, \ldots, a_n, 0)$$
$$+ P_{n+1}(a_1, a_2, \ldots, a_n - 1, 1). \tag{25}$$

It is easy to check that (19) does indeed satisfy this equation, but Kingman [12], [13] raised the more general question, how can partition structures, whose probabilities must satisfy (25), be characterized? Define X as the set of all sequences $x_1, x_2, x_3, \ldots$ such that

$$x_i \geq 0, \qquad x_i \geq x_{i+1}, \qquad \Sigma x_i \leq 1,$$

and put $x_0 = 1 - \Sigma x_i$. Given any such sequence $x = (x_1, x_2, \ldots)$, suppose that in an infinite population a fraction x_i are painted with color C_i, while a fraction x_0 are painted in different colors, where each such color is different from every C_i. Let $P(\mathbf{a}; \mathbf{x}, n)$ be the probability that a sample of n genes from this population has the color partition $\mathbf{a}$. Then Kingman showed that any $P(\mathbf{a})$ satisfying (25) is of the form

$$P(\mathbf{a}) = \int_X P(\mathbf{a}; \mathbf{x}, n) \rho(dx),$$

where ρ is a probability measure on X. Further ρ, the so-called representing measure, is unique, and for the partition probability formula (19) turns out to be the Poission–Dirichlet distribution previously defined. These observations allowed Kingman to obtain a powerful theory of partition structures whose applicability extends beyond the sphere of mathematical genetics in which it had its origins.

A second broad class of Kingman's results concerns coalescent processes. The derivation of (19) used an argument working in a "forward" time direction. However, it is often simpler and more natural in genetics to work backwards in time: one can be certain of having two parents, four grandparents, and so on, while the number of one's descendants in any generation is random. In any model of biological evolution the lines of descent of any set of genes, when traced backwards in time, ultimately coalesce in a common ancestor. Kingman [14] was able to show that (19) arises as a natural property of the structure of any such process, together with the constant mutation of genes from one allelic type to another. Thus (19) has a natural "biological" flavor to it, and may almost be taken as characterizing processes with reproduction, mutation and a random number of offspring for each individual. Further properties of population processes of this type, often found by working backwards in time and thus also considering novel features such as the ages of the alleles currently existing in the population, have been found by Donnelly and Tavaré [1], Griffiths [5], [6], Kelly [7], [8], Saunders et al. [16] Tavaré [17] Watterson, [18], and others.

5. Concluding Remarks

My own research interests for the past several years have been in areas other than those described above, mainly in human population genetics and the problems of inferring the genetic basis of diseases using family and other data. This is, of course, in line with the inductive side of population genetics described above, although it has no connection with the sampling formula. I have been asked to describe what conclusions, or what moral, might be drawn from my experiences as a probabilist. So far as the events described above are concerned, it is hard to draw any but the obvious ones: I had great luck in arriving at the sampling formula, and better luck in having colleagues who were able to carry the analysis of its properties far beyond what I could do myself. My most memorable experience was the Stanford seminar series of 1964; perhaps the two main things I learned from this were how varied are the talents that different people have, and how a hard-headed recognition of one's own skills, together with an open sharing of ideas with colleagues of different abilities, leads to the happiest and most productive research career.

Furthermore, having decided that one's skills and interests lie in the applications of probability rather than in the pure theory, I would recommend that one consider seriously spending a significant portion of one's time working in surroundings appropriate to those applications. I have spent as much time, in the last 15 years, as a member of a biology department as of a mathematics department, since I feel that only in this way can I understand the approach to real problems as perceived by a geneticist (rather than a mathematician). The two perceptions of these problems are different, and having suffered through many a statistics seminar in which "applied" problems were used merely as a vehicle for unimportant mathematical elaborations, I prefer the approach I have suggested. It is not an easy approach since one needs the contact with mathematicians and their modes of thought also, but if one is interested in applications, it is one worth trying.

References

[1] Donnelly, P. and Tavaré, S. (1986) The ages of alleles and a coalescent. *Adv. Appl. Prob.* 18, 1–19.

[2] Ewens, W. J. (1972) The sampling theory of selectively neutral alleles. *Theoret. Popn Biol.* 3, 87–112.

[3] Feller, W. (1951) Diffusion processes in genetics. *Proc. 2nd Berkeley Symp. Math. Statist. Prob.*, 227–246.

[4] Fisher, R. A. (1958) *The Genetical Theory of Natural Selection*. Dover, New York.

[5] Griffiths, R. C. (1980) Lines of descent in the diffusion approximation of neutral Wright–Fisher models. *Theoret. Popn Biol.* 17, 37–50.

[6] Griffiths, R. C. (1985) Asymptotic line-of-descent distributions. *J. Math. Biol.*

[7] Kelly, F. P. (1977) Exact results for the Moran neutral allele model. *Adv. Appl. Prob.* 9, 197–201.

[8] Kelly, F. P. (1979) *Reversibility and Stochastic Networks*. Wiley, New York.

[9] Kimura, M. (1968) Evolutionary rate of the molecular level. *Nature* 217, 624–626.

[10] Kimura, M. and Crow, J. F. (1964) The number of alleles that can be maintained in a finite population. *Genetics* 49, 725–738.

[11] Kingman, J. F. C. (1975) Random discrete distributions. *J. R. Statist. Soc.* B 37, 1–22.

[12] Kingman, J. F. C. (1975) The population structure associated with the Ewens sampling formula. *Theoret. Popn Biol.* 11, 274–283.

[13] Kingman, J. F. C. (1978) The representation of partition structures. *J. Lond. Math. Soc.* 18, 374–380.

[14] Kingman, J. F. C. (1982) On the genealogy of large populations. *J. Appl. Prob.* 19 A, 27–43.

[15] Kolmogorov, A. N. (1959) The transition of branching processes into diffusion processes and associated problems of genetics (in Russian). *Teor. Veroyatnost. i Primenen.* 4, 233–236.

[16] Saunders, I. E. Tavaré, S. and Watterson, G. A. (1984) On the genealogy of nested subsamples from a haploid population. *Adv. Appl. Prob.* 16, 471–491.

[17] Tavaré, S. (1984) Line-of-descent and genealogical processes and their applications in population genetics models. *Theoret. Popn Biol.* 26, 119–164.

[18] Watterson, G. A. (1984) Lines of descent and the coalescent. *Theoret. Popn Biol.* 26, 77–92.

[19] Wright, S. (1960) On the number of self-sterility alleles maintained in equilibrium by a given mutation rate in a population of given size: a re-examination. *Biometrics* 16, 61–85.

R. L. Tweedie

Richard Tweedie was born in the Australian country town of Leeton, New South Wales, on 22 August 1947. He was educated in Leeton, and in 1965 was awarded a National Undergraduate Scholarship to The Australian National University, Canberra, where he gained a BA with first class honours and the University Prize in 1968, followed in 1969 by an MA with a thesis on random walk theory.

In October 1969 he commenced a Ph.D. in probability theory at Cambridge University under David Kendall, and completed a thesis on Markov chain theory in 1972. He returned in 1972 to a postdoctoral fellowship at the Institute of Advanced Studies, The Australian National University. In 1974 he joined the Division of Mathematics and Statistics, CSIRO, as part of the expansion of that division under Dr Joe Gani. He spent 1978 as associate professor at the University of Western Australia, and returned to CSIRO as senior regional officer of the Division of Mathematics and Statistics in Melbourne. In 1981 he joined Siromath Pty Ltd as general manager, and was appointed managing director in 1983, the post which he currently holds.

His major research has been in the structure and application of Markov chains and related areas. He has carried out the bulk of his consulting work in biological modelling, and has published over 60 papers in both mathematical and nonmathematical journals. He was awarded a D. Sc. from The Australian National University in 1986 for this work.

He has been active in promoting the use of probability and statistics in the general community in Australia, and he is the first statistician from the private sector to be elected president of the Statistical Society of Australia.

Richard Tweedie married Catherine Robertson in 1971 and has a daughter, Marianne. He has travelled widely, particularly when working with colleagues in the USA and Finland. For relaxation he reads science fiction and detective fiction, having an extensive collection of the type of No. 1 world bestsellers acquired for reading on aeroplanes. When he had time, he used to enjoy woodwork, cooking and playing tennis and squash.

In and Out of Applied Probability in Australia

1. Preface and Apologies

When I was asked to contribute to this volume I was both flattered and depressed: flattered, to find myself in the company of many applied probabilists who were famous names in the field before I ever entered it; depressed, because I feel too young to be asked to write even this brief autobiographical note with all the intimations of a completed career that such an activity implies.

I am reminded in fact of similar feelings of depression in my early mathematical career. It was then (and probably is now) a popular theory that all mathematicians of distinction had made their contributions before they were 20; and we students frequently despaired, not because our research was not up to the standard of Abel and Galois, but because we hadn't even *begun* research. Perhaps the most famous mathematician of that period, Tom Lehrer, encapsulated the feeling when he said, "When Mozart was my age, he had been dead for three years!"

An autobiographical article usually describes a significant, and preferably also a completed, life's work. While I am unable to provide the brilliant insights of an Abel of applied probability, or the considered reflections of a lifetime spent in the area, this is nonetheless not an unreasonable time for me to attempt to describe my personal view of the craft of an applied probabilist. I have spent the 15 years of my academic and research career in applied probability; now I am becoming more concerned with research and consulting management, as well as statistical analysis rather than modelling. Although, regrettably, I see myself as being on the way out of applied probability, I will attempt to describe at least the following:

— how I entered applied probability, and what aspects of the craft attracted and continue to attract me;
— my own view of the levels at which probability can be "applied", ranging from the use of structural mathematical results to the direct application of probabilistic thinking to data analysis;
— where I personally believe applied probability is going, and how it relates to the industrial mathematical consultancy in which I am now involved.

There are several points which my own varied experience has led me to feel are worth making, and which I will expand on below. They are that:

— the mathematics carried out by *real* applied probabilists should have some connection to nonmathematical questions; the reason for the mathematics should thus be understandable at some nontechnical level;
— applied probabilists need to understand the structural aspects of their models, in order to evaluate the qualitative applicability of the mathematics to particular contexts;

— the mathematical models formulated by applied probabilists should be motivated by a consideration of *real* data;
— an applied probabilist must also become a statistician, in order to associate models in practice with data, to estimate model parameters and test their goodness of fit;
— applied probabilists will flourish only in projects involving major contact and collaboration with nonmathematicians;
— an applied probabilist must be able to work with computers, as most applicable models are too complex for useful analytic solution.

Ron Pyke [10] has analysed the aspects of probability that differentiate between real applied probability and the bulk of the material published as applied probability. I believe one other major discriminating factor is the type of person doing the work, and the six points above describe a personality as much as a particular type of output.

There has been increasing pressure on academic research to take place in areas of relevance. My ideas are based on an unusually mixed career in academic research, pure research and commercial consulting groups, and I hope that this article will contribute something of value to applied probabilists who are examining their relevance in the world today.

2. Entering Applied Probability in Australia and England

My own entry into the subject contains no particularly useful lessons on what makes an applied probabilist, except that it illustrates the importance of chance phenomena. I was born in Leeton, a country town in Australia, and all my primary and secondary education took place there. Chance favoured my school with good teachers; the quality of teaching in Australian country areas was variable, but in both mathematics and English I was well-taught and encouraged. At the same time the country-school syllabus was patchy in physics and chemistry, and I found my background and motivation quite inadequate in these subjects at university; thus I never became an unemployed physicist, as did many of the more talented students of my time. With hindsight this deficiency in my education or my talents proved to be lucky. Even luckier was a chance visit, on a field trip with a group of geography students, of the master of one of the residential colleges at The Australian National University (ANU). He suggested I apply for a National Undergraduate Scholarship which the ANU then awarded to encourage good undergraduate students. I won this scholarship, went to ANU rather than to the University of Sydney (in whose catchment area my school traditionally lay), and at that point my introduction to applied probability became a much more likely event.

For a country of its size, Australia has produced a surprisingly large number of statisticians and an even more surprisingly large number of probabilists. The statisticians can be traced to a number of sources, but my impression is that the Australian probabilistic school owes its importance almost entirely to the influence of P. A. P. Moran, first professor of statistics at ANU. His students, many of whom staffed the research and the teaching departments at the university when I attended it, had included E. J. Hannan, J. Gani, C. R. Heathcote, P. J. Brockwell, C. C. Heyde, E. Seneta and W. J. Ewens, all of whom have had a significant effect on my own career. I was taught by Ewens, Heathcote and Seneta in applied probability; they all aroused my enthusiasm and introduced me to a subject which I not only enjoyed mathematically, but whose context in the nonmathematical world I could also understand.

This supports my first point in Section 1 above; for me the interest in the probability I learned was that it really was *applied*, and hence the mathematics involved was, at least to some extent, connected to nonmathematical questions.

From this standpoint, much that appears under the label of applied probability in the academic literature is no such thing. It is, if you like, pure probability theory, and should be treated as such. Pure probability theory is an honourable pursuit: it is certainly a respectable branch of pure mathematics, it is usually far harder technically than applied probability and it may contain the basis for later applied probabilistic results. It differs from applied probability in its motivation: as pure mathematics it generates its own internal problems and development, whereas applied probability should (again, at some level) be driven by externally generated questions.

I have never felt technically qualified to work with facility in the deepest areas of pure probability; but with Eugene Seneta introducing Markov chains and branching processes, and Chip Heathcote discussing random walks, queues and dams at ANU, I had a very early entry into subjects on which I have worked since.

When I finished my undergraduate work in 1968, I followed the standard Australian practice of the time and went overseas for my graduate work. Chance favoured me again: ANU provided two scholarships to its graduates each year to study anywhere in the world, and I gained one of them. By this time, I was convinced I wanted to study applied probability, although I then thought of it simply as probability theory, having done very little classical pure probability theory. David Vere-Jones, with whom I have had a long, spasmodic but highly productive collaboration, was at ANU at the time, and he recommended I try to study with David Kendall, who had supervised his own D.Phil. David Kendall agreed to take me; I filled in the time between the end of the southern academic year in December and the beginning of the northern academic year in September doing research with Chip Heathcote on random walks; and in 1969 I commenced research in Markov chain theory in Cambridge.

I worked in three areas. I completed the random walk research which I had begun in Canberra; I worked on approximations to stationary measures, extending some results of Seneta; and I worked on the R-theory of Markov chains originated by Kendall and Vere-Jones 10 years earlier. All of this work could be (and has been!) published as applied probability; but only the second area really qualifies as applied, even in an indirect way, by the criteria I have suggested in this paper.

Nevertheless, by the time I returned to Australia in 1972, I had obtained some nice results, acquired some useful technical abilities and, in particular, begun to work on the structural aspects of general probabilistic models. I will describe some of these in the next section, for two reasons. First, because I welcome this chance to place my own work in a general context; but second, because I wish to illustrate the value of theory as well as data analysis.

3. The Structure of Markov Processes: Arguably Applied Probability?

I recall a delightful "applied probability" paper which gave a *very* nonmathematical description of a fascinating problem, and then went on to say, essentially, "We shall model this as follows. Let (Ω, F, P) be a probability space admitting a Hausdorff topology..." The application was then decently buried. Is this applied probability? At one level, as I will argue below, it may well be: a first, fundamental main question is the way in which the results, no matter how theoretical, *can* be used.

The majority of my own papers in "applied probability" begin as follows: "Let $\{X_n\}$ be a Markov chain on a general measure space E, homogeneous in time, with n-step transition probabilities for $X \in E$, $A \subseteq E$, given by

$$P^n(x, A) = \mathbb{P}(X_n \in A | X_0 = x)."$$

These processes are pleasant in many ways. Because they occur as a chain of values in discrete time, many of the major mathematical problems of continuous-time stochastic processes disappear. They may become duller for mathematicians; but they become easier to understand, and I think they still model many problems worth modelling. Moreover, discrete-time Markov chains are often embedded in many much more complex processes and provide a key to their analysis, so that these relatively simple processes repay understanding in a wide variety of models.

The simplest cases for study are when the values taken by $\{X_n\}$ are in a countable set, say $E = \{0, 1, \ldots\}$. For applied purposes there is usually no loss in generality in assuming *irreducibility*: that is, for all pairs x, y, there exists n such that $P^n(x, y) > 0$. With this assumption, a classical result which is both elegant and important is that the chain is of one of only three types:

— *positive recurrent*, which means that if $\tau_x = \inf(n: X_n = x)$, then

$$\mathbb{E}(\tau_x | X_0 = y) < \infty \quad \text{for all } x, y;$$

— *null recurrent*, a marginal case where

$$\mathbb{E}(\tau_x | X_0 = y) = \infty \quad \text{for all } x, y$$

but
$$\mathbb{P}(\tau_x < \infty | X_0 = y) = 1 \quad \text{for all } x, y;$$

— *transient*, where
$$\mathbb{P}(\tau_x < \infty | X_0 = x) < 1 \quad \text{for all } x.$$

Most real-life situations modelled by Markov chains have a structure depending on a few fundamental parameters; which of these types describes the chain depends on the parameter values rather than the overall structure.

With some significant exceptions (such as branching-process models and epidemic models which are not irreducible), models of reality need to be of the positive-recurrent type, since, in particular, a unique stationary probability distribution Π then exists satisfying the equation

$$\Pi(x) = \sum_y \Pi(y) P(y, x) \quad \text{for all } x, \tag{1}$$

and the n-step transition probabilities in the aperiodic case converge to Π, i.e.

$$P^n(x, y) \to \Pi(y) \quad \text{for all } x, y \tag{2}$$

in some suitable mode of convergence. For models where (2) does not hold, relevance to real situations is much harder to argue.

The fact that such a classification occurs is now a simple classical result. The main questions I have worked on in probability theory revolve around these classification results, and can be summarized as follows:

— how much of this clear structure remains on an arbitrary (noncountable, nonstructured) space? This is usually a question in pure probability theory (and one studied very widely by very many people); but it has substantial applied motivation, and any answer to it helps delineate the range of applicability of applied models;
— when does a given model, defined by the (often intuitively derived) one-step probabilities $P(x, A)$, actually fall in the useful class of positive recurrent chains? This is a real applied probability question in my opinion, and in view of the expanding role of computers (of which more below), one of the theoretical questions of substantial applied importance.

The answers to these questions are surprisingly nontechnical and general.

First, given the appropriate definition of irreducibility, the three-way classification of Markov chains into positive-recurrent/null-recurrent/transient types extends to a virtually arbitrary state space. Suppose E is any space, restricted only by needing to support a σ-field of sets with a σ-finite measure ϕ on E. A chain $\{X_n\}$ on E is ϕ-irreducible if for all x

$$\mathbb{P}(X_n \in A \text{ for some } n | X_0 = x) > 0$$

whenever $\phi(A) > 0$; that is, the chain has some possibility of reaching A from any starting point if A is nontrivial (as measured by ϕ). The structural results for Markov chains on the integers extend to any ϕ-irreducible chain almost completely (see [17]); here, "almost completely" means that only a few ϕ-null sets may have "wrong" properties, and although this is distressing for pure probabilists (and I have spent many hours of only occasionally fruitful work trying to delete such null sets, as all pure probabilists do!), it is of little practical consequence. Other more analytic approaches to nondiscrete state spaces (such as in [11]) do not give such complete analogues of the three-way classification; mathematically, they give interesting results and pose difficult problems, but for applications they seem much less useful.

The power of the seemingly innocuous ϕ-irreducibility concept was first exploited by Harris [3] and later a group based at the University of Minnesota (see Orey [9]). More recently, the real reason for this power has been discovered by Nummelin [8], who made the remarkable observation that ϕ-irreducibility enables the construction of an augmented state space and chain for which a single point is an atom (i.e. can be reached with positive probability). Consequently, renewal theory, which underpins most results on the integers, also underpins ϕ-irreducible chains on even the most general space.

Is this elegant work applied probability? Clearly at one level it is not; but at a deeper level I would argue that it is. I have myself often used such structural results to guarantee that actual applied examples could be studied in their natural settings: for example, I have considered a queueing model for the times of cooling of pig iron, set by the original modeller in an artificial integer-valued state space but equally easily analysed in the natural real line space once it could be shown that the structural results applied there.

For me personally, the value of studying general-state-space chains has always been that they removed the need to use the often very artificial integer-valued structure of the space. Moreover, I have always been delighted by the fact that the leap from integer-valued chains to completely general-valued ϕ-irreducible chains was so easy, and hence for genuine applications on, say, the real line or Euclidean space, no extra theory exploiting specific Euclidean properties seemed necessary. Since ϕ-irreducibility solves real problems and usually, for such real problems, is quite a respectable assumption, I have had little inclination to explore the more pure-mathematical intricacies of non-ϕ-irreducible chains.

Given then that a ϕ-irreducible chain *may* be positive recurrent, how do we decide that it is? Again, the classical integer-valued result extends to general spaces; if $P(x, A) = \mathbb{P}(X_n \in A | X_{n-1} = x)$ is the one-step distribution governing transitions from x, then the chain is positive recurrent if there exists $\varepsilon > 0$, a function $g > 0$, and a "suitable" set C such that

$$\int P(x, dy) g(y) \leq g(x) - \varepsilon, \qquad x \notin C \qquad (3)$$

$$\int P(x, dy) g(y) \quad \text{bounded}, \qquad x \in C. \qquad (4)$$

Here a "suitable" set is a set in the "centre" of the space; a finite set of the integers, or, usually, a compact set on Euclidean space (see [18] for details).

Is this applied probability? I would argue that it is, and much more definitely so than the structural classification result itself. It gives checkable conditions for positive recurrence and has been easily used to explore conditions under which a range of frequently constructed models have stationary distributions (see e.g. [5]).

From this area of my own craft, I would stress the second point in Section 1: an understanding of the theoretical aspects of a general class of models is necessary for an applied probabilist. I have often been able to use such knowledge in deciding when a given model was of real relevance. My own favourite example of this is in Vere-Jones and Ogata's paper [9], where I am particularly pleased with the acknowledgement!

I should not leave the reader with the impression that I think one can argue that all Markov chain theory is applied probability. My first piece of research provides an excellent example of when it is not. I considered conditions under which a solution to (1) existed in the case when $\{X_n\}$ was a random walk on $\{0, 1, \ldots\}$; that is, when for some distribution $\{a(x)\}$, $P(x, y) = a(y - x)$, $y > 0$, with the boundary condition

$$P(x, 0) = \sum_{-\infty}^{-x} a(y).$$

If $\{X_n\}$ is positive recurrent this is applied probability; one can use (3) and (4) with $g(x) = x$ to deduce that $\{X_n\}$ is positive recurrent provided

$$\sum_{-\infty}^{\infty} xa(x) < 0.$$

However, I considered the existence of a (nonsummable) solution to (1) in the *transient* case, and showed that at most one solution exists, and that the necessary and sufficient condition for a solution is that, for some $\delta \neq 1$,

$$\sum_{-\infty}^{\infty} a(x)\delta^x = 1.$$

I still consider this result elegant; but I cannot think of any application or interpretation for it.

4. Consulting in CSIRO: Real Applied Probability?

I returned to Australia to the Department of Statistics at the ANU Institute of Advanced Studies; with no formal teaching responsibilities I was able to commence lines of work, such as those above, which I still continue to follow. Despite the many applied contributions of members of that department, I must confess that my own introduction to genuine applications of mathematics only came when I joined the CSIRO Division of Mathematics and Statistics (DMS) in 1974, when Joe Gani became its chief.

CSIRO, is, I believe, a unique scientific organization. Other countries fund research by grants to institutions such as universities; in addition to an effective university system for teaching and research, Australia has had for some 60 years a parallel research system, government funded, working in principle on problems of importance to Australia. Its practical achievements include shrinkproofing of wool, development of myxomatosis infection methods for rabbit control, and invention of the Interscan aircraft landing system. It is currently under great pressure to produce ever more relevant results with ever-decreasing funding; but when I joined, the economy was booming and, with the arrival of Joe Gani, DMS was receiving strong support for research and consulting.

The role of DMS was, and is, twofold. Its scientists are to develop statistical and mathematical techniques of relevance; they are also to work in collaboration with other CSIRO scientists in the application of such techniques.

I did not initially take easily to the latter role. I was a relatively pure applied probabilist and wanted to continue in that vein; to a large degree I did, but slowly I became involved in consulting and collaborative research to an increasingly great extent.

Such collaborative work gradually led me to a strong belief in the following thesis: for applied probability to be *real* applied probability, *the mathematical models must be associated with real data.*

And by real data, I do not usually mean data found in a reference of 10 or 30 years ago and already over-analyzed because no one has obtained any other data. I mean data needing a new method of analysis, where the applied probabilist has been involved closely with the other scientists who are helping, or directing, the framing of scientific questions.

I believe G. E. P. Box once said "95% of statistics is framing the correct questions". The same should be true of applied probability. The consideration of previously analysed data sets to initiate new methodology has always seemed to me to be somewhat sterile, although the use of such data to test new methods can of course be illuminating; and in principle any data-based justification seems better than none at all. (In this context it is instructive, if depressing, to scan probability journals seeking data-sets.)

In CSIRO I found, as did several of my colleagues, real opportunities to apply probability theory to externally-motivated problems: that is, to problems in other disciplines. The more I saw of such applications the more I developed the belief that *a real applied probabilist must become a statistician.* Data-free modelling needs no statistics; but when data appear, methods of inference relevant to the model, and in particular methods of estimation of model parameters, become of paramount importance. This is scarcely an original viewpoint—one sees Maurice Bartlett remarking on the link between statistics and applied probability virtually as the subjects were being born— but it is often more honoured in the breach than in the observance.

As an open research field I recommend inference in applied probability: it is difficult, data-sets are usually too sparse for comfort, and methods of inference may have to be very model-specific to be applicable. However, once an inference method is found for a model, the work on it is much more likely to become real applied probability.

My own work in this area has been almost exclusively on methods of point estimation for models I was working with. I have taken the no doubt contentious view that estimation of error bounds is a secondary problem, because in my own experience point estimation has always been hard enough. I very much hope that the next generation of probabilists will succeed in supplying much stronger statistical methods for use with applied probability.

During my seven years at CSIRO, I continued working on a number of the probability problems in the last section; but I also became involved in a number of real-world problems, of which I shall describe a selection in the hope that they will illustrate the points made above.

4.1. Estimation of Polynesian Founder Populations

A major question of Pacific anthropology concerns the size of founder populations. Could a small expedition such as that of Thor Heyerdal have founded a viable population, or does one need an expedition with hundreds of demographically suitable pioneers? Ian Saunders and I studied this problem with Norma McArthur, a demographer and prehistorian at ANU [7], [13]. The results were fascinating, especially to anthropologists: the work was certainly real applied probability.

It illustrates three aspects of real probabilistic modelling (and indeed real applied statistics) which I believe can be widely generalized:

— the probabilistic model was simple mathematically (although at least one clever idea was involved which afterwards seemed very obvious!);
— the analytic solution proved nonetheless impossible; a solution by simulation was, however, easy and effective;
— the major contribution of a probabilist turned out to be the mode of approach and the questions addressed; these are often quite different from those considered by deterministic modellers.

And for those who do not wish to read [7] or [13], I should reveal that, at least in a computer model, one has a very good chance of settling a Polynesian island successfully with a very small initial population indeed, provided one's initial companions are chosen with care!

4.2. Myxomatosis models

Epidemic models are widely studied, in great depth and with very complex tools. I know of only a few cases where they have been applied. One such occurred when Ian Saunders was given data on the spread of myxomatosis in rabbits by the CSIRO Division of Wildlife Research. He showed that the usual homogeneous mixing assumptions, on which those complex results depended, were rather inappropriate; but he was able to generalize the infection rate model appropriately and fit it to the existing data [12]. Again, this was real applied probability, and it helped to decide which strain of virus to use to control the rabbit population. Two major lessons from this work are:

— a solid theoretical understanding of model structure is needed to apply previous work;
— real data is a different guide to the validity of model assumptions than is mathematical convenience.

4.3. Muttonbird Growth

I had one year's leave from CSIRO in 1978, to take up the position of associate professor at the University of Western Australia. In Western Australia I worked with John Henstridge on modelling the growth of muttonbird chicks. This was the oldest data I have worked with, collected by a pioneer Australian ornithologist, Dom Serventy, in Tasmania in 1954 and left unanalysed owing to lack of suitable methods. The main aspect of the growth of such chicks is a series of spectacular weight gains at increasingly separated time points, between which chicks lose weight as their parents are absent collecting the next meal; thus, normal growth curves seem inapplicable. We proposed a reservoir-like model and fitted it in pieces; inputs as independent variables, weight-loss between feeds as exponential decay, input times as a Poisson-like process [4]. The major general points this brought home to me are:

— the independence inherent in many real applied probability models can usually be exploited for inference;
— basic applied probability models are applicable in very different contexts (despite my comments on their need to be data-driven).

4.4. Laplace Transform Estimation Methods

In conjunction with Hans-Jurgen Schuh, Paul Feigin and Byron Morgan (with much advice from others, not least Ian Saunders again) I have been involved in developing estimation techniques based on the fact that many probabilistic models are simpler to analyse when time is integrated out to give a Laplace transform (see [14], [2], [20], [6] for examples and applications). Much of this work is still mathematically quite unsatisfactory although it has proved of considerable practical use in several fields: it illustrates that:

— rudimentary estimates are better than none at all; this research was motivated by a need to fit models to data which we could not fit otherwise;
— the difficulty of more detailed inference in applied probability.

These examples have no particular unifying theme, and are only some of the high points of a much greater number of CSIRO examples of applied probability in which I have been involved. Without further comment, however, I would point out that all of them illustrate my fifth point in Section 1, that *real applied probability projects should involve major contact and collaboration with a nonmathematician*.

Finally, in describing my own evolution as an applied probabilist, I must mention the influence of the stream of visitors, both Australian and overseas, encouraged to spend time at DMS by Joe Gani. The resulting exposure to a wide variety of problems, techniques and general approaches to the application of results was of profound benefit to me and to many others. It is only after the event that the effect of many of these contacts can be seen; but for many years, DMS undoubtedly provided an environment for mathematical education unparallelled in Australia.

5. The Advent of Computers

My own work in applied probability has suffered from my relative inability to use modern computing power to best advantage. This partly reflects my education. I was in the first class of statisticans ever taught computing at ANU. We had at least a two-week turnaround on batch jobs; and, with an average of one mistake per job control card (*those* were the days!), a ten-week course allowed me to correct only the first five control cards in my first program. I do not recall actually computing anything at all in that course, and I fear the experience left me with no desire to compute anything ever again.

However, my irrational tendency to avoid computing whenever possible also partly reflects my own desire, and I am sure the desire of almost every other probabilist of my generation, to provide *analytic* solutions to tractable mathematical problems.

This is a hard goal to abandon. Nonetheless, I believe it is at best only a partial accomplishment for modern applied probabilists, and that the ability to compute well has become a necessary, and possibly primary, accomplishment if one is to acquire the craft of probabilistic modelling.

The simplest probability models are combinatorial, and susceptible to counting methods. There is an intermediate range of complexity where methods may be analytic and models may be interesting (queueing models are an example, as are some epidemic models). I now tend to believe that *realistic probabilistic models are usually too complicated to analyse save by simulation or computerized numerical approximation*.

This is often taken by computer scientists or applied mathematicians to mean that the major areas of probabilistic research, such as asymptotic results or approximations, have become redundant. This may, of course, be true for some specific areas. In general, though, I believe that computer results, which are themselves approximate, must be enhanced by the best possible use of analytic results before computing begins.

I also believe that, although it may initially seem paradoxical, the increasing ability to obtain solutions to models by computational methods means that the general structural results, examples of which I have given in Section 2 above, will become even more important: they are often the only guarantee that a result exists at all, and a simulation may then lead to its details. The need for mathematics to underpin models will not disappear; but much of the research on solving models by analytic means may well do.

Clearly this is a personal view, and I suspect it may be an *ad hoc* justification of the areas I have worked in, as opposed to those I have never touched. I can give two examples to support my belief in the importance of structural underpinning, although I shall not try to justify further my growing disbelief in the usefulness of some of the analytic methods of solution studied by so many in so much detail.

5.1. Truncation Methods

Suppose we have any Markovian model and we mean to calculate Π as in (1) above. Typically in practice this is not possible analytically. One of my own early interests [17] was to work on methods of estimating Π using $_{(n)}P$, the $n \times n$ 'north-west corner' truncation of P. It is possible to construct approximations $_{(n)}\Pi$ based only on $_{(n)}P$ such that

$$_{(n)}\Pi/_{(n)}\Pi(0) \to \Pi/\Pi(0).$$

Eugene Seneta [15] gives many results concerning this approach for positive recurrent chains, and he shows that in *all* cases when $\{X_n\}$ is irreducible, $_{(n)}\Pi/_{(n)}\Pi(0)$ tends to *some* limit. Convergence does not guarantee convergence to a quantity of interest. We do need a theoretical result, such as the criterion of (4) to guarantee the existence of Π, to which convergence then proceeds.

5.2. A Model for Genetic Frequency

In CSIRO DMS again, de Hoog *et al.* [1] used quite sophisticated numerical analysis to solve the stationary equations

$$\Pi(A) = \int_E \Pi(dy) P(y, A)$$

for a Markov model of genetic frequency with $E = [0, 1]$ as state space. The numerical result is an extraordinary distribution, multimodal in the extreme. Is it an approximation to a real solution to the integral equation? Again, use of structural theoretical results showed that one and only one solution Π existed, and hence the theory of numerical methods shows that it is in fact approximated by the numerical solution.

None of the examples given above, other than the Polynesian model in [7], illustrate my view that any realistic model can be studied best, and often only, by simulation methods, although many other researchers could provide instances of this type. Marcel Neuts, for example, has devoted considerable efforts over the past few years to attacking queueing theory and related applied probability models with the goal of setting up models susceptible to computer solution. I do feel that the need for such computational skills is worth stressing, and the implications for the teaching of future applied probabilists is clear.

There is, however, a real danger to the future of the craft of applied probability modelling in the advent of cheap and effective computing. This danger seems to me to have already overtaken much "applied" statistics research where many papers pose problems of marginal interest, fail to solve them in any analytic way, but include a wealth of simulation output as an alternative solution. I have argued that this may be the best, and indeed only, way to solve many applied probability problems; the danger is that it enables some kind of answer to be supplied to almost any supposedly applied problems, and leaves the real craft of modelling in peril, as all crafts are, of being swamped by mass-produced products of poor workmanship.

The only solution I can see to this is that journal editors must be very much on their guard against accepting computer solutions to problems unless they judge the problem to be sufficiently interesting that such a solution, with its inherent problems and flaws, is still a worthwhile scientific contribution. Applied probability modelling has been, like any branch of mathematics, unwilling to admit heuristic arguments as proofs; only the best modellers have enough insight to provide correct heuristics consistently. I have no doubt that these standards should be kept very strongly in mind as the tide of simulation "proofs" rises about the already somewhat swamped journals of our craft.

6. Out of Applied Probability

In 1981, I left CSIRO to take up the newly created position of General Manager of Siromath Pty Ltd, a novel venture by CSIRO in conjunction with the Australian Mineral Development Laboratories and a partnership of consulting actuaries. Siromath is a consulting company, providing statistical, mathematical, operations research and computing consulting to business, industry and government.

Since joining Siromath, I have gained considerable managerial knowledge of both the needs and the desires for mathematical skills in business and industry. I have also carried out personally several major statistical consulting projects and directed very many more. Yet I have rarely, if ever, used applied probability models in this commercial environment. Siromath has had only a few projects involving "applied probability" modelling, out of a total of some hundreds of problems. This is despite Siromath having from its own staff, or via access to CSIRO staff, the ability to put together a group of applied probabilists of world class.

This might force one to conclude that, at least in Australia, the only viable venues for applied probability research and consulting are academic ones. This has certainly not been a uniform, worldwide, experience. Bell, IBM, Philips and the like have clearly felt that an applied probability approach (queueing-theory models, for example, or network models) have sufficient value to justify quite large groups working in the area. Reliability theory has earned major support from Boeing, amongst others. In principle, areas such as networking, reliability, quality control, stochastic sampling theory, inventory theory all seem to have great potential for increasing commercial efficiency. Why then have we found, so far, little call from industry or government in a country such as Australia?

I can see a number of reasons why the commercial use of applied probability in my current context may be difficult. Among them are:

— the high cost of data-gathering for fitting models. Probabilistic models come into their own in complex situations where, say, regression methods or analysis of variance or other essentially linear techniques are inadequate. However, fitting any nonlinear model demands detailed data and it is often hard to justify collection costs except in a large-scale concern.

— the ill-understood conclusions available from models. Probabilistic models give probabilistic answers. These are not usually single-value decision-prompting answers, but a quantification of the range of outcomes. Industrial and commercial decision-makers usually do not respond well to such answers, because of the inherent difficulty in adjusting to probabilistic thinking.

— the cost of a single piece of applied probabilistic or indeed any real mathematical research. The Siromath experience so far is that an external consultancy in a country such as Australia may carry out major projects involving the application of known techniques to large problems; but projects developing new methods are only cost-effective if the problems are really on a larger scale again. I regret that it appears to me that only a very few pieces of applied probability research solve major new problems; most results in the literature would have a very poor cost–benefit profile, except perhaps for groups such as telephone companies, where, essentially, the potential bulk usage of a new piece of research warrants the funding of that research.

Despite these concerns, I do see applied probabilists in Siromath or similar organisations having some areas of great potential. I would encourage anyone developing skills in the craft of applied probability modelling to work in areas such as these:

— commodity sampling methods, currently much understudied and overdependent on simple statistical rather than appropriate applied probabilistic models;

— reliability theory for complex processes such as power plants;
— inference from simulation modelling;
— projection and prediction methods based on official statistics and large-scale samples.

These areas are connected through an emphasis on the interaction of different components in complex systems. Standard statistical methods seem to me to handle this simplistically and poorly. Applied probability modelling should be able to grapple with at least some of the nonlinear and unparametrizable interactions usually present, and perhaps give qualitative results which identify questions which can be answered and pinpoint data needed to answer them.

7. The Future in Australia for Probability

I may appear pessimistic about the actual use of probability in my current environment, despite the potential I see in the areas of application at the end of the previous section.

I am perhaps conditioned by the local environment. Australia is a relatively small country. Larger economies may be able to afford more applied probability even in a very project-oriented environment such as that in which I now work. I do believe, however, that many of the lessons learned at Siromath and briefly described above will be relevant elsewhere.

It does seem to me that the luxury of carrying out the versions of "applied probability" which do not have direct applicability (or indeed working on any other form of pure mathematics) in academic seclusion will disappear soon, for all but the very best, as the postwar economic boom conditions continue to evaporate. Those of us who, as aspiring applied probabilists, realize we are not Galois or Abel, Kolmogorov or Feller, Kendall or Moran, must decide whether to take up statistics (promising workable answers to necessarily oversimplified problems) or whether to try to improve the real applicability of our probabilistic modelling methods.

Unless we soon produce an applied probabilist or two of the calibre of Kolmogorov and Feller, who can substantially redirect probability theory towards applications as, say, Fisher and some of his generation directed statistics, applied probability may well become a branch of pure mathematics, because the problems in making true probability applicable often seem very hard indeed, and beyond the capacity of most working in it today. I would regret this. I have found probability theory enjoyable, interesting and even exciting because it is not only mathematically elegant but also, at various levels, genuinely of external importance.

I hope we can persuade potential users, be they commercial clients, university colleagues or funding groups, that applied probabilistic answers can be directly useful. I hope that we continue developing the craft of applied probability modelling, drawing on the varied skills of mathematicians, statisticians and computer scientists; and I hope this volume aids in the increasing refinement, and particularly the use, of that craft.

Acknowledgements

An autobiographical paper is a good place for acknowledgements. I should express my thanks to Chip Heathcote, Eugene Seneta and David Kendall for bringing me into the subject; to David Vere-Jones in New Zealand and Elja Arjas in Finland for introducing a number of problems and people to me at critical times; to Joe Gani for introducing me to the potential wider relevance of all mathematical research; and to Jim Boen for giving me the basis of a consulting philosophy. More generally, I owe debts to a large number of collaborators, including notably Ian Saunders, Pekka Tuominen and Paul Feigin in probabilistic ventures, but also including many friends from other fields, for introducing me to so many interesting problems and to such fascinating solutions.

References

[1] de Hoog, F., Brown, A. D. H., Saunders, I. W. and Westcott, M. (1985) Numerical calculation of the stationary distribution of a Markov chain in genetics *J. Math. Anal. Appl.* To appear.

[2] Feigin, P., Tweedie, R. L. and Belyea, C. (1983) Weighted area techniques for explicit parameter estimation in hierarchical models. *Austral. J. Statist.* 25, 1–16.

[3] Harris, T. E. (1956) The existence of stationary measures for certain Markov processes. *Proc. 3rd Berkeley Symp. Math. Statist. Prob.* 2, 113–124.

[4] Henstridge, J. D. and Tweedie, R. L. (1984) A model for muttonbird growth. *Biometrics* 40, 117–124.

[5] Laslett, G. M., Pollard, D. B. and Tweedie, R. L. (1978) Techniques for establishing ergodic properties of continuous-valued Markov chains. *Nav. Res. Log. Quart.* 25, 455–472.

[6] Leedow, M. I. and Tweedie, R. L. (1983) Weighted area techniques for the estimation of the parameters of a curve. *Austral. J. Statist.* 25, 310–320.

[7] McArthur, N. A., Saunders, I. W. and Tweedie, R. L. (1976) Small population isolates: a simulation study. *J. Polynesian Soc.* 85, 307–326.

[8] Nummelin, E. (1978) A splitting technique for Harris recurrent Markov chains. *Z. Wahrscheinlichkeitsth.* 43, 309–318.

[9] Orey, S. (1971) *Lecture Notes on Limit Theorems for Markov Chain Transition Probabilities.* Van Nostrand, London.

[10] Pyke, R. (1975) Applied probability: an editor's dilemma, in *Proc. Conf. Dir. Math. Stats. (Supplement to Adv. Appl. Prob.)*, 17–37.

[11] Revuz, D. (1975) *Markov Chains.* North-Holland, Amsterdam.

[12] Saunders, I. W. (1980) A model for myxomatosis. *Math. Biosci.* 48, 1–15.

[13] Saunders, I. W. and Tweedie, R. L. (1976) The settlement of Polynesia by CYBER 76. *Math. Scientist* 1, 15–26.

[14] Schuh, H.-J. and Tweedie, R. L. (1979) The estimation of parameters in hierarchical models using Laplace transform estimation. *Math. Biosci.* 45, 37–67.

[15] Seneta, E. (1980) *Non-negative Matrices and Markov Chains.* Springer-Verlag, New York.

[16] Tweedie, R. L. (1971) Truncation procedures for non-negative matrices. *J. Appl. Prob.* 8, 28–46.

[17] Tweedie, R. L. (1976) Criteria for classifying general Markov chains. *Adv. Appl. Prob.* 8, 737–771.

[18] Tweedie, R. L. (1982) Criteria for rates of convergence of Markov chains, with applications to queueing and storage theory. In *Papers in Probability, Statistics and Analysis*, ed. J. F. C. Kingman and G. E. H. Reuter, Cambridge University Press, Cambridge.

[19] Vere-Jones, D. and Ogata, Y. (1984) On the moments of a self-correcting process. *J. Appl. Prob.* 21, 335–342.

[20] Young, R. R., Anderson, N., Overend, D., Tweedie, R. L., Malafant, K. W. J. and Preston, G. A. N. (1980) The effect of temperature on times to hatching of eggs of the nematode *Ostertagia circumcincta*. *Parasitology* 81, 477–491.

Index

Achievement testing   16
Adapted process   245
Adaptive management   82–4, 86
Additive processes   173
Age-distributions   4
Aitken, A. C.   224, 231
Algorithmic mathematics   217
Algorithmic probability   219–221
Ancillaries, relevant   7
Annihilation process   245
Anthropology, computerized   300
Archaeological seriation   5, 6
Architecture, statistics in   69–70
Astronomy   6–7
Australia, conditions in   35, 37, 38, 305–7
Australian National University   34, 48, 229–30, 243, 293, 298

Balance, partial   100, 191
Balanced systems   206
Ballistics   13, 14
Ballot theory   147, 148
Bandit problems   191–2
Barbu, Ion   253
Barnard, G.   7, 8, 113
Bartlett, M. S.   viii, 68, 72, 84, 128, 187, 190, 193, 228, 299
Bay, Z.   143–5
Bellman, R.   169, 216
Berkeley Symposia   16, 19, 22, 23, 73, 156, 159, 173, 286
Bernoulli Society   3, 116, 132
Binomial moments   144
Biomathematics   73–4, 78–82
Biometry   71–3, 162–3
Bird navigation   6
Birmingham, University of   173

Birth-and-death processes   95, 100, 134, 176
Bombay, University of   127, 243
Boundary-value problems   106, 174
Box, G. E. P.   50, 299
Braşov conferences   260, 263
Brooklyn College   167–8
Brownian motion   96, 136, 144–5

Cambridge, University of   47, 65–6, 67–9, 190–2, 267, 294
Campbell, J. T.   224, 230
Centre of Mathematical Statistics, Bucharest   258–60
Certainty equivalence principle   192
Chain-binomial epidemic models   69, 79, 84
Chess   89, 167, 171
City College, New York   11–12
Clustering problem   22
Coalescent processes   289
Cochran, W. G.   46, 199
Cohen, J. W.   114, 115, 117, 177, 215
Cold Spring Harbor   49, 159
Collaboration with non-mathematicians   5–6, 8, 17, 26, 39, 49, 176, 258, 260, 271–2, 289–90, 302
Columbia University   12–14, 16–17, 26
Combinatorics   95, 133, 135
Commutation rules   246
Compensation techniques   121
Computers, importance of   22, 35, 56, 85, 92, 96, 103–4, 133–4, 207, 214–216, 302–4
Congestion theory, *see* Teletraffic
Consulting work   6–8, 17, 50, 175, 200, 202, 206, 232, 293, 298–302
Continued-fraction expansion   263

309

Continuous tensor products   242
Continuum mechanics   92, 106
Control, stochastic   130, 190, 192, 208
Conveyor belts   200
Convolution semigroup   240
Cornell University   73–4, 129
Corruptions, random   7
Coverage and packing problems   23
Creation process   245
Crow, J. F.   158, 281, 283, 286
CSIRO   35, 46, 298–302, 303

Dams, *see* Storage models
Daniels, H. E.   47, 172, 173, 174
Darboux functions   255–7
Data analysis   299
Delft, Technological University of   91, 102–3
Denjoy, A.   254, 265
Density estimation   6
Differential equations   93
Differential geometry   35, 37
Diffusion equation   161, 170
Diffusion models in population genetics   160, 163, 279
Diophantine equations   141
Dirichlet problem   120–1
Disease control   82–4
Distribution approximations   20–21
Doob, J. L.   5, 243
DSIR   187–9, 224, 227
Dynamic programming   191
Dynamic reversibility, *see* Reversibility

Earthquakes, *see* Seismology
Econometrics   33, 36
Editorial experience   40, 54, 132, 177–8, 208, 270–1, 304
Electrical noise   170, 171
Electron emission, secondary   144
Entropy   239
Eötvös Competition   142–3
Epidemic models   4, 69, 73, 188
Ergodic theory   113
Erlang, A. K.   96, 146
Ewens, W. J.   294
Expectation   191, 241

Feller, W.   viii, 5, 50, 95, 114, 156, 160, 172
Fermat's last theorem   266
Fermionic stochastic calculus   246

Fibonacci numbers   142
Finch, P.   278
Finney, D.   71, 75
First-order cocycles   242
Fisher, R. A.   22, 48, 49, 68, 80, 152, 154, 155, 163, 237, 240, 278, 279
Fluctuation theory   101, 131
Fokker–Planck equation   155, 156, 170

Game theory   133, 267
Gani, J.   48, 75, 113, 128, 132, 191, 215, 242, 243, 278–9, 294, 298
Gaussian processes   39, 40
Gegenbauer polynomials   159
Gelfand–Neumark–Segal construction   238
Genetic frequency   303–4
Genetics, Mendelian   163
Geometrical probability   23–4
Geometry   54–5, 90, 92
Geostatistics   54
Gittins, J.   191–2
Gnedenko, B. V.   118, 226, 243
Grad–Solomon tables   19–20
Great Depression   11, 89, 167
Green's function methods   170, 174
Grenander, U.   187, 216
Group screening   50
Group theory   95

Haldane, J. B. S.   152, 154, 163
Halmos, P. R.   238, 239
Hannan, E. J.   48, 52, 190, 278, 294
Harmonic analysis   146
Harvard University   169–71
Heisenberg uncertainty principle   246
Heyde, C. C.   116, 294
Hope-Simpson, R. E.   70, 81
Hospitals, function and design of   69
Hudson, R.   245, 246
Hungarian Academy of Sciences   145

Immunology   45
Indecomposable measures   240
Indeterminism   130
India, conditions in   246–7
Indian culture   236
Indian Institute of Technology, Delhi   243–4
Indian Statistical Institute   236–42, 244–5
Infinite allele model   160, 283, 284

Information conditions 39
Information theory 94–5, 113, 239
Insensitivity 97, 98, 104
Institute of Mathematical Statistics 159
Insurance risk problem 135
International Mathematical Olympiads 252
International Teletraffic Congresses 101, 113, 115–16
Inventory analysis 21
Ionescu Tulcea, C. T. 253
Iowa State College, Ames 12, 157–8
Iterated logarithm, law of 38
Itô calculus 6, 245

János Bolyai Mathematical Society 146
Johns Hopkins University 51, 72, 197–9, 201
Jordan, C. 143–5
Jurimetrics 25

Kac, M. viii, 5, 130, 148, 193
Kalman filter 35, 190
Karlin, S. 215, 278, 281, 286
Keilson, J. 116, 117, 121, 215
Kempthorne, O. 158, 281, 282
Kendall, D. G. 56, 101, 128, 134, 148, 224–5, 259, 267, 294
Khintchine, A. Ya. viii, 102, 148, 238, 243
Kihara, H. 153–4
Kimura, M. 281–3
Kimura–Crow model 160
Kinetic gas theory 96, 100
Kingman, J. F. C. 5, 162, 286, 287, 289
Kolmogorov, A. N. viii, 95, 148, 239, 240, 243
Kolmogorov equations 155, 156, 160, 279

Ladder phenomena 136
Lancaster, O. 45, 230
Laplace transform methods 145, 179, 301
Law, statistics in 20, 24–6
Lebesgue's theorem 254
Lévy, P. viii, 95, 146, 240, 243, 264
Lévy–Khinchine representations 240
Lévy process 135
Lie groups 35, 241, 243

Likelihood ratio 37, 39
Limit theorems 7, 8
Lincoln Laboratory, MIT 171
Linear innovations 36
Linear operators 131
Loránd Eötvös University, Budapest 143, 146
Lush, J. L. 157, 158

Mahalanobis, P. C. 237, 240
Malaria control 75, 82
Manchester, University of 128, 190, 242–3
Manifold 37, 38
Manuscripts, copying of 4
Mapping function 153
Markov chains 5, 120, 131, 134, 217, 224, 261, 266, 279, 294–8
Markov processes 114, 117, 136, 207, 261
Markov renewal processes 135, 208
Markovian models 110, 119, 123
Martingales 40, 117, 131, 207
Maryland, University of 114, 117
Matching problem 143
Mathematical education 122–3, 133, 218–19, 232–3, 269–70
Mathematical programming 133
Mathematics, new applications of 104, 302
Matrix-geometric vector 217
Maximum likelihood estimator 36, 37
McGregor, J. 278, 281, 286
Measles, epidemiology of 69, 70, 79, 84
Medical informatics 77
Melbourne, University of 33, 45, 47, 277
Method of phases 134
Mihoc, G. 252, 258, 260
Miklós Schweitzer competition 145
Modelling, biomathematical 78–82, 301
Modelling, craft of 3–8, 91, 95, 107, 129–30, 219–221
Modelling, Markovian 110, 123
Modelling, probabilistic 67, 76, 84, 95, 134–6, 202–4, 260
Modelling, stochastic 227–9
Moment problem 238
Monte Carlo simulations 200
Moran, P. A. P. 34, 36, 48, 56, 128, 190, 229, 278, 294
Moscow University 226

Multiple equation systems 36
Multivariate analysis 20, 21–22
Mutation, stepwise 161
Myxomatosis modelling 300–1

National Institute of Genetics,
    Mishima 154–5
Neutral theory of molecular
    evolution 160, 163, 283
Neuts, M. 121, 304
New Zealand, conditions in 230–3
Neyman, J. 5, 23, 130, 237
Non-Bayesian assumption 285, 286
Normal distribution 19, 21
North Carolina, University of 46–7
Nuffield Foundation 69–70, 72
Number theory, probabilistic 147, 149, 263–6

Observables 241
Office of Naval Research 13, 15–16, 26
Ohta–Kimura model 161–2
Operational research 66–7, 95–6
Optimization 190
Order statistics 147
Ornstein–Uhlenbeck process 174, 176
Oxford, University of 71–3, 224–5
Ozone, stratospheric 53

Palaeomagnetism 48
Palásti conjecture 23
Palm, C. 101, 102, 114
Parasitic disease 76, 82–4
Partial differential equations 94, 189
Particle counters 144, 145
Partitions, random 288
Passage times 119, 120
Percolation process 189
Perturbations, random lateral 7
Phase-type distribution 121
Phi-mixing sequences 262–3
Pinball, legality of 25
Point processes 97, 136, 145–6, 207, 227–8
Poisson–Dirichlet distribution 287, 288
Poisson equation 94, 121
Poisson flows 136
Poisson processes 21, 97, 135

Poland, changes in 111–13, 117–18
Pollaczek, F. 100, 114, 118–19
Polymerization 189, 191, 192
Positive-definite kernels 242
Potential theory 119
Prabhu, N. U. 110, 116, 117, 177
Predicate calculus 94
Prediction 34, 36, 40, 94
Probabilistic modelling, *see* Modelling, probabilistic
Probability, applied 130–4, 293–302
Problem-solving 89
Product form 102
Propeller blades 92
Propositional calculus 94
Publishing problems 132, 147, 177, 208, 244–5

Quality control 13–14, 18–19
Queueing networks 103–4, 136, 191, 200–1, 205–7
Queueing theory 5, 69, 96, 101, 103, 114–15, 134–6, 146–8, 172, 179, 200

Radar design 171
Radio direction-finding 66–7
Random drift, genetic 158, 159, 161
Random system with complete connections 261–6
Random walks 131, 135
Ranga Rao, R. 238, 240
Rao, C. R. 237–40, 244
Rational transfer function systems 36
Regeneration points 101
Relaxation method 92
Relaxation time 104
Reliability theory 178–9
Repairman problem 146
Research environment 34–5, 129–131, 147, 175, 289
Research funding 133, 147, 175
Research Triangle Institute 50–1
Reuter, G. E. H. 5, 128
Reversibility 98, 189
Revue Roumaine 254, 270
Riemann hypothesis 149, 265
Riesz decomposition 121
Robustness of least squares 50
Rochester, University of 175–81
Russian school of mathematics 226–7, 241

Sample path analysis   207
Sampling formula   284–289
Sampling techniques   13–14, 18, 20
Scaling, multidimensional   5
Schistosomiasis (bilharziasis)   82–3
Schottky's formula   146
Schrödinger equation   246
Seismology   227–8
Selective neutrality   286
Self-sterility   279
Semi-Markov process   146
Seneta, E.   229, 294, 303
Sequential analysis   13–14, 190
Serial correlation   48
Server workload   101
Service systems   205
Shannon, C.   94, 113, 239
Shape, statistics of   6
Shells, helicoidal   93
Shifting balance theory   163
Signal detection   170, 172
Simulation models   69, 304
Sinai, Ya. G.   239, 241
Siromath Pty. Ltd.   304–6
Sloan–Kettering Institute   73–4
Social psychology   17–18, 20
Southwell's relaxation method   92, 94
Spatial models   188
Spatial processes   187
Spectral density   39
Spline techniques   6
Stanford University   14–15, 19–21, 26, 72, 278–9, 281, 289
State-space approach   35, 36, 172, 241
Stationary system   36
Statistical equilibrium   96, 99
Statistics, status of   41, 56, 231–2
Steklov Institute, Moscow   240–1
Stepping-stone model   156
Stochastic control, *see* Control, stochastic
Stochastic differential equations   246
Stochastic equations   117
Stochastic modelling, *see* Modelling, stochastic
Stochastic ordering   136
Stochastic processes   114, 145–6, 162
Stochastic Processes and their Applications   viii, 116, 132, 177
Stochastic Processes and their Applications (Conferences)   116, 177
Stochastic programming   267
Storage models   48, 128, 134–6

Structure, mathematical   110, 122
Supplementary variable   100
Switching algebra   94
Sylvania Electronic Systems   172–4

Taboo probabilities   120
Takács, L.   101, 114, 117, 215
Tata Institute, Bombay   243
Tauberian theorems   145
Tăutu, P.   259–60
Telephone systems   89, 94, 95
Teletraffic   95–6, 103, 113, 115, 146
Temperature, atmospheric   53
Theoretical physics   169–70
Threshold model   188
Time change, random   120, 121
Time series analysis   12, 36, 37, 187, 190
Tracer kinetics   176
Traffic flow   24
Traffic handling capacity   95
Transition totals, distribution of   189
Triangles, random   55
Truncation methods   303
Tukey, J.   35, 50, 53
Tungsram Research Laboratory   143–5
Tweedie, R.   229

Uniformity, tests of   51
University of Wisconsin, Madison   158, 159
USSR   225–6, 240–1
Utrecht, University of   105

Varadarajan, V. S.   238, 241, 242
Vere-Jones, D.   294
Von Mises, R.   92, 106

Waiting lines, *see* Queueing
Walker, A. M.   5, 71
Watson, G. S.   6, 35
Wellington, Victoria University   224, 230, 231
Whittle, P.   224, 231
Wiener, N.   94, 113, 117
Wiener–Hopf technique   100, 131, 135, 136, 174
Wilks, S.   5, 50
Wishart, D. M. G.   172–4
World Health Organization   74–7
Wright, S.   152–4, 155, 159, 163, 279